Wolfram Schommers

Expansion of Physics Through Nanoscience

Also of interest

Nanomaterials.
Volume 1: Electronic Properties
Ghatak, Mitra, 2018
ISBN 978-3-11-060922-6, e-ISBN 978-3-11-061081-9

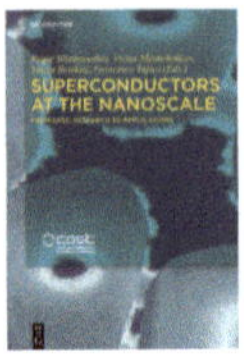

Superconductors at the Nanoscale.
From Basic Research to Applications
Wördenweber, Moshchalkov, Bending, Tafuri (Eds.), 2017
ISBN 978-3-11-045620-2, e-ISBN 978-3-11-045680-6

Nanoscience and Nanotechnology.
Advances and Developments in Nano-Sized Materials
Van de Voorde (Ed.), 2018
ISBN 978-3-11-054720-7, e-ISBN 978-3-11-054722-1

Electrons in Solids.
Mesoscopics, Photonics, Quantum Computing, Correlations, Topology
Bluhm, Brückel, Morgenstern, von Plessen, Stampfer, 2019
ISBN 978-3-11-043831-4, e-ISBN 978-3-11-043832-1

Wolfram Schommers

Expansion of Physics Through Nanoscience

What Is Time at the Basic Level?

DE GRUYTER

Author
Prof. Dr. Wolfram Schommers
Department of Materials Science
and Engineering
University of Texas at Arlington
19031 Arlington
TX 76019
USA
wolfram.schommers@partner.kit.edu

ISBN 978-3-11-052460-4
e-ISBN (E-BOOK) 978-3-11-052559-5
e-ISBN (EPUB) 978-3-11-052466-6

Library of Congress Control Number: 2020938292

Bibliographic information published by the Deutsche Nationalbibliothek
The Deutsche Nationalbibliothek lists this publication in the Deutsche Nationalbibliografie; detailed bibliographic data are available on the Internet at http://dnb.dnb.de.

Cover image: Sakkmesterke/iStock/Getty Images
Typesetting: Integra Software Services Pvt. Ltd.
Printing and binding: CPI books GmbH, Leck

www.degruyter.com

Preface

The impact of nanoscience (nanotechnology) will be tremendous. The manipulation of atoms and molecules will allow to construct new scientific and technological worlds and will bring fundamentally new perspectives in all fields, for example, in the development of new electronic devices and also in the field of medicine. In the case of nanobiotechnology, big changes are expected. It has been speculated that through nanoscience the bodies of human beings will be transformed into undecaying systems, that is, an infinite life is not excluded through the prospects in nanoscience.

On the other hand, the threats in nanotechnological applications are large; in connection with self-organizing processes, uncontrolled processes and strong destructions may appear. It has been speculated that even the entire earth could be transformed into a system hostile toward life. In other words, biological individuality, that is, life, would no longer be possible on the Earth.

To put it exaggeratedly, in nanoscience (nanotechnology) we move between two limiting cases, between two poles, and these poles are given by "total destruction" and "infinite life."

We have to be careful; it must be a challenge for science to understand and to describe all these processes in order to avoid disasters. On the other hand, realistic theories open the door for fantastic new possibilities, just in connection with medicine.

This situation requires reliable theoretical conceptions for the understanding of the various effects in nanoscience. In this book we discuss critically traditional methods, and the necessity of new aspects is underlined.

•

In nanoscience and nanotechnology, there are various disciplines working with principally different systems. Examples for typical directions are functional nanomaterials, food chemistry, medicine with brain research, quantum and molecular computing, bioinformatics and nanoelectronics. All these disciplines have their basis at the nanolevel. Why?

In nanoscience, we work at the ultimate level. Here, the properties of "all" kind of condensed matter have its origin. In fact, nanoscience (nanotechnology) reflects the smallest level at which the properties of the physical reality emerge and at which biological individuality comes into existence. This material world is defined in "absolute" terms. Daniel and Mark Ratner formulated this feature in their book *Nanotechnology and Homeland Security* as follows: "[. . .] through nanotechnology we can make materials whose amazing properties can be defined in absolute terms: This is not only the strongest material ever made, this is the strongest material it will ever be possible to make." In other words, in nanoscience (nanotechnology) scientists work at the ultimate level.

https://doi.org/10.1515/9783110525595-202

The biological structures also have their origin at the nanolevel, building up of molecule by molecule, macroscopic biological systems. That is, biological individuality has its origin at the nanolevel too and is expressed in its ultimate form at this level.

•

The development of the scanning microscope was the essential step toward nanoscience. This device made it possible to move single atoms in a controlled manner from one position to another. To do that in a "controlled" way is important here.

It became possible to manipulate matter at its ultimate level. This is the basis for technological developments. It now makes sense to engineer physically real systems at the nanometer scale, just at the level where the basic features are physically defined. In other words, technology has reached the level of science. Nanoscience and nanotechnology are basically undistinguishable.

We have one theory for all disciplines in nanoscience, and this theory is given through the basic laws of theoretical physics. In traditional technologies (micro- and macrotechniques), engineers do not work at the ultimate level. They use more or less phenomenological descriptions which, in general, cannot be derived from the laws of theoretical physics, that is, each technological direction has its own description.

The topics in nanoscience are more basic than in micro- and macrotechniques and we have here to work with the conceptions of theoretical physics. But it has to be underlined that also theoretical physics enters a new realm when it is applied to nanoscience. New forms of conceptions and novel tools are obviously necessary. The class of self-organizing processes, which belong to the heart of nanoscience, is an important and typical example. This field is analyzed in this monograph.

Self-organizing processes are "time sensitive," that is, the theory must have a realistic conception for the phenomenon "time." Does traditional physics treat the time in a satisfying way? This is not the case, but becomes particularly necessary when we enter the nanorealm. This field is also analyzed in this monograph.

•

What is time? We cannot put a piece of time on the table and detect it. Nevertheless, human beings experience the phenomenon "time." There is one type of time in traditional physics and this is defined by our clocks. But within science and technology, this type of time is used and applied too, without changing its principal definition.

No doubt, outstanding theories have been introduced in physics, but the notion "time" remained unchanged, despite the developments in the theory of relativity. Scientists developed quantum theory, but a quantum aspect of time is missing. That what we define as time is an external parameter and is not a property of a physically real system (object). But this is a deficit! This is an important topic and we analyze the situation in this book.

The external time is not coupled to physical reality. This type of time is used in the description of physically real processes but, on the other hand, time itself is not produced through physically real processes. In other words, this type of time is not coupled to the physically real world.

We need a more sophisticated type of time in physics and nanoscience, which is not external in character, but is coupled to the systems under investigation. That is, a "system-specific time" is needed.

The external time, displayed by our clocks, reflects rather a naive point of view. This is clearly expressed through a statement by Julius Fraser. He gave the following comment in his book *TIME, the Familiar Strange*:

> The time is like the cable of the cable railway of San Francisco. The cable is driven by a far and invisible machine, is however hidden. We know that it moves, because the carriages are connected and are carried with it. Completely similarly we usually see the time in everyday life as a universal, cosmic movement of the present, of the now; they are propelled by natural or divine forces and matter, life, humans and society are connected with them and are propelled and along-moved for a while.

The San Francisco cable railway is of course a metaphor but it exactly reflects the situation of how we experience the phenomenon "time." However, we avoid giving a concrete conception. For example, Albert Einstein remarked the following: "Time is that what we measure with a clock." However, time is obviously more than that: a human not only looks at the clock, but he/she has a certain "time feeling." Einstein told us how time behaves but not what time actually is.

The clock time is, as we already stated, merely an "external" time and has nothing to do with the objects under investigation and it is in particular not the reason for the time feeling of human beings. This external time cannot be rationally explained by the conceptions of traditional physics. It is a metaphysical element, so to speak. The metaphor with the San Francisco cable railway underlines that. Where is such a time-producing machine hidden?

There is not "one" system existing, singular in character, which produces what we call time. But time has to be based on physically real systems. In other words, not "one" singular system produces the time, but "each" physically real system (object) does it. This exactly means "system specific." This is the conception.

A lot of physically real systems (objects) are in the universe but none of them is preferred and singular. Each of them must therefore have its own time structure which is, in other words, system specific. This peculiarity is compatible with the individual time feeling of human beings.

In fact, we have to assume that the time behaves system specifically and does not reflect a singular feature as is the case in traditional physics. All that indicates that the time is not comprehensively treated enough in traditional physics. This is particularly accentuated by the fact that there is no quantum aspect for the time. An operator for the time does not exist in traditional quantum theory.

Erwin Schrödinger tried to introduce such an operator within the formalism of traditional quantum theory but without success. The philosopher Mario Bunge underlined that the role of time in the usual form of quantum theory is not acceptable. In fact, during the last decades, the quantum aspect for time became more and more a relevant item. In this monograph, an overview is given and its relevance for nanoscience is constructively discussed.

•

No doubt, theoretical physics is important for nanoscience. On the other hand, nanoscience has an influence on the developments in theoretical physics. New aspects in nanoscience are often challenges for basic physics and problems come up, which were not considered before. In those cases, we have an expansion of theoretical physics through nanoscience (nanotechnology). Here the "nature of time" is a prominent example.

With nanoscience, we enter a higher level of reality, higher and deeper than that on which traditional physics is based. This step is required when we want to describe the typical features of nanosystems realistically. For the understanding of the effects at the nanolevel, basic conceptions have to be deepened and expanded. In particular, the nature of space and time plays an essential role in this connection. Here Mach's principle comes into play and has to be fulfilled. We discuss these basic features in all chapters of this book.

The effects in nanoscience are essentially quantum phenomena and are time sensitive. It is argued in this monograph that the external time of traditional physics is not sufficient for the analytical treatment of nanosystems and we obviously need here the conception of the system-specific time. In Chapters 4–6, we discuss this point in detail, also in connection with mathematical formulations. In Chapter 7, we underline that the existence of the system-specific time is of general interest and can probably solve the problem that science has with respect to the cosmological constant.

Nanomedicine is of particular interest. Here, the nanoengineering of brain functions is an often discussed topic. In this connection, an overlap of matter states with mind states is inescapable. In particular, the interrelation between basic reality, mind, brain and matter has to be considered. We discuss this point particularly in Chapters 4 and 5. The principles of evolution and the methods of behavior research deliver essential arguments.

Wolfram Schommers
University of Texas, Arlington, USA
and
Karlsruhe Institute of Technology

Contents

Chapter 1
Challenges in nanoscience

With nanoscience, we enter a new field not only in science but also in technology. Science and technology grow together and become undistinguishable. In other words, technology has reached the level of science.

The door for completely new perspectives will open. In nanoscience, we work at the ultimate level where the properties of matter and biological individuality emerge. Can we transfer the basic laws of physics, used so far within the frame the traditional technologies, to the realm of nanoscience? New possibilities create specific challenges, just in connection with theoretical descriptions. Therefore, the realistic theoretical treatment of future nanoscientific problems demand new and extended physical conceptions, which were, so far, not relevant.

In fact, the more we penetrate into the details of nature, the more we need realistic conceptions for their description, that is, we need conceptions with increasing sophistication and reliability. When we enter the realm of nanoscience, we penetrate deeper into the details of nature. In such cases, the basic physical laws will be more disclosed and we have to respect this situation.

It is an illusion to believe that the traditional laws of theoretical physics can be transferred without modifications to the realm of nanoscience (nanotechnology). In this chapter, some relevant points are quoted.

1.1 Nanoscience: basic questions

Nanoscience will be the dominant direction for technological developments in the future. This new science will influence our lives to a large extent. Specific manipulations of matter in the atomic realm, for example in materials science and medicine, will open completely new perspectives on all scientific and technological disciplines. The impact of nanotechnology will be far-reaching: Brain functions will be influenceable and we may be able to manipulate intelligence through nanotechnological operations. The potential advantages gained through nanotechnology are tremendous, but there are serious threats as well, which will be underlined further.

For the production of optimal nanosystems with tailor-made properties, it is necessary to analyze and construct such systems by adequate theoretical and computational methods. What theoretical tools does a future nanoscientist (nanoengineer) need? At the nanolevel, the properties of matter emerge and even biological individuality comes at this level into existence. This situation makes nanoscience to a discipline with “ultimate” character [1]. The theoretical methods have to be selected and developed accordingly.

https://doi.org/10.1515/9783110525595-001

1.1.1 Technology has reached the level of science

Again, in nanoscience and nanotechnology, we work at the "ultimate level" where the properties of materials and also biological individuality emerge. The complete characteristics of nanosystems are established at this level, and there is no scientific level above it. This means the term "ultimate level." Here, technology has reached the level of science.

In this situation, we have to apply the physical laws in their basic form and these are identical with the laws of theoretical physics. Although we have a lot of directions and disciplines in nanoscience (nanotechnology), which are partly very different in character, we are nevertheless working here at the same theoretical footing – in contrast to the traditional technologies.

In nanoscience, we have "one" theory for all the disciplines and all nanophenomena, and this is given by the basic laws of theoretical physics. Why? The theoretical laws of nanoscience are basic; there is no level above the ultimate nanolevel. That is, at the nanolevel, the theoretical laws of physics are responsible for understanding the phenomena of the development of devices. They represent the most basic frame of natural science. Table (1.1) summarizes this fact:

$$\left\{\begin{matrix} \text{Nanoelectronics} \\ \text{Nanostructured materials} \\ \text{Nanomedicine} \\ \text{Bioinformatics} \\ \vdots \end{matrix}\right\} \Leftarrow \text{Laws of theoretical physics} \qquad (1.1)$$

In fact, all the disciplines quoted in table (1.1) are governed by the structure and dynamics of atoms (molecules), and the same theoretical conceptions with various boundary conditions have to be applied.

The situation is different in traditional technologies. Within these disciplines, that is, in micro- and macrotechnique, engineers do not work at the ultimate level, but they more or less use phenomenological descriptions, that cannot be deduced from the basic physical laws and each discipline of micro- and macrotechnique has its own description.

As we already remarked, in nanoscience there is only "one" theoretical frame for all directions and phenomena, and this is given at this ultimate level by the basic laws of theoretical physics. That is, at the nanolevel, scheme (1.2) is characteristic in connection with scheme (1.1):

$$\begin{matrix} \text{Ultimate level} \\ \Uparrow \\ \text{Laws of theoretical physics} \end{matrix} \qquad (1.2)$$

The following questions arise. Are the present basic laws of theoretical physics tailor-made to the ultimate level of nanoscience? Are they adapted to the ultimate level of nanoscience? Do they "completely" describe the essential features of it? Can we completely understand the entities and the processes at the nanolevel through the tools of theoretical physics developed so far? We do not, and we know why. There was simply no need for new perspectives. We normally develop physical theories if there is a need for them. Even the notion "nanoscience" was widely unknown before certain path-breaking experiments were done.

1.1.2 The invention of the scanning tunneling microscope

The invention of the scanning tunneling microscope was the starting point for nanoscience. In fact, with the development of this instrument, approximately 40 years ago, nanoscience became an important scientific and technological discipline, since for the first time single atoms could be moved in a controlled manner from one position to another, and we learned to manipulate matter at its ultimate level.

The processes at the ultimate level, where the properties of matter have their origin and where biological individuality comes into existence, became controllable. That is, at this level, the systems can be changed systematically in order to create experimentally new possibilities in an artificial way, for example, novel devices and to cure cancers.

1.1.3 Self-organizing processes

In nanoscience and nanotechnology, self-organizing processes are of particular relevance. They belong to the heart of nanoscience. A self-organizing process can be characterized as follows: We prepare the initial system A and put this system into a given environment. Then, system A develops to a new state, state B. The experimentalist, in general, does not know the final state B at the beginning when the self-organizing process starts. In other words, the new system, the outcome remains, in general, unknown until the self-organizing process is finished.

This must be avoided because the "threats" are too large [2] and we must precisely know what we want to develop, and we must know the final state B "before" the self-organizing process is initiated by the experimentalist.

Is that what we want to develop identical with that of what nature creates? Since the threats within the frame of nanoscience can be tremendous [2], we have to predict the outcome, that is, we have to predict the final state B on the basis of reliable theoretical laws. This is in fact necessary. We must know what we do.

1.1.4 New phenomena require new theoretical developments

We have the following situation: The laws of theoretical physics are not tailor-made to the ultimate level of nanoscience. These laws are basic because there is no level above the ultimate nanolevel.

Again, this branch became first conscious through the path-breaking development of the scanning tunneling microscope. Just the "controlled" manipulation of single atoms brought the nanolevel into existence. The new field of nanoscience (nanotechnology) came into the focus of the scientific community. At this level, new peculiarities came into play that were not considered or recognized before. However, these new peculiarities require new sophisticated theoretical conceptions. Since we are working here at the basic level, the new theoretical conceptions expand the usual frame of theoretical physics through the effects at the nanolevel.

Just the phenomenon "time" needs a more realistic basis. The reason is obvious. So far, within the frame of traditional physics, "stationary" processes are in the foreground, as already underlined by Ilya Prigogine [3]. However, self-organizing processes at the nanolevel require more than that. Self-organizing processes reflect "nonstationary" processes. In the next section, it will be outlined that the present time conception is not sufficient for the treatment of self-organizing processes, particularly at the nanolevel.

In conclusion, the laws of theoretical physics have to be extended when we want to describe the processes at the ultimate nanolevel. In other words, the theoretical laws are not yet adapted to the nanolevel.

1.2 Time

The time that we use in everyday life is shown by our clocks. In the following, we will mark it by the Greek letter τ, instead of t and call it "clock time." Only this type of time exists in traditional physics. The characteristics of time τ were introduced by Newton more than 350 years ago, but its attributes remained up to the present day. The time τ is basically classical.

1.2.1 No quantum aspect of time!

In fact, the character of the clock time τ is not changed during the developments of the various disciplines in theoretical physics (classical mechanics, theory of relativity, quantum theory). Clearly, within the theory of relativity, the value of τ changes when the clock moves with a certain velocity or when the clock is positioned in a gravitational field, but only the value of τ is changed but not its character. Even modern quantum theory uses the clock time, that is, modern quantum theory is based on the classical time parameter τ. There is no quantum aspect of time so far.

The clock time τ appears in Schrödinger's equation without to modify its classical character. The coordinates x, y, z behave statistically in traditional quantum theory but not the time τ. This is important when we deal with self-organizing processes. A nanosystem develops in nature from state A to state B. We have to assume that this is a full quantum-mechanical transition. If such a transition (the self-organizing process) is described by Schrödinger's equation, we work with a classical time τ. In other words, the time-dependent transition is described classically, which cannot be realistic because the nanosystem reflects in general a quantum system. The self-organizing process has to be considered as a "nonstationary" process and, if it is described within Schrödinger's theory, the potential $U(\mathbf{r})$ within Schrödinger's equation has to be replaced by $U(\mathbf{r}, \tau)$.

In conclusion, only the classical time τ appears in the quantum-theoretical description of traditional quantum theory; no other type of time is defined or introduced here. Already, Erwin Schrödinger and Wolfgang Pauli strongly argued that we need a quantum aspect of time. Ilya Prigogine worked in this field as well; he strongly emphasized that we need a quantum picture for time.

All authors (Schrödinger, Pauli, Prigogine and others) argued that a quantum aspect of time should be reflected through an "operator" for the time, reflected by an operator description of time. In other words, the time should be no longer a simple external parameter (given by τ defined through an external clock) as in classical mechanics and traditional (conventional) quantum theory. For this purpose, Schrödinger analyzed this point on a general footing. He tried to introduce an operator for the time which however does not exist within the frame of traditional quantum theory. Let us give a few historical details.

Does there exist a time operator within traditional quantum theory? The answer is negative! An operator for the time is definitely not definable within traditional quantum theory. We can only overbear the classical character of time, expressed by the clock time τ within Schrödinger's equation, by introducing an operator for the time. This is however not possible as Schrödinger himself demonstrated. This must have major implications. Schrödinger's equation can, therefore, only be considered as a limited tool for describing quantum phenomena, and this type of description is at best restricted to "stationary" phenomena [3].

In conclusion, a quantum aspect of time cannot be introduced within traditional quantum theory without changing the structure of the physical laws themselves, that is, without changing the structure of Schrödinger's equation itself. It must be emphasized that Schrödinger's equation cannot be deduced. There is no consistent physical frame for that. But what has to be changed? Obviously, the most basic features of the principal conception have to be changed. Let us briefly outline two conceptions (views) that are principally different from each other.

The space as container

The physical basic frame, on which Schrödinger's equation is based, is obviously not realistic enough. It is the so-called container model [4–6]. Let us repeat the main characteristics of the container model. Within the frame of this model, the entire material entities are assumed to be embedded at each time τ in space. Not only the sun, the moon and everything else, but also the physically real entities of quantum field theory are equally located in space.

This imagination is based on what we have directly before our eyes in everyday life. We are firmly convinced that we observe at each clock time τ, certain space structures are made of physically real (material) entities. This is, however, merely a superficial impression, but the situation is not so simple. We have to be careful!

What is the container made of within this model? The container is identical with what we call space. This model is naive but it works well in many cases. Nevertheless, the container principle is not realistic because within this frame the nature of the space–time is assessed not precisely enough. Details are given in Ref. [4–6], but we will explain the situation in more detail later.

Projection principle

We may state that the details of the container model are not realistic and the model has to be considered, at best, as an approximation. A more realistic basis for the physical phenomena in the world is desired. The quantum aspect of time is the only problem that the container model is obviously not able to solve. We will discuss in this and in the following chapters that the container model is not developable.

Within this situation, the so-called projection principle gets a particular relevance [4, 5]. Here, the physically real world is not embedded in space and time, but it is projected onto space and time. This conception is adapted to the basic features of space and time and is therefore more realistic from the very first.

In fact, the projection principle exactly delivers what is desired in connection with the "quantum aspect of time." We not only get an operator for the time but also each system (nanosystem) now contains a time variable, t, which is system specific and behaves quantum mechanically as well. In other words, the projection principle offers the essential items we need for the understanding of the self-organizing processes belonging to the heart of nanoscience (nanotechnology). Let us quote the relevant results here.

In the momentum–energy representation, the main differences to the traditional form of quantum theory can be depicted easily. New elements appear within this development. Characteristic variables as well as novel operators come into play. In other words, it is an expansion of physics initiated through nanoscience. But the basic view, on which the new conception is based, had to be changed: We go from the "container model" to the "projection principle."

1.2.2 Container model and projection principle: the main characteristics

Within the container model, the material objects are embedded at each clock time τ in space. It is assumed that there is nothing outside of space and time. This is, in fact, what we have directly before our eyes and what we experience.

In the case of the projection principle, the information about the physically real (material) objects is projected onto space and time. We get a picture of reality, and the objects themselves appear at each clock time τ as geometrical structures within the picture. That is, within the frame of the projection principle, it is assumed that the impressions before our eyes are geometrical structures.

1.2.3 Quantum aspect of time within the projection principle

Within the projection principle, the tools for the description of self-organizing processes are given by an operator for the time, which takes in projection theory the form

$$\hat{C}(\alpha, E) \tag{1.3}$$

and we have a novel expression for the Hamiltonian having the specific form

$$\hat{H}\left(\alpha, -i\hbar\frac{\partial}{\partial E}\right) \tag{1.4}$$

The parameter α sums up coordinates in operator form and the momentums. The variable E is the energy. The time operator $\hat{C}$ is not known in traditional quantum theory, and also not the Hamiltonian $\hat{H}$ in this form. The operator $-i\hbar\partial/\partial E$, which appears in $\hat{H}$, is the operator for the time coordinate; $-i\hbar\partial/\partial E$ is not defined in traditional quantum theory.

The operators $\hat{H}$ and $\hat{C}$ belong together, which is reflected by the fact that the operator $-i\hbar\partial/\partial E$ in $\hat{H}$ and the variable E in $\hat{C}$ form a commutator of the form

$$\left[E, -i\hbar\frac{\partial}{\partial E}\right] = i\hbar \tag{1.5}$$

Equation (1.5) is similar to the relations we have in usual quantum theory for the momentum and the coordinate [4, 6].

The entity $\hat{C}(\alpha, E)$ is the time operator for the system under investigation, that is, it is system dependent. The operator $-i\hbar\partial/\partial E$ is valid for each system.

All three expressions (1.3)–(1.5) do not exist in traditional quantum theory. The Hamiltonian is defined in traditional quantum theory but without the operator $-i\hbar\partial/\partial E$.

These three expressions change the situation in connection with self-organizing processes completely. They represent the situation in the momentum–energy representation. The equivalent expressions exist in space and time. Instead of the operator $-i\hbar\partial/\partial E$ for the time coordinate, we get the system-specific time t, and the variable t reflects the quantum time as well and exists simultaneously to $-i\hbar\partial/\partial E$.

All laws and expressions deduced within this frame are independent of the clock time τ, but τ remains an essential element. τ has nothing to do with the system under investigation and reflects the reference time for human observers.

Relations (1.3)–(1.5) are the basis for the description of self-organizing processes relevant in connection with nanosystems. It solves the problem with the cosmological constant, which we will discuss in Chapter 7. In other words, the existence of a system-specific time is of principal interest.

In this way we get a quantum time t, which is system specific. The time t reflects the quantum aspect of time together with $\hat{C}(\alpha, E)$ and $\hat{H}(\alpha, -i\hbar\partial/\partial E)$. Just this possibility is not given in traditional quantum theory, where we have merely an external parameter τ for the time without any quantum aspect. Nevertheless, the clock time τ is also in connection with the projection principle an essential entity. As we have remarked previously, τ serves here as reference time.

1.2.4 Descriptions

The self-organizing processes in nanoscience are in general quantum-mechanically organized. Here, we have to distinguish between three cases: The process in "nature," the situation in "traditional quantum theory" and the assessment within the frame of the "projection principle." The various possibilities are summarized in scheme (1.6):

$$\begin{aligned} &A \rightarrow B \\ &A \rightarrow B_\tau \\ &A \rightarrow B_t \end{aligned} \tag{1.6}$$

The details will be explained in the following sections:

The processes in nature

The processes in "nature" start at the state A and end at state B as outcome [see scheme (1.6)], are quantum processes. The nanosystem leaves the state A at clock time τ_A and reaches the final state B at clock time τ_B. For times $\tau > \tau_B$, the system is permanently in the stationary state B. The self-organizing process takes place for $\tau_A \leq \tau \leq \tau_B$ and is time dependent.

What does "time dependence" here mean? It does not refer to the clock time τ because τ is an external time and has nothing to do with the system under investigation and, furthermore, τ is a classical element introduced by Isaac Newton within the frame of classical mechanics.

But the term "time dependence" demands an internal quantum time for the understanding of the transition from A to B. Clearly, in all three cases [scheme (1.6)], human beings observe with respect to clock time τ, but the observation process itself is not responsible for the self-organizing process from A to B. Before we discuss this point in more detail, we would like to analyze the situation in traditional quantum theory.

Traditional quantum theory

The time dependence of the self-organizing process within traditional quantum theory is based on the classical clock time τ, although the process itself is quantum mechanical in character. In Schrödinger's equation for nonstationary processes, only the clock time τ appears. That is, the outcome B is described τ dependent, which we want to denote by B_τ [see scheme (1.6)].

For the transition from A to B in the time interval $\tau_A \leq \tau \leq \tau_B$, the time dependence of the internal process is relevant. Within the frame of traditional quantum theory, we have, therefore, to consider this transition as a "classical" process, that is, the nanosystem behaves classically within $\tau_A \leq \tau \leq \tau_B$. On the other hand, for $\tau > \tau_B$ the system is in a stationary state and the time is not of particular relevance here and we may consider the behavior of the nanosystem as "quantum mechanical" in character. This is of course only a rough classification, but it underlines the present tendencies.

We may state that the overall situation in traditional quantum theory is hardly acceptable. There is no operator for the time, and there are problems in connection with the uncertainty relations as well when we consider the time and the energy [4]. As we have already remarked previously, the coordinates x, y, z behave statistically in traditional quantum theory but not the time τ. We can, therefore, not expect that the state B_τ is close to the real state B of nature; it will even have no similarity with it. Thus, we have

$$B_\tau \neq B \tag{1.7}$$

Equation (1.7) is not due to the unrealistic choice of models, but it reflects the principal limits of traditional quantum theory.

The following has to be underlined once again: The clock time τ appears in Schrödinger's equation without changing its character; τ remains a classical time, which is external in character and comes exclusively into play through our clocks that we use in everyday life. This situation is reflected in the fact that an "operator" for the time is not importable within Schrödinger's theory (Section 1.2.1). The details are difficult to analyze because Schrödinger's equation cannot be deduced; there is no consistent physical frame for that. Schrödinger's equations are guessed and used within the frame of the container model.

Projection theory

There is no doubt that the physical conception behind the projection principle is more sophisticated than that of traditional quantum theory. We will analyze the details in this chapter. Here, we have a system-specific time t, which is able to describe the self-organizing processes in nanoscience, at least in principle. The variable t is quantum mechanical in character and, therefore, the transition from A to B becomes a fully quantum-theoretical description. This is an advantage against traditional quantum theory where the system-specific time t is not known (defined) but merely the clock time τ.

An operator for the time is defined in projection theory and this is expressed through eq. (1.3). Furthermore, the Hamiltonian is reformulated through an operator for the time coordinate $-i\hbar\partial/\partial E$ [see expression (1.4)]. The details concerning the analytical treatment of the projection effects will be discussed in Chapter 6.

All of that means that, not only the coordinates x, y, z behave statistically within the projection principle, but the time t as well. Here, we have the following situation: At clock time τ, the coordinates x, y, z and the system-specific time t behave statistically.

So, the self-organizing processes, which take place in the realm of nanoscience, are fully describable at the quantum level and the quantum aspect of time is given through the existence of the system-specific time t. The outcome B of nature [see relation (1.6)], denoted within the frame of the projection principle by B_t, fulfills

$$B_t = B \tag{1.8}$$

at least in principle.

Relations (1.7) and (1.8) express certain features: In principle, it is not possible to describe self-organizing processes within the frame of traditional quantum theory, and this is expressed by eq. (1.7). It is however possible to describe self-organizing processes through the projection principle, which is formulated by eq. (1.8).

1.3 The basic features of nanoscience

The quantum aspect of time could only be introduced on the basis of the projection principle. The container model is obviously not able to construct such a quantum time when we work on the basis of traditional quantum theory. All attempts failed (Section 1.2).

Both conceptions are based on what we have directly before us in everyday life, but they are assessed differently. Within the container model it is assumed that the objects before our eyes are physically real (material) in character, whereas within the projection principle it is assumed that the object before us are "geometrical" structures. There is no doubt that the projection principle reflects a realistic viewpoint because the features of space and time are treated with this conception more realistically. We will justify this statement in this section.

1.3.1 What do we experience?

What is the most direct experience of human beings? We touch an object of the world outside with our fingers and feel it directly. Therefore, we conclude that the object before our eyes is material in character and the container model is assumed to be valid. But the projection principle delivers exactly the same impression. Here, there is a direct contact between geometrical structures, the geometrical structures of the fingers and the object, and this situation takes place simultaneously in reality, where the physically real objects (material things) are located. There is a strict correlation between the geometrical structure situation, which is directly before our eyes, and that of the material situation in the world outside, which is not directly before us. In conclusion, the world outside is felt through the container model and the projection principle in the same way, but the mechanisms are assessed differently [7].

The container model is still the standard conception of traditional technologies (macro- and microtechnique). There is no doubt that the container model could also be applied with success in nanoscience. Also, stationary processes at the nanolevel seem to be successfully described by the container model. However, in the case of nonstationary processes, which are necessary for the description of self-organizing at the nanolevel, the container model has to be replaced by the projection principle (Section 1.2). There is no doubt that nanoscience stimulates to go new ways at the basic level of theoretical physics.

1.3.2 General remarks

A human being is faced in his life with various materials, functional natural objects, a diversity of food and so on. He is confronted with a lot of other important phenomena relevant for life, in particular, within the realm of medicine. All that is experienced likewise by all human beings.

These phenomena directly influence our actions and thoughts in daily life. All these things have their origin at the nanolevel. The nanolevel is the "ultimate level" at which the characteristic properties of our everyday life emerge. Furthermore, at the nanolevel individual existence has its origin. Our direct impressions have their origin at the nanolevel and there is no level above it. There is no scientific view, which could reveal deeper insights and there are no further properties in connection with condensed matter. In other words, we have reached the ultimate level.

The development of the scanning tunneling microscope was an important step in nanoscience and this instrument made it possible to manipulate matter at this basic level where the properties of matter have their origin. This situation makes it necessary to apply the physical laws in their basic form.

The properties of nanosystems are given through an ensemble consisting of a few hundred/thousands atoms, that is, the "ensemble" emerges the ultimate properties. The behavior of such an ensemble and the sort of material are in the foreground here and emerges basic features.

There is no doubt that the atoms themselves are defined, but possibly only approximately. The use of an ensemble of atoms (molecules), responsible for the primary features of nanosystems, is an intuitive recipe in analogy to what we have in front of us in everyday life, and these are objects in space like sun, moon and trees.

However, we have to be careful in connection with such analogies. In principle, the nanoensemble could be composed in another way, for example, by a "unified whole." There are, in fact, indications for that, first proposed by Ilya Prigogine. In the case of a unified whole, the manipulation of an atom would reflect a change of the entire unified whole. Ilya Prigogine remarked that even elementary particle could not be really basic, but an outcome of the second law of thermodynamics. The choosing of the second law of thermodynamics as the fundamental conception, instead of object structures, is an example. Also, the projection principle opens the possibility for such views.

1.4 The levels of physical science

Science penetrates deeper and deeper into physical reality, and with nanoscience, we reach the ultimate level where the properties of ordinary matter emerge, but also the basic characteristics of biological systems (human beings, animals, etc.). Nanoscience works directly at this ultimate level. There is no scientific level, which could reveal deeper insights and there are no further properties in connection with condensed matter.

1.4.1 On the selection of theoretical views

The deeper we go into physical reality, all the more we have to select the physical laws with care, that is, the theoretical views have to be chosen with increasing sophistication. At the highest, ultimate level (here the nanolevel is meant), the properties are particularly sensitive to small variations in the notions and the theoretical laws. Thus, nanoscience is more sensitive against variations of the basic laws of theoretical physics than in micro- and microtechnology.

At the ultimate level of nanoscience, the laws of "traditional" theoretical physics are in specific cases are not sufficient enough for an adequate understanding of the nanophenomena. In such cases, we have to extend the usual theoretical frame. The "nature of time" is an example for that (Section 1.2).

The ultimate level is definitely defined, but is it really expressed in final form through an ensemble of atoms or molecules? Is the ultimate level really described in the form of atomic entities when we try to understand nanophenomena? We have to be careful, just in connection with brain functions, which are obviously coupled to the states of the mind. Is the traditional view adequate at all? It is of course a good approximation, but is it really the final basis?

1.4.2 Qualitative new effects at the nanolevel

We experience this world, which we call "macroscopic," in a direct way. But how do we experience the world at the nanolevel? The following question arises: What imaginations can be transferred from the macroscopic world to the nanorealm? Here, we have to be careful. At the nanolevel, qualitative new effects appear that are unknown in the macroscopic world of everyday life for which our intuition is tailor-made.

In nanoscience, the behavior of ensembles consisting of atoms and molecules are in the center. We usually base the theoretical treatment in macro-, micro- and nanotechnology on our intuitive imaginations, which we experience in everyday life. These are assumed to be material objects in space when we base our statements on the container model. Once again, through the transition from the macroworld to the nanorealm, qualitatively, new phenomena emerge, and these new effects need a reliable theoretical basis. Can we simply transfer the laws of macro- and microphysics to nanoscience? No, we cannot, as we have demonstrated in Section 1.2 with respect to the problem of quantum time.

In all our usual considerations, we assume that the material world is embedded in space. This is, as we know, the container model, which can also be characterized as an object-in-space conception. This object-in-space conception is assumed in traditional physics to be valid at "all" levels of description (Figure 1.1), that is, when we go scientifically from the macroscopic world around us to the nanolevel. We outlined in Section 1.2 that the container model has to be replaced by the projection principle, even then, when the container model works well and this is in fact the case for some space–time regions.

Improvements in connection with the physical descriptions at the nanolevel have to be developed if it turns out that the laws are not sufficient and not precise enough. Since we are working in nanoscience at the ultimate level, where the properties of matter come into existence, such improved and new conceptions have to be developed in terms of basic arguments. Can nanoscience stimulate to go new ways at the basic level? Yes, it can and we already did it, and we demonstrated that with respect to the nature of time (Section 1.2).

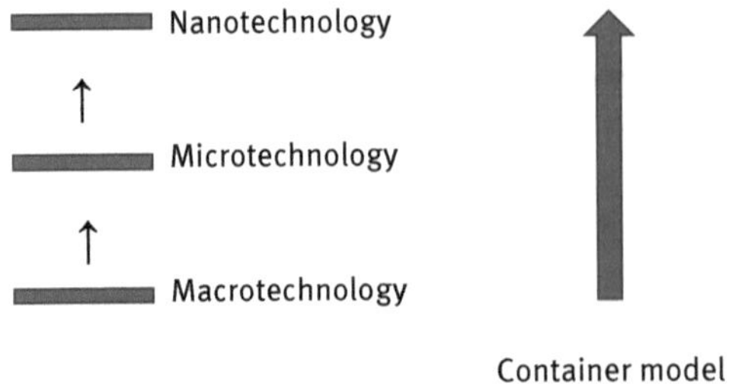

Figure 1.1: The theoretical treatments that are applied on the various technological levels (macro-, micro- and nanotechnology) are essentially influenced by our intuitive imaginations. These intuitive imaginations are mainly based on facts that we experience in everyday life: We assume throughout that there are material objects in space. That is, it is assumed that the container model is valid. However, through the transition from the macroworld to the nanorealm, qualitatively, new phenomena emerge, but it is supposed in traditional physics that the container model is not influenced through this transition, and it is used at all levels. Furthermore, the conception of an external time, shown by our clocks, is also applied without any modification in nanoscience. As we have outlined in Section 1.2, this is doubtful. Nanoscience stimulates to go new ways at the basic level of theoretical physics.

1.5 What is outside?

A human being or another biological creature observes the environment. The information of the world outside is given in the form of physically real (material) objects. This information is transferred by our sense organs into the mind (brain) of the human being. (The differences between mind and brain are not essential for our statement.) The mind processes this outside information and produces a "picture" of the world outside. This picture represents an inside version of the world outside.

In other words, we experience the world by our sense organs, that is, the human being interacts with the physically real world and information flows via our senses into the mind. The mind of the human being processes this outside facts and produces a picture of the physically real world outside. It is a picture in space and time. It is a space structure at clock time τ. This is the most direct information that a human being has about the world outside; on this direct information, the theoretical views are primarily based.

1.5.1 One-to-one correspondence

Within this view, we have a world outside and an inner image. The inner world represents a picture of the world. In this connection, it is essential to note that we are usually firmly convinced that the space structures at clock time τ of the inner picture are identical with that of reality outside. This is the fundamental assumption on which the container model is based. It is throughout assumed that there is a one-to-one correspondence between the inner picture and the world outside. For example, the well-known psychologist C. G. Jung wrote [8]: “When one thinks about what consciousness really is, one is deeply impressed of the wonderful fact that an event that takes place in the cosmos outside, produces an inner picture, that the event also takes place inside . . . .”

This statement represents the typical view and reflects the usual conception that human beings have about the whole scenario.

The statement by C. G. Jung suggests just what we have already pointed out: The image as well as the real world outside are embedded in space and time and are assumed to have exactly the same structures. These are space structures at clock time τ.

However, there is a serious objection. Why should events in nature occur twice, once outside of us and once again in the form of a picture? This would be against the “principles of evolution” and the “principle of usefulness,” respectively. Nature always works in accordance with these principles.

Nevertheless, C. G. Jung’s statement with respect to “outside world” and “inside world” is correct, but the assumption that an event takes place twice is obviously not realistic. We have to assume that there is no “one-to-one correspondence” between the reality outside and the inner picture of it. A one-to-one correspondence is far from being realistic.

Clearly, the world in front of us is the inside world and is not the direct material reality as the container model suggest. But, this is not the essential point. The critical point here is to assume that the structure of the inside world is identical with the structure of the world outside.

Through this one-to-one correspondence, we may in fact always assume (consciously or unconsciously) that we have directly the physically real world outside in front of us. We maintain this view in this book when we talk about the usual conception of the world, which is given by the container principle. However, we already remarked that this container model with its one-to-one correspondence is not realistic and we come to the so-called projection principle. For example, the quantum aspect of time could not be formulated within the frame of the container model and the projection principle had to be applied (Section 1.2). The deeper reason for that is given through the features of space and time, which will be discussed in the next section.

1.5.2 Impressions

We have to always keep in mind that the container model is an “impression” of human beings and is not a direct fact. It is an impression that human beings have in everyday life. That is, we only have the impression that all the material objects in the world outside are positioned in a space before us. Persons, cars, planes, the sun, moon and stars are pictures of reality in our mind; we have only the impression that all these things are located outside us.

This conclusion is supported in Ref. [9]:

> We have devices in the cerebral cortex which – comparable with a television screen – produce "pictures" in our awareness from the nerve-excitations coming from the retina. It is characteristic for the sight-process that our awareness does not register the picture of a candle on the retina inside the eye, but we have the impression that we are standing opposite the candlelight which is located in the space outside, not standing on its head but upright. We see "real objects" in front of us and around us. Within this act of perception, the eye, the optic nerves and the brain work together. To see without the brain is as impossible as to see without eyes.

This statement underlines that the impressions before us reflect the inner world. This in particular means that the spontaneous impressions, which appear unconsciously within so-called assumptionless observations in everyday life in front us, reflect "observer-dependent" facts and are primarily not the absolute truth. Here, we have to take into account that space and time are entities with specific features.

Within the container model, space and time also belong to the world outside. Is it possible at all that space and time are entities of the reality outside in which the physically real (material) objects are embedded? The answer to this question is essential when we assess the world outside from the physical point of view. In the next section (Section 1.6), we will analyze the situation.

1.5.3 Goggles with inverting glasses: an important experiment

From the discussion in Section 1.5.2, we have to conclude that the impressions in front of us are observer dependent. The observations within the assumptionless observations of everyday life come into existence through our mind. We need information from the world outside, but the direct impression in front of us is essentially a product of mind and is not directly the material world outside. This impression is identical with the space structure at clock time.

In other words, all we have before our eyes is essentially influenced by the mind; it is an invention of the human observer's mind. It reflects "merely" a picture of reality but it is not the material reality itself. The goggle experiment with inverting glasses demonstrates this fact, which has been described by Thomas Kuhn as follows [10]:

> An experimental subject who puts on goggles with inverting lenses initially sees the entire world upside down. At the start his perceptual apparatus functions as it had been trained to function in the absence of goggles, and the result is extreme disorientation, an acute personal crisis. But after the subject has begun to learn to deal with the new world, the entire visual field flips over, usually after an inverting period in which vision is simply confused. Thereafter, objects are again seen as they had been before the goggles were put on. The assimilation of a

> previously anomalous visual field has reacted upon and changed the field itself. Literally as well metaphorically, the man accustomed the inverting lenses has undergone a revolutionary transformation of vision.

Due to the goggles with inverting glasses, a visual transformation took place spontaneously, that is, without the conscious action of the concerned human being. It is an experiment within the frame of assumptionless observations of everyday life. In conclusion, the spontaneous impressions, which human beings experience, are influenced by the mind and, therefore, they are not observer independent. It particularly demonstrates that the world before our eyes is a picture of reality. Let us still analyze the various steps in connection with the goggle experiment.

Goggle experiment: the various steps

Let us consider a human being, denoted here by the letter A, who puts on at time τ_A goggles with inverting glasses. For times

$$\tau < \tau_A \tag{1.9}$$

observer A has a certain space structure P_τ before his eyes. Does this space structure reflect the world outside (reality) or the inner world (picture of reality)? This question is relatively easy to answer.

After human being A has put on the goggles, he sees the entire world within a certain time interval $\Delta\tau$ upside down, that is, instead of the state P_τ. The space structure is now in the state $P_{\tau_A+\Delta\tau}$, which is different from P_τ: $P_{\tau_A+\Delta\tau} \neq P_\tau$. In other words, at $\tau = \tau_A$ a transition from P_τ to $P_{\tau_A+\Delta\tau}$ takes place:

$$P_\tau \rightarrow P_{\tau_A+\Delta\tau} \tag{1.10}$$

The transition is effectuated by the goggles. The state $P_{\tau_A+\Delta\tau}$ remains conserved during the time interval $\Delta\tau$.

This is the case for times τ with

$$\tau_A < \tau < \tau_A + \Delta\tau \tag{1.11}$$

At time

$$\tau = \tau_A + \Delta\tau \tag{1.12}$$

the entire field flips over and we again get the space structure P_τ, that is, we have the reverse transition

$$P_{\tau_A+\Delta\tau} \rightarrow P_\tau \tag{1.13}$$

Thereafter, objects are again seen as they have been before, without goggles. There is no intervention through human being A at $\tau = \tau_A + \Delta\tau$. That is, the goggles are still present in unaltered position. This state P_τ remains for all times τ larger than $\tau_A + \Delta\tau$:

$$\tau > \tau_A + \Delta\tau \tag{1.14}$$

Again, the goggles are still present in the case of eq. (1.14); the flipping over of the entire visual field takes place unconsciously.

Transition (1.13) is due to an inner process and is carried into execution through the mind–brain complex of human being A. Since the mind–brain complex belongs to the inner world of A, it cannot influence the world outside, that is, the mind–brain complex cannot remove the goggles at time $\tau = \tau_A + \Delta\tau$ [eq. (1.12)], and they are still present for $\tau > \tau_A + \Delta\tau$ [eq. (1.14)]. That is, there are no changes in the world outside.

Then, the space structure P_τ before the eyes of human being A for $\tau > \tau_A + \Delta\tau$ can only be a picture of reality (the inner world). The mind–brain complex operates on the basis of the space structure $P_{\tau_A+\Delta\tau}$ at time $\tau = \tau_A + \Delta\tau$ [eq. (1.12)] and which is present for $\tau_A < \tau < \tau_A + \Delta\tau$ [eq. (1.11)]. Thus, the space structure $P_{\tau_A+\Delta\tau}$ is nothing else than a picture of reality.

In other words, the space structures P_τ and $P_{\tau_A+\Delta\tau}$ are states of the mind. They are not the material world; the physically real (material) world exists but not before the eyes of human being A. How are the entities "reality, matter, mind, brain and the picture of reality" ordered relative to each other? This situation is discussed in Chapters 4 and 5.

Conclusion

Without doubt, the experiment with inverting glasses is basic with respect to the nature of space and the role of the observer. In particular, it demonstrates convincingly that the world in front of us is actually inside the head.

But what is outside? How is the world outside (reality) structured? The answers on these questions are tightly linked with the nature of space and time. We will analyze the situation in Section 1.6. But we may already quote here that the real physical process outside remains of course unchanged, only the kind of perception is concerned through the goggles with inverting glasses. The impression, which we feel to be outside, is actually an image within the mind of the observer. This is convincingly confirmed by the goggle experiment, but this experiment demonstrates more than that. This point is discussed in the next section.

1.5.4 The reality-picture unit

From all statements given before, we have to conclude that there is a "reality" (world outside) and a "picture of reality" (inner world). The world before our eyes, which we experience spontaneously in everyday life, can only be the picture of realty and not the reality itself. The direct observation exhibits a space structure at time τ consisting of geometrical positions and not of material entities.

The mind–brain complex

It is demonstrated by the experiment with the goggles that we do not change the world outside but the inside world. The picture is created by the mind–brain complex on the basis of the information from the world outside. The mind–brain complex is of particular importance. In Chapters 4 and 5, we will describe a possible conception for the functionality of the mind–brain complex. In the case of a nanotechnological impact on the mind–brain complex, we have to know what we do and this requires at least a minimum knowledge about the functionality of such a complex.

In the case of the goggle experiment with the inverting glasses, we do not change the world outside but the world inside is changed. The outcome of the experiment can only be interpreted in this sense. The mind–brain complex of the observer is in action and manipulates the picture of reality that appears spontaneously before the observer's eyes.

Our actions in everyday life are based on this picture of reality. Manipulations with respect to the picture of reality can be dangerous and can lead to misjudgments. This is independent of the true structure in the world outside, that is, if there is a one-to-one correspondence between reality and its picture or not.

The invention of nature

The observations and activities of human beings are based on the reality-picture unit. But this conception is not an invention developed by the human being but it is an invention of nature itself. There is a definite order behind the reality-picture unit, that is, nature has developed a conclusive system, after that a human being can reliably do the observations and activities. The goggles with inverting glasses are ignored by the mind–brain complex after a brief time interval, and this is because they disturb this reality-picture unit, which a human being needs for his daily actions in the original form developed by nature in the course of evolution. The picture is adapted to reality. Both, the reality and its picture, are compatible to each other. This adapted state is disturbed by the effect of the goggles with inverting glasses. But the mind–brain complex developed a counteraction in order to compensate the inverting-glasses effect.

Such counteractions will possibly not take effect if there are far-reaching changes in connection with the mind–brain complex itself. Nanotechnological changes with respect to the mind–brain complex cannot be possibly compensated through the mind–brain complex itself.

The picture is changed but not in reality (world outside). The picture is no longer compatible with reality. In other words, the information in the picture is no longer in accordance with the structure of reality outside. This can lead to a catastrophe. That is, we have to be careful when we change the mind–brain complex nanotechnologically.

1.6 The basics of space and time

In Section 1.5, we talked about the relationship between reality (world outside) and the picture of reality (inner world). Is there a one-to-one correspondence between the structures of reality and those of its picture? In order to answer this question, we have to say something about the basic nature of space and time.

1.6.1 The experiences in everyday life

Can space and can time be considered as physically real entities outside a human being? Can the physically real world be located in space and time? In Section 1.2, we discussed the quantum aspect of time, but we did not ask about the most basic features of time.

The time is defined in basic form through our experiences in everyday life. There is no other possibility to search for the items of time. The problems with respect to time are strictly accompanied with the problems of space. Space and time belong somehow together. This connection follows directly from the similarities they have with respect to their common features.

Space is usually characterized by the coordinates x, y, z and the time by the letter τ. In what form do they appear in nature? Let us first consider the time τ, which we measure with our usual clocks. Clearly, a clock does not produce τ. But what is the mechanism behind τ? Here, the metaphor with the San Francisco cable railway is a possibility to understand τ.

The external time, displayed by our clocks, reflects rather a naïve point of view. This is clearly expressed within a statement by Julius Fraser. He gave the following comment:

> The time is like the cable of the cable railway of San Francisco. The cable is driven by a far and invisible machine, is however hidden. We know that it moves, because the carriages are connected and are carried with it. Completely similarly we usually see the time in everyday life as a universal, cosmic movement of the present, of the now; they are propelled by natural or divine forces and matter, life, humans and society are connected with them and are propelled and along-moved for a while [11].

The San Francisco cable railway is of course a metaphor, but it exactly reflects the situation on how we experience the phenomenon "time." Our time feeling corresponds to such kind of process and it is perceived as external in character. It is, however, a singular, metaphysical process of which the entire world has equally embraced. From the scientific point of view, such a metaphysical conception is not acceptable at all. It can simply not be used within serious physical descriptions, even when it meets the situation well, but it is too naive and not realistic.

The cable railway metaphor reflects an “external” time, and this behavior is exactly expressed through our clocks, that is, the cable railway defines qualitatively what we called “clock time.” However, the time τ cannot be grasped in this way.

1.6.2 Can space and time exist in isolated form?

What is the space made of and what is the time made of? Define space and time physically real entities? In traditional, classical physics, space and time are considered as physically real quantity, which effectuate the phenomenon of inertia. It should be a “physically real something,” but different from matter. What can we say about this situation?

Are space and time accessible to empirical tests?

The basic characterization of space is given by points. Our three-dimensional space consists of three real numbers x, y, z that we call coordinates. On the other hand, time is also characterized by points, and each time point is given by one number, which we have marked previously by the Greek letter τ which is the clock time.

If space and time would be physically real quantities, these basic quantities, that is, x, y, z and τ have to be accessible to empirical tests. This is definitely not possible. The following facts show that and demonstrate that there is no possibility to measure the basic points x, y, z and τ of space and time.

The single elements x, y, z and τ of space and time cannot be experienced with the five senses. We definitely cannot see, hear, smell or taste space and time, that is, the basic elements of space and time, characterized by x, y, z and τ, are not accessible to experimental tests. This is quite independent of the character of space (space–time). In principle, they could be absolute but nonabsolute as well. Also, measuring instruments are not definable for the experimental determination of the space–time points x, y, z and τ, and such kind of devices are even not thinkable.

One might object that we experience the phenomenon “space” incessantly in everyday life, and this is because at each clock time τ we have space structures, that is,, x, y, zstructures, before our eyes. However, we do not experience the basic elements x, y, z at τ, but we observe “objects” in such cases and this has nothing to do with the observation of the single elements x, y, z and τ.

What do we actually observe in connection with the space–time elements x, y, z and τ? Here, the following facts are relevant: We never observe single elements x, y, z and τ, but we are only able to state the following:

We observe distances in connection with material bodies (masses),

and

we observe time intervals in connection with physically real processes.

Only what is observable can be considered as physically real. Since the basic elements of space and time can principally not be observed (measured), the elements x, y, z and τ of space and time may not be considered as physically real entities like matter.

This in particular means that space and time should never be the source for physically real effects, for example, inertia. "Nonobservable" here means "nonexistence" in the form of a physically real entity. The scientific standpoint requires such a viewpoint.

Space and time do not reflect physically real quantities, that is, the basic elements x, y, z and τ do not reflect physically real quantities. Thus, they may not be considered as the source for physically real effects.

The elements x, y, z and τ cannot be measured. An empty space–time cannot exist as physically real unit. From the scientific point of view, only those entities that are observable can lead to physically real effects. Nobody can cut out a piece of space or a piece of time and put them on the table. Space and time in isolated form, that is, without material objects, do not exist. Nobody can invent an experimental

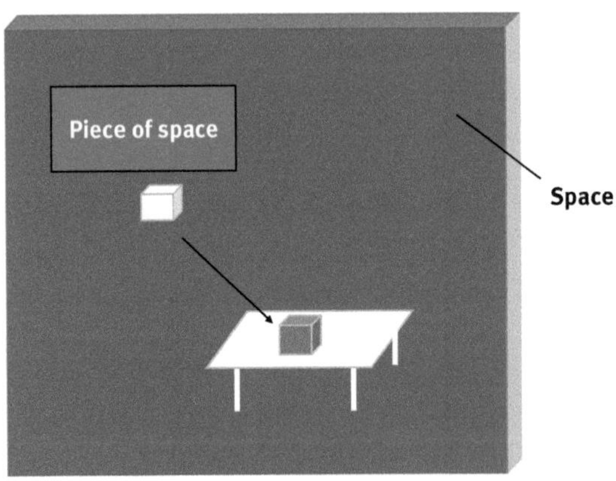

Figure 1.2: What is the "space" made of and what is the "time" made of? Can we cut out a piece of space and can we put this piece of space on the table or similarly, can we treat a piece of time in this way? Nobody is able to put a piece of space or a piece of time on the table, and nobody can invent a procedure for that. The question is simply ridiculous. Pieces of space and pieces of time cannot be isolated and cannot be treated as independent entities.

method for that. In fact, to discuss such a possibility reflects a ridiculous situation, and this is because no one can imagine that (see also the illustration in Figure 1.2). In conclusion, the basic elements of space and time, that is, x, y, z and τ, cannot be identified with a real something in analogy to matter.

Mach's principle

Already Ernst Mach argued in this way. Also for him, the basic elements of space and time, characterized by x, y, z and τ, cannot be considered as a "physically real something" to be capable of physically real effects. The effect of inertia is such a space phenomenon within Newton's theory. Ernst Mach rejected Newton's absolute space radically.

In conclusion, also after Mach, the space (space–time) can never be the source for physically real (material) effects, that is, it can never act on material objects giving them certain physically real properties. The effect of inertia is an example that Mach's principle is entirely based on the previously discussed fact. After that, we can never observe space and time because its elements (the coordinates x, y, z and the time τ) are in principle not observable or measurable. We can only say something about "distances" in connection with masses, and "time intervals" in connection with physical processes.

Once again, space and time can never be the origin of physically real effects. The space–time continuum is not a physically real entity like ordinary matter. However, Mach's principle goes not far enough. This principle does not explicitly forbid that matter can be embedded within the space–time block.

We will use the term "Mach's principle" in connection with the following fact: Space and time can never be the source of physically real processes and, furthermore, physically real systems can never be embedded into a space (space–time). We cannot embed physically real (material) bodies into a space or a space–time, which itself is not physically real.

As far as we know, Mach never forbade explicitly the possibility to embed matter into space and time, although this peculiarity should follow from Mach's principle. Let us repeat: How can a physically real something (matter) be embedded within a system that is not a physically real entity (space, space–time)? Such a construction is perhaps pragmatic but it has no scientific background.

Such a matter-in-space conception may be suggested when we base our knowledge on what we have directly in front of us. But, as we have recognized, these everyday life impressions do not reflect material reality itself, but "only" a picture of it. The goggle experiment, discussed in Section 1.5.3, demonstrates that. A physically real something (matter) cannot be embedded in space and time. Instead of the container principle, we, therefore, discuss in this monograph, together with the container principle, a more realistic physical principle, the "projection principle." This principle fulfills Mach's principle.

1.6.3 The mind creates space and time!

Space and time are not physical real elements and, therefore, they cannot belong to the reality outside because this reality contains exclusively physically real entities when we judge the situation from the observer's point of view. It makes no sense to assume that space and time are entities of reality outside since they are not observable.

The observer is only able to observe with his five senses and the measuring instruments physically real entities. Thus, space and time can only belong to the mind of the observer. They are created by the mind. Then, we come inevitably to the projection principle: Space and time do not belong to the physical reality. On the other hand, human beings are caught in space and time and do the observation in space and time. This means that the physical reality is not directly observable, but it exists. We perceive geometrical structures in space and time, which are projections of facts of physical reality outside. In this way, we obtain a picture of physical reality. However, the physical reality itself remains hidden in its basic form. The situation is summarized in Figure 1.3. The figure reflects the reality-observation principle.

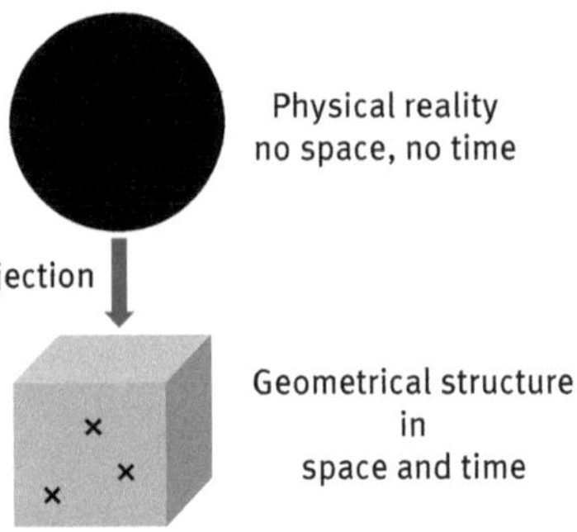

Figure 1.3: The physical reality does not contain space and time. On the other hand, human beings do the observation in space and time. Thus, the physical reality is not directly observable. We perceive geometrical structures in space and time, which are projections of facts of physical reality outside. In this way we obtain a picture of realty. The physical reality remains hidden in its basic form. The figure reflects the reality-observation principle.

The information about the physically real objects in reality outside is projected onto space and time. The physically real information is processed by the mind and with the help of space and time, created by the mind, it is represented as picture (picture of reality). The notion "picture of reality" has often been used in the text before.

Due to the fact that space and time are not physically real, the container model has to be replaced by the projection principle:

$$\begin{gathered}\text{Container model}\\ \Downarrow\\ \text{Projection principle}\end{gathered} \tag{1.15}$$

Already, the philosopher Immanuel Kant came to the conclusion that space and time are elements of the mind. Kant's contribution is important and we will give more details in Section 1.9.

1.6.4 The unconscious perception of space and time

The features, which we have discussed in the sections before, are irrevocable facts that have to be considered when we treat and analyze space and time scientifically. But what about the space feeling and the time feeling of human beings or other biological creatures? How does a human being experience "space" and how does he experience "time" in everyday life, that is, unconsciously?

Both effects appear spontaneously without thinking and are unconscious phenomena and take place exclusively at the level of everyday life. Situations of this kind define our elementary feelings with respect to space and time. There is no other level which could mediate such elementary feelings.

Space feeling

What do we understand under space feeling? What kind of observation creates space feeling? The structures before us at time τ reflect objects in geometrical form, which are embedded in space having the elements x, y, z and τ. What are the facts on which the elements x, y, z and τ are based?

We observe, primarily, objects and their extensions. For the perception of two objects, a certain "extension" is introduced by nature. Since space and time are elements of the mind (Section 1.6.3), the notion "nature" is identical with "mind." Nature invented extensions in order to give human beings the possibility to order the world before his eyes. Here, extension has to be considered as a basic notion and we should not try to analyze it further, that is, we should not try to explain the notion "extension" by more basic notions, more basic than the notion extension itself [5]. The effect of extension is just that what conveys "space feeling."

The notion extension reflects a qualitative effect. We meet it at first in this basic form just in connection with our assumptionless observations in everyday life. Again, the effect of extension appears spontaneously without thinking and it has to be classified as a qualitative phenomenon. This assumptionless space impression is nothing else than an inner image and is positioned within the mind of the observer.

The image before us contains, in general, a lot of objects and, therefore, we have a lot of extensions. The mind organizes this "ensemble of extensions" as one phenomenon which we call "space." But we may also express this effect as follows: There are various distances for various objects that are located in the same space with "one" characteristic global space extension.

That is all what we can say about space within our assumptionless everyday impressions. Once again, this is a qualitative impression. But we need more than that when we want to analyze the physical processes in the world theoretically.

From the picture before us we get the coordinates at a certain clock time when put a grid over the image [5]. The grid is a fictitious net. The coordinates x, y, z at time τ are the elements of the "fictitious net," which the observer intellectually puts over the image in front of him [5].

Time feeling

Concerning the phenomenon "time," let us repeat the following as essential point: We never observe single time values τ. We define them only! We define a time value τ as a clock hand but, as we have stated in Section 1.6.1, the value τ does not exist without clock. Our clocks are constructions that are based on that what is called "time feeling."

How comes this time feeling into existence? It is a phenomenon through "structure changes" that appear unconsciously in the space before us. The structure changes have in general a certain tendency to proceed in a certain direction. Such situations (structure changes) qualitatively create a new effect, that we experience as "time feeling." In this way, the phenomenon "time" comes into play. How is the time feeling connected to the time values τ, which we called "clock time" in the section before? This is a simple procedure: We formally relate to each object structure in space a time value τ (position of a clock hand).

1.7 The system-specific time

There are distances in connection with space structures and there are time intervals with respect to space structure changes. These space structures and their changes are only definable with a "real something" and this something is given by the physically real objects and systems, respectively, which are located in the world outside.

1.7.1 The metaphysical aspect

Human beings are caught in space and time. Material objects (systems) do not exist without space and time (x, y, z and τ) and, on the other hand, space and time do not exist without physically real objects or systems. The mind organizes the world in this way, if we work within the frame of the projection principle. The details will be discussed in Chapters 4 and 5.

The transition from an external time to an internal time notion

The cable railway metaphor reflects an "external" time, and this behavior is exactly expressed through our clocks. Also, here, the time must be connected to a real something, which is given through the San Francisco cable railway. In other words, the real something is nothing else than a metaphor for an external time machine. As we already stated in Section 1.6.1, such kind of time machine is, from the scientific point of view, not acceptable. Such a time machine has never been observed and we do not know even where its position in space is. Thus, the external time can only be judged as a metaphysical notion. We may define an external time by a clock, but this clock is not a counterpart of an actually existing external time machine.

If we leave this conception of a metaphysical, "external time," we come inevitably to an "internal" time, defined by the physically real systems (objects) themselves, that is, the time becomes then a system-specific entity. This is an important step: To go from an external to an internal time notion. Then, the metaphysical aspect of time is eliminated, as we will recognize in the following subsections.

Within such a conception, time is connected to a physically real system. This is a fact and is dictated by the experience. "One" time (one time structure), for example the clock time τ, means "one" system, that is, no other system defines what we call as time. For example, the San Francisco cable railway, which is assumed to produce τ, and this type of time is used for the description of "all" the other systems in the universe.

A singular system

Such a lone system is singular in character and can only be realized by a machine, which exists independently of the other physically real systems in the universe. In fact, that is an unphysical and nonscientific situation. As we remarked previously, a singular, metaphysical time machine has never been observed. It is therefore not existing from the scientific point of view.

Since the time is a really existing phenomenon, we come to the following conclusion: Each physically real system must have a time structure. This is a requirement and can be fulfilled with the frame of the projection principle. The time within such a conception is individual in character and it does not reflect singular features, that is, it is system specific without the existence of a singular time machine.

1.7.2 The time spectrum

In Section 1.6.4, we talked about space feeling and also about time feeling of human beings. Let us briefly repeat: The structures before us at time τ reflect objects in geometrical form, which are embedded in space, having the elements x, y, z.

Primarily, we observe objects and their extensions. The effect of extension is just what conveys "space feeling."

On the other hand, time feeling is a phenomenon through "structure changes" that appear unconsciously in the space before us. These space structure changes have, in general, a certain tendency to proceed in a certain direction. Such situations (space structure changes) create qualitatively a new effect, which we experience as "time feeling."

The features of the internal time

We work within the frame of an internal time view. That is, each physically real (material) system or object defines its own time structure that we called system-specific time. The cause for the time feeling is space structure changes. This is our basic conception for the phenomenon time.

The system under investigation changes its space structure in the course of clock time τ. At time τ, the system has a certain space structure $\{x, y, z\}_{\text{system}}$. The clock time τ, which is the reference time, moves monotonically from small to large values, that is, it moves strictly in one direction:

$$\tau_1, \tau_2, \ldots, \tau_i, \ldots \tag{1.16}$$

with

$$\tau_1 < \tau_2 < \cdots < \tau_i < \cdots \tag{1.17}$$

The space structure of the system varies and is, in general, different for different values τ_i. The system can be a single object or an ensemble of objects. Therefore, we have the following allocation:

$$\begin{aligned} \tau_1 &\rightarrow \{x, y, z\}^1_{\text{system}} \\ \tau_2 &\rightarrow \{x, y, z\}^2_{\text{system}} \\ &\vdots \\ \tau_i &\rightarrow \{x, y, z\}^i_{\text{system}} \\ &\vdots \end{aligned} \tag{1.18}$$

The variation of space structures means that there exists a time feeling existent, or we may say that, through the space structure changes, a time effect comes into existence. It is a time effect with respect to the system itself, that is, it is a system-specific time effect. In the case of a constant time interval $\Delta\tau = \tau_m - \tau_n$, the time effect varies with varying space structure. This is of course only a qualitative statement. How is this phenomenon expressed in a more quantitative manner?

Let us choose for the system-specific time the letter t. In other words, the system-specific time effect is characterized by t. The variable t is responsible for the time behavior of the system under investigation. We have space structure changes and to each space structure, we have to allocate a certain time characteristic, for example, the letter t_i. Then, we get an ensemble of system-specific times:

$$t_1, t_2, \ldots, t_i, \ldots \tag{1.19}$$

To each of these time values belongs a certain space structure, that is, we have

$$\begin{gathered}
\{x, y, z\}^1_{\text{system}} \rightarrow t_1 \\
\{x, y, z\}^2_{\text{system}} \rightarrow t_2 \\
\vdots \\
\{x, y, z\}^i_{\text{system}} \rightarrow t_i \\
\vdots
\end{gathered} \tag{1.20}$$

Due to eqs. (1.18) and (1.20), each space structure is correlated to two time values. For example, space structure $\{x, y, z\}^i_{\text{system}}$ in eqs. (1.18) and (1.20) is characterized by τ_i and t_i.

1.7.3 The standard system

The reference system is based on a certain time feeling. The time feeling may be different for different individuals and often develops in connection with systematic and constant changes in the environment of the individuals. For example, the time feeling can be based on the movement of the sun. However, the introduction of a reference time structure only makes sense if it is equally used by all individuals. It is therefore a standard system.

The timescale of such a standard can be chosen arbitrarily. The usual standard reference system is defined through our clocks. The clock hands of all clocks show the time τ. That is, there exists one τ scale and this has been chosen arbitrarily.

1.7.4 Timescales

The system-specific time t is different for different systems (objects). But here, the timescales of all systems should be identical and, furthermore, they have also to be identical with the reference τ scale. Otherwise, we would not be able to relate the various time structures to each other and there would be no relation to the reference time τ. That is, the τ scales of the physically real systems can be chosen arbitrarily as well, but have to be identical with the τ scale.

However, there are other features of the various system-specific times that cannot be chosen arbitrarily. The system-specific time t is a "quantum time" (Chapter 6) and it is not only characterized by a τ scale, but also through the τ ranges as well, which are different for different systems. The quantum aspect of the system-specific time t is expressed by probability densities with respect to the variable t.

The analysis given in Section 1.7.2 reflects a classical treatment and the system-specific time t is given in the classical approximation. This is particularly expressed through the allocations in eq. (1.20). But we do not want to deepen this point in this chapter and we refer to Chapter 6.

In conclusion, the system-specific timescales (t scales) and the reference timescale (τ scale) have to be chosen identical, that is, we have

$$t\ scale = \tau\ scale \tag{1.21}$$

Each space structure $\{x, y, z\}^i_{\text{system}}$ in eqs. (1.18) and (1.20) is characterized by the two time values τ_i and t_i. Furthermore, since both timescales are identical [scheme (1.21)], with eqs. (1.18) and (1.20), we get the following allocations:

$$\begin{aligned} t_1 &= \tau_1 \\ t_2 &= \tau_2 \\ &\vdots \\ t_i &= \tau_i \\ &\vdots \end{aligned} \tag{1.22}$$

The system under investigation moves through an ensemble of time values

$$t_1, t_2, \cdots, t_i, \cdots \tag{1.23}$$

and we want to express this ensemble (spectrum) by the symbol $\{t\}_{\text{system}}$:

$$\{t\}_{\text{system}} = t_1, t_2, \ldots, t_i, \ldots \tag{1.24}$$

With eqs. (1.18), (1.20) and (1.22), we have

$$\begin{aligned} \tau_1 = t_1 &\rightarrow \{x, y, z\}^1_{\text{system}} \\ \tau_2 = t_2 &\rightarrow \{x, y, z\}^2_{\text{system}} \\ &\vdots \\ \tau_i = t_i &\rightarrow \{x, y, z\}^i_{\text{system}} \\ &\vdots \end{aligned} \tag{1.25}$$

Relation (1.25) has to be seen as a selection process; it defines a selection process. A space structure is observed through selection at a certain clock time τ. This is in fact the situation in everyday life and reflects the most basic information a human being can have.

1.7.5 Observation through selection

Constant time intervals

Let us suppose that a human being observes two systems (objects), denoted by m and k. He registers at clock time τ_1 the space structures of the two systems and, corresponding to eq. (1.18), we have

$$\tau_1 \rightarrow \{x,y,z\}^1_{\text{system } m}, \{x,y,z\}^1_{\text{system } k} \tag{1.26}$$

The human observer registers, on the other hand, at clock time τ_2 the changed space structures of the two systems m and k, which is expressed by

$$\tau_2 \rightarrow \{x,y,z\}^2_{\text{system } m}, \{x,y,z\}^2_{\text{system } k} \tag{1.27}$$

Also, here, we have space structure changes and to each space structure, we have to allocate a certain time characteristic as in the case of eq. (1.20). Since we have at τ_1 and τ_2 four space structure [see eqs. (1.26) and (1.27)], we must have four system-specific time values: $t_{1m}, t_{1k}, t_{2m}, t_{2k}$. These values are connected to the space structures given in eqs. (1.26) and (1.27), and we have

$$\begin{aligned} &\{x,y,z\}^1_{\text{system } m} \rightarrow t_{1m} \\ &\{x,y,z\}^1_{\text{system } k} \rightarrow t_{1k} \\ &\{x,y,z\}^2_{\text{system } m} \rightarrow t_{2m} \\ &\{x,y,z\}^2_{\text{system } k} \rightarrow t_{2k} \end{aligned} \tag{1.28}$$

When we go from τ_1 to τ_2 we have space structure changes with respect to system m and with respect to system k. With eqs. (1.26) and (1.27), we immediately get for the space structure changes $\Delta\{x,y,z\}_{\text{system } m}$ and $\Delta\{x,y,z\}_{\text{system } k}$, the relations

$$\begin{gathered} \Delta\{x,y,z\}_{\text{system } m} = \{x,y,z\}^2_{\text{system } m} - \{x,y,z\}^1_{\text{system } m} \\ \downarrow \\ \Delta t_m = t_{2m} - t_{1m}, \Delta\tau_{12} = \tau_2 - \tau_1 \end{gathered} \tag{1.29}$$

and

$$\begin{gathered} \Delta\{x,y,z\}_{\text{system } k} = \{x,y,z\}^2_{\text{system } k} - \{x,y,z\}^1_{\text{system } k} \\ \downarrow \\ \Delta t_k = t_{2k} - t_{1k}, \Delta\tau_{12} = \tau_2 - \tau_1 \end{gathered} \tag{1.30}$$

The human being compares the space structure changes $\Delta\{x,y,z\}_{\text{system } k}$ and $\Delta\{x,y,z\}_{\text{system } m}$ with each other, and these space structure changes correspond to time effects $\Delta t_m = t_{2m} - t_{1m}$ and $\Delta t_k = t_{2k} - t_{1k}$.

If we have

$$\Delta\{x,y,z\}_{\text{system } m} > \Delta\{x,y,z\}_{\text{system } k} \tag{1.31}$$

the space structure changes of system m are larger than the space structure changes of system k. Then, the processes with respect to system m last longer than the processes in connection with system k. This is, in fact, what we experience as time feeling.

In the case of eq. (1.31), we work with constant time intervals. From eqs. (1.29) and (1.30), we obtain in fact the constant time intervals

$$\Delta t_m = \Delta t_k = \Delta\tau_{12} \tag{1.32}$$

All that is judged relative to human observer's own time feeling, which is expressed through the clock by $\Delta\tau_{12} = \tau_2 - \tau_1$.

Constant space intervals

Let us consider again the two systems m and k, and let us suppose that the space structure of both systems change monotonically and continuously. A human being observes m and k at clock time τ_1. At clock time τ_2 he registers system m and at clock time τ_3 he registers system k. Then, we have the following situation:

$$\begin{aligned} \tau_1 &\rightarrow \{x,y,z\}^1_{\text{system } m}, \{x,y,z\}^1_{\text{system } k} \\ \tau_2 &\rightarrow \{x,y,z\}^2_{\text{system } m} \\ \tau_3 &\rightarrow \{x,y,z\}^3_{\text{system } k} \end{aligned} \tag{1.33}$$

The observations are done under the condition that the space structure changes are equal for both systems:

$$\Delta\{x,y,z\}_{\text{system } m} = \Delta\{x,y,z\}_{\text{system } k} \tag{1.34}$$

In analogy to eqs. (1.29) and (1.30), we have with eq. (1.34)

$$\begin{gathered} \Delta\{x,y,z\}_{\text{system } m} = \{x,y,z\}^2_{\text{system } m} - \{x,y,z\}^1_{\text{system } m} = \Delta\{x,y,z\}_{\text{system } k} \\ \downarrow \\ \Delta t_m = t_{2m} - t_{1m}, \Delta\tau_{12} = \tau_2 - \tau_1 \end{gathered} \tag{1.35}$$

and

$$\begin{gathered} \Delta\{x,y,z\}_{\text{system } k} = \{x,y,z\}^3_{\text{system } k} - \{x,y,z\}^1_{\text{system } k} = \Delta\{x,y,z\}_{\text{system } m} \\ \downarrow \\ \Delta t_k = t_{3k} - t_{1k}, \Delta\tau_{13} = \tau_3 - \tau_1 \end{gathered} \tag{1.36}$$

The systems m and k are different from each other and condition (1.34) is fulfilled. In particular, we have

$$\begin{aligned} \Delta t_k &\neq \Delta t_m \\ \Delta\tau_{13} &\neq \Delta\tau_{12} \end{aligned} \tag{1.37}$$

and

$$\begin{aligned} \Delta t_m &= \Delta\tau_{12} \\ \Delta t_k &= \Delta\tau_{13} \end{aligned} \tag{1.38}$$

If we have

$$\Delta t_k > \Delta t_m \tag{1.39}$$

the structural changes of system m in the time interval Δt_m are identical with the structural changes of system k in the time interval Δt_k. Then, the processes inside system m are quicker than the processes inside system k.

In the case of eq. (1.39), we work with constant space structures. From eqs. (1.35) and (1.36), we obtain the constant space structure intervals, expressed by eq. (1.34).

All that is judged relative to the human observer's own time feeling, which is expressed through his clock. The reference time intervals $\Delta\tau_{12}$ and $\Delta\tau_{13}$ refer to system m and system k. We have $\Delta\tau_{13} > \Delta\tau_{12}$ in the case of $\Delta t_k > \Delta t_m$, that is, the time interval registered by the clock is larger in the case of system k than the registered time interval in the case of system m. Again, both systems move through exactly the same space structures but in different clock-time intervals.

For human beings, the basic information about the world outside is given by the unconscious experiences in everyday life. This information in the form of a space structure $\{x, y, z\}_{\text{world}}$ is at clock time τ directly before the eyes of human beings.

This procedure does not reflect the system-specific time structure $\{t\}_{\text{world}}$, which is not directly observed at clock time τ, but the space structure $\{x, y, z\}_{\text{world}}$. Nevertheless, we can make statements about the system-specific time because there is a coupling between τ and t. This coupling is expressed by eq. (1.25). However, we already mentioned before that the system-specific time t reflects basically a quantum variable, which does not allow a formulation like eq. (1.25).

The observation of N systems

In summary, the introduction of the system-specific time through the projection principle makes it possible to eliminate metaphysical time machines like the San Francisco cable railway that is used as metaphor here. Within the frame of the projection principle, we are able to compare the time structures of system m with that of system k and human observer is also able to relate these timescales relative to his own time feeling, which is expressed through the clock.

So far, we have studied not only one system but also two with different space structures. We want to extend this view from two to N physically real systems. They are observed in connection with the reference system on the basis of the τ scale. We observe the N systems successively at the clock times $\tau_1, \tau_2, \ldots, \tau_i, \ldots$. That is, the N systems are observed at each time τ_i simultaneously. Instead of eqs. (1.18) and (1.20), we get

$$
\begin{aligned}
&\tau_1 \rightarrow \{x,y,z\}^1_{\text{system }1}, \{x,y,z\}^1_{\text{system }2}, \ldots, \{x,y,z\}^1_{\text{system }N} \\
&\tau_2 \rightarrow \{x,y,z\}^2_{\text{system }n}, \{x,y,z\}^2_{\text{system }2}, \ldots, \{x,y,z\}^2_{\text{system }N} \\
&\quad\vdots \\
&\tau_i \rightarrow \{x,y,z\}^i_{\text{system }n}, \{x,y,z\}^i_{\text{system }2}, \ldots, \{x,y,z\}^i_{\text{system }N} \\
&\quad\vdots
\end{aligned}
\tag{1.40}
$$

and

$$
\begin{aligned}
&\{x,y,z\}^1_{\text{system }1} \rightarrow t_{11} \\
&\{x,y,z\}^1_{\text{system }2} \rightarrow t_{12} \\
&\qquad\vdots \\
&\{x,y,z\}^i_{\text{system }m} \rightarrow t_{im} \\
&\qquad\vdots
\end{aligned}
\tag{1.41}
$$

At each time τ we have N space structures and N system-specific time values.

Instead of eq. (1.22), we obtain, in the case of N physically real systems, the more general relation

$$
\begin{aligned}
&\tau_1 = t_{1,1} = t_{1,2} = \cdots = t_{1,N} \\
&\tau_2 = t_{2,1} = t_{2,2} = \cdots = t_{2,N} \\
&\vdots \\
&\tau_i = t_{i,1} = t_{i,2} = \cdots = t_{i,N} \\
&\vdots
\end{aligned}
\tag{1.42}
$$

Like eq. (1.22), relations (1.42) have to be seen as selection processes, where each of the times $t_{11}, t_{12}, \ldots, t_{1m}, \ldots$ belongs to one of the N systems (objects). These are the observations through selection processes.

The system-specific time t is identical with the clock time τ, but t is not τ. This is valid for each of the N systems: At τ we have N system-specific time values.

1.7.6 Some characteristic features

The phenomenon "time" is based on space structure changes. Here, space structure changes do not mean that specific space structures have to be known explicitly, but that space structures can be arbitrary and only their changes are relevant for the phenomenon of time.

In other words, we experience the time feeling in connection with physically real systems (objects) via space structure changes. Thus, we read the time from a physically real system; they define what we call "time." That is, the phenomenon time is system specific. The systems make the time and not a hidden global time machine like the San Francisco cable railway.

With the choice of τ, we adjust the clock time to the time feeling of human beings. The τ scale reflects the standard system (Section 1.7.3). The choice of the τ scale of the clock is independent on any physically real system. As we outlined in Section 1.7.3, the timescales of individual systems have to be identical with the timescale of the standard system (clock),

But we perceive only those systems, whose individual time values are in accordance with τ. If the values t of a certain physically real system (object), having the time spectrum $\{t\}$, are all different from τ, the system is not observable. This is the case for all values t of the spectrum $\{t\}$ if $\tau \neq t$.

We set the time structure

$$\{t\} \rightarrow t_a, t_1, t_2, \ldots, t_b \tag{1.43}$$

of a physically real system (object) in relation to the reference time τ, which reflects, in analogy to eq. (1.42), a time structure as well:

$$\tau \Rightarrow \{\tau\} \rightarrow \tau_1, \tau_2, \ldots, \tau_i, \ldots \tag{1.44}$$

with

$$\tau_1 < \tau_2 < \cdots < \tau_i < \cdots \tag{1.45}$$

Applying eq. (1.22), we get

$$t_1 < t_2 < \cdots < t_i < \cdots \tag{1.46}$$

and

$$\tau_1 = t_1 < \tau_2 = t_2 < \cdots < \tau_i = t_i < \cdots \tag{1.47}$$

Again, the reference time spectrum (1.44) can in principle be chosen arbitrarily. However, it is realized through our clocks that are based on the time feeling of human beings.

From the correspondence between τ and t [eq. (1.22)], we can deduce certain observation laws. Let the values t_a and t_b be the lower and the upper limits of the t spectrum $\{t\}_{\text{system}}$ of a certain physically real system (object). Then, the system is observable if the observation time τ is within the range of $t_a \leq \tau \leq t_b$. On the other hand, the system cannot be recognized by the human observer if $\tau < t_a$ or $\tau > t_b$. To each value t of the t spectrum, belongs a characteristic space structure of the material system in the world outside, and the human being just registers the space structure that belongs to $t = \tau$.

The situation corresponds to what we observe in everyday life: We have at each clock time τ a certain space structure before our eyes. The situation is immediately understandable and does not need further explanation. Again, according to usual clocks, which reflect our time feeling, the time τ moves evenly from low to large values. In this way, the space structure of the system is observed through selection.

This step-by-step observation through selection is obviously an invention of evolution and, to observe the system at once, it unburdens the individual to master everyday life.

1.7.7 Time spectra and selections

To each physically real system or object belongs a system-specific time spectrum $\{t\}_{\text{system}}$. In the case of N systems, we have N time spectra.

$$\{t\}_{\text{system}\,n}, \quad n = 1, 2, \ldots, N \tag{1.48}$$

At clock time τ_i, one of the t values of each time-spectrum is selected. The reason is obvious: We are caught in space and time and a human being can only observe the information of the system under investigation at clock time τ, that is, only one t value of the spectrum $\{t\}_{\text{system}}$ is observable and this value is given by $t = \tau$. In the case of N systems, we have N t values at clock time τ. This is expressed by

$$\{t\}_{\text{system }n} \xrightarrow{\text{selection}} t^i_{\text{system }n} = \tau_i, \quad n = 1, 2, \ldots, N \tag{1.49}$$

All N systems have the same time value at τ_i. Equation (1.49) describes a selection process. Systems for which a t value is not defined at τ_i, are not observable, but can exist. This is fulfilled within the conception described in Sections 1.7.1–1.7.5.

We compare time structures $\{t\}^k_{\text{system}\,n}$, $n = 1, 2, \ldots, N$ with the reference time structure $\{\tau\}$ [see relation (1.44)]. The variations of space structures (space structure changes) reflect the behavior of what is defined as time, which is a system-specific entity. Within this conception, we can say something about the time behavior of physically real systems (objects).

Since "all" physically real systems are equally observed by human beings and selected through the reference time τ, we get the impression that "all" systems are described by the reference time τ, that is, their own system-specific time t is suppressed and is shifted into the background. But this situation is embossed through observation processes performed by human beings. In fact, the observer is here in the center, and the observers are caught in space and time, that is, t values with $t \neq \tau$ are not observable.

Remark

Clearly, a nanostructure formed by atoms and/or molecules is such a space structure at time τ. Thus, nanosystems too reflect system-specific time spectra $\{t\}_{nano}$ for each system. There is no doubt that this view represents a basic feature for systems at the nanolevel. For the description of nanosystems, this basic view should be taken into account, just in connection with self-organizing processes.

1.7.8 Space structure changes and observation features: overview

The phenomenon "time" is an objective fact. We experience it at the macroscopic level of everyday life. This is the most direct and basic experience a human being can have in connection with time and time feeling, respectively. The phenomenon time is born at the level of everyday life. The origin of the so-called time feeling comes into existence through space structures and their changes. These are space structures of material objects. This was our basic assumption in Sections 1.7.1–1.7.5. Furthermore, we based our statements with respect to time on the fact that human beings are caught in space and time. We are particularly caught in the time τ, which is expressed by our clocks. We do not experience the physically real (material) world outside τ. Observations for times smaller than τ and for times larger than τ are not possible.

Physics developed from classical mechanics to quantum theory, and the conceptions of physics were essentially further developed by quantum theory. However, the notion "time" remained classical in character in the transition from the classical description to the methods of quantum theory.

In fact, the clock time τ is not changed during the developments of the various disciplines in theoretical physics. Clearly, within the theory of relativity, the value of the clock time τ changes when the clock moves with a certain velocity or when the clock is positioned in a gravitational field. However, only the value of τ is changed but not its character.

As we remarked several times, Erwin Schrödinger tried to introduce a quantum time within the frame of usual quantum theory, but without success. We have obviously to analyze the basics, on which the time is based, in more detail. First of all, we have to eliminate the "external" time notion. We did that in this section and introduced an "internal" system-specific time. The main items for the description of a nonexternal time notion can be summarized as follows:

1. The time feeling has its origin in space structure changes of the material environment.

2. Human observers are caught in the clock time τ. There are no observations possible outside τ. Only system-specific times with $t=\tau$ are observable.
3. It is a classical description in the macroscopic realm, that is, quantum-theoretical characteristics are not yet considered, but in Chapter 6.

In this way, an external time machine in the sense of the San Francisco cable- railway could be avoided. Again, the system-specific time, introduced in Sections 1.7.1–1.7.5, is still classical in character without any quantum-theoretical characteristic. The quantum aspect of time, which we discuss in Chapter 6, may not be in contradiction to the macroscopic, classical treatment given in this section. In fact, we will recognize that the quantum-theoretical formulation of the phenomenon "time" is compatible with the classical view of time given here.

The quantum aspect of time t (Chapter 6) effectuates that t behaves statistically and fluctuates, and there can be t values at clock time τ with $t \neq \tau$, that is, t may fluctuate between the past and the future.

1.7.9 Space–time feeling: elementary steps

In Sections 1.6.4 and 1.7.8, we indicated the essence of space and time within projection theory. In this section, we want to continue this discussion on the basis of specific considerations.

Space feeling

The separate perception of objects comes into existence through extensions of space. These space extensions generate the human being's space feeling. For the perception of two objects, a certain "extension" is introduced through nature. Nature invented extensions in order to give human beings the possibility to order the world (environment) before his eyes. As we already remarked in Section 1.6.4, here extension has to be considered as a basic notion.

It defines what we call as the "space," assessed from the qualitative point of view. For the quantitative treatment of the space phenomenon, we put a coordinate system over the space impression. In the case of two objects, we have x_1, y_1, z_1 for object 1 and x_2, y_2, z_2 for object 2. The space extension itself is then expressed by the distance of the two objects:

$$\Delta x = x_2 - x_1, \quad \Delta y = y_2 - y_1, \quad \Delta z = z_2 - z_1 \tag{1.50}$$

The space feeling itself is not dependent on the size of the distances Δx, Δy, Δz, but it comes into existence through the existence of Δx, Δy, Δz. The magnitude of Δx, Δy, Δz is not relevant here. We have "one" space feeling for an ensemble of physically real objects before us, which are located in one space. There are various distances for various objects that are located in the same space with “one” characteristic space extension. Therefore, we have only “one” type of space feeling. In other words, the space feeling cannot be dependent on the distances expressed by Δx, Δy, Δz for each pair of the object ensemble.

Time feeling

While the phenomenon of space feeling is due to an existing space extension, time feeling comes into existence through “space structure changes,” which create a qualitatively new effect that we connect with the notion “time.” The time feeling itself is an attribute of the human being’s mind. As we remarked in Section 1.6.4, the space structure changes have, in general, a certain tendency to proceed in a certain direction.

The quantitative treatment of the phenomenon “time” is in analogy to the treatment with respect to space. Then, the space structure changes are expressed by the differences of the corresponding time values t_1 for space structure 1 and t_2 for space structure 2. The space structure change itself is then expressed by the time interval of the two space structures: $\Delta t = t_2 - t_1$.

The time feeling (the time effect) comes into existence through the existence of Δt, but also through the magnitude of Δt, that is, in contrast to space feeling, which is not dependent on the distances Δx, Δy, Δz, the time feeling is dependent on the time-distance Δt. This feature reflects the fact that human beings have a certain sense for dynamical processes. In this way, human beings get the ability to study and influence the structure and dynamics of the physically real objects in the environment.

Conclusion

In summary, we may state the separate perception of material objects through space extensions, the space feeling is created, whereby the space extensions effectuate the separations. On the other hand, the perception of physically real space structure changes creates the time feeling. In this case, we have two space structures, and both structures create the same space feeling, but the time feeling comes into existence through the changes.

1.7.10 The time aspect in traditional physics

What is time? Augustinus von Hippo gave more than 1,600 years ago the following answer to this question: “If no one asks me, I know what it is. If I wish to explain it to him who asks, I do not know.” This is exactly the situation up to the present day. Even Albert Einstein gave the question “What is time?” no definite answer. He said: “Time is that what a clock show.”

In fact, we cannot say much more about the phenomenon “time” in traditional physics. Isaac Newton postulated an absolute time, which is identical with our clock time τ. After Newton, the absolute time can even exist without any matter or other entities. The situation concerning time has not essentially changed. Clearly, the clocks show another time if they move or if they are put into a gravitational field, but the Newton's basic definition remains, also in the usual form of quantum theory.

Projection theory opens the door for a more sophisticated view. We need in fact a more realistic time conception since the new technologies (nanoscience and nanotechnology) emerge new effects, and for their understanding, a more sophisticated view for the time is required.

The container model works with one space and one time. Within the frame of the projection principle, there is possibility for various spaces and various time structures, and this is because space and time are not physically real entities. Projection theory requires the existence of a system-specific time t. We will recognize in Chapter 6 that the existence of both, the momentum $\mathbf{p} = (p_x, p_y, p_z)$ and the energy E, can only exist if a system-specific time t exists. The clock time τ remains in the form of a reference time.

1.8 Other types of biological individuals

The experiment with inverting glasses underlines the fact that the world (picture) before our eyes is observer dependent. We have to assume that all members of a certain species have the same impression. This in particular means that the members of other species have possibly other pictures of the world outside before their eyes. This is, in fact, confirmed within the frame of behavior research. In this connection, the so-called chick experiment is relevant.

In other words, we have to conclude that the picture of reality must be species-dependent. We have to assume that the actions of other biological systems are, in general, based on a picture of reality that is different from that of human beings. How can this statement be verified?

1.8.1 The chick experiment

Within the so-called chick experiment, the behavior of a turkey is studied. The participants in this experiment are a turkey, its chick and a weasel. The study has been done by Wolfgang Schleidt and the topic belongs to the realm of behavior research.

The chick experiment is fundamental. From the behavior of human beings in everyday life, we draw our fundamental information about the world outside. This information is basic and is the starting point for scientific developments, starting with classical mechanics essentially developed by Isaac Newton.

Once again, Wolfgang Schleidt used for his studies a turkey, its chick and a weasel. Using these three animals, he investigated the behavior of the turkey and the weasel in certain situations. Such types of experiments are important because they could influence our world's view basically, that is, the importance is thoroughly comparable with specific key experiments in physics.

The experiments have been done at the macroscopic level, that is, at the level of everyday life. What are the details?

The weasel is the deadly natural enemy of turkeys. A weasel, which approaches the nest of a turkey, will be attacked by the turkey with violent pecks.

What did Schleidt do? His experiments were very simple and convincing. He worked with methods that are located at a level used more or less in daily life, but the implications are far-reaching because the very basis is concerned here. Let us take the following instructive text from Ref. [12]:

> It is known that a turkey sitting on its just-hatched chick attacks everything which approaches its nest. This is of course not true in the case of one of its own chicks which has for any reason left the nest. In order to protect the chick, it will steer back the squeaking the little bird with calming calls into the nest. All this seems to be nothing more than normal; indeed, the turkey actually shows almost human behavior.
>
> The fact is however that the perception apparatus of turkeys must be quite different from that of humans. This can be demonstrated by means of two simple experiments.
>
> 1. Schleidt blocked the ears of the turkey so that it could not hear anything. After a certain period of pacification one of her chicks approached the nest and a serious disaster happened: Without hesitation the turkey strongly pecked the chick with its beak until it was dead.
>
>   The turkey saw its chick approaching but did not identify it. Everything that is "unknown" and that approaches its nest is attacked.
> 2. Schleidt implanted into the body of a stuffed weasel, the deadly enemy of the turkey, a little loudspeaker which emitted the sound of a squeaking little chick. By means of a hidden device he moved the stuffed weasel up to the nest. In this case also something happened which was quite unexpected: The turkey saw the weasel coming but did not identify it; after some hesitation, it even allowed the weasel into the nest and gave it protection.
>
>   These dramatic and unexpected results lead to the conclusion that the turkey must experience the world optically quite differently from the way that we do, even though the eyes of the turkey are quite similar to ours. There is obviously no similarity between what the turkey experiences and what a human being sees in the same situation! [12]

The experiments are reproducible, that is, all turkeys experience the world in this way. In other words, the experiments reflect a general scientific fact and are not only an individual instance.

It is a nonphysical experiment but the outcome could be important for physics. It is particularly not based on the notions and the experimental methods of traditional physics.

1.8.2 Assessment

Schleidt worked with methods of everyday life. However, his experiments have to be considered as fundamental and are obviously relevant for the developments in scientific research. Schleidt demonstrated convincingly that the perception apparatus of a turkey must be quite different from that of human beings. These experiments led to dramatic and unexpected results. They particularly demonstrate that the turkey must experience the world optically in a quite different way than human beings do, although the eyes of the turkey are quite similar to ours. There is obviously no similarity between the pictures of reality of a turkey and a human being in the same situation.

Both systems, human being and turkey, react correctly in the normal case because both species are able to exist in the world. This can only be possible from the point of view of the modern principles of evolution if their particular views of the world are correct.

Although the conceptions of the world of humans and turkeys are on the one hand different from each other, they are on the other hand correct in each case.

This means that neither of these two conceptions of the world can be true in the sense that they are a faithful reproduction of nature: Objective reality (basic reality) must be different from the pictures, which biological systems construct from it.

In summary, the experiments by Wolfgang Schleidt deliver essential contributions for our understanding of what we call reality-picture principle. They can help to learn something about the relationship between reality and any kind of observer (human being, turkey, etc.). However, we can only say that the perception apparatus of the turkey is different from that of humans; the details are not accessible in this way but we do need more information for the answering of such principal questions. We even do not know if such biological systems (turkeys) experience "their world" within the framework of space and time. A more detailed analysis is given in the next section.

The chick experiment offers a general conception, because there is no reason to believe that turkeys have to be considered as an exception.

1.8.3 Chick experiment and projection effects

The result of the chick experiment is intuitively judged within the frame of the container model. This is simply a matter of habit. But this way obviously leads to conclusive statements, even when the outcome of this experiment is hardly understandable within the container model. On the other hand, the projection principle offers the possibility to explain the chick experiment straightforward. Nothing has to be changed when we consult the basic items of this principle for the explanation of Schleid's experiments. In Figure 1.3, the main characteristics of the projection principle are given.

The world, which biological systems experience, is observer dependent but species-dependent as well. The species dependency is due to the principles of evolution, where the principle of usefulness (Section 1.5.1) is in the foreground in the biological development of the members of a species. After this principle only that information is pulled from the reality outside (basic reality), which is useful for daily life and which is essential for survival for the individuals. In other words, this information is species dependent.

The mind–brain complex processes this information, at least in the case of a human being. But we may suppose that such kind of procedure is also a characteristic in the case of turkeys, if such a mind–brain complex actually exists for them. However, we have to assume that the mind–brain complex is species dependent as well, and we have

$$[\text{mind–brain complex}]_{\text{turkey}} \neq [\text{mind–brain complex}]_{\text{human being}} \tag{1.51}$$

That is, not only the pulled information from basic reality is species dependent but the mind–brain complex as well, which processes the pulled information. These processes lead to the pictures of reality. Thus, the picture of reality of a turkey must be different from the picture of reality of a human being:

$$[\text{picture of reality}]_{\text{human being}} \neq [\text{picture of reality}]_{\text{turkey}} \tag{1.52}$$

This is just what the chick experiment has taught us.

The picture of reality consists of a frame of representation on which the (species-dependent) information is projected. In the case of human beings, the frame of representation is given by space and time. In the case of turkeys, the frame of representation is unknown. We get

$$[\text{frame of reference}]_{\text{human being}} = \text{space and time} \tag{1.53}$$

$$[\text{frame of reference}]_{\text{turkey}} = ? \tag{1.54}$$

The situation is summarized in Figure 1.4.

1.8.4 The impact of nanoscience

Other types of biological individuals obviously live in their own world. The principles of evolution, in particular the principle of usefulness, dictated the peculiarities of each of these worlds. For example, human beings and turkeys live in parallel worlds, as was demonstrated through the chick experiment. We may state quite generally that the world outside is observer dependent for all types of biological individuals (Section 1.5.2), but it is, in addition, species dependent as well. The chick experiment underlines that.

The level at which the particular peculiarities emerge is the nanolevel. That is, within the frame of nanotechnological changes we vary the basic features of biological individuals. If a human being wants to change the characteristics of another type of individual nanotechnologically, he must know the "basics" of the

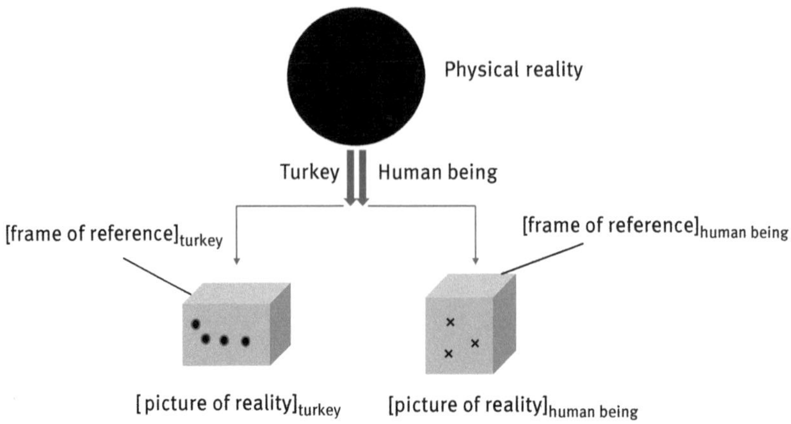

Figure 1.4: The world, which biological individuals experience, is observer dependent and also species dependent. The species dependency is due to the principle of usefulness and is in the foreground in the biological development of the members of a species. After this principle, only that information is pulled from basic reality, which is useful for daily life and for survival. The mind brain complex processes this information. We may suppose that such kind of procedure is also a characteristic in the case of turkeys, if such a mind–brain complex actually exists for them. In any case, we have to assume that the mind–brain complex is species dependent as well. These processes lead to the pictures of reality. Thus, the picture of reality of a turkey must be different from the picture of reality of a human being. This is just what the chick experiment has taught us. The picture of reality consists of a frame of representation on which the (species-dependent) information is projected. In the case of human beings, the frame of representation is given by space and time. In the case of turkeys, the frame of representation is not known.

characteristic features of this type of individual. Otherwise, the outcome of the nanotechnological impact can lead to a catastrophe, as we have seen in the case of the chick experiment.

If the reality-observation principle of projection theory is the realistic background for the assessment of such situations, it is important to know how the mind–brain complex and the frame of representation work with respect to the other type of biological individual. Do they exist at all or in the form as human beings imagine?

1.9 Immanuel Kant and his thoughts about space and time

Nanoscience requires a reliable view about the world outside. Concerning statements about reality outside, we need statements about space and time. The reason is obvious: Human beings are caught in space and time and master their lives exclusively within the frame of space and time.

Within the conception of the projection principle, space and time are not physically real entities but serve exclusively for the representation of facts in the world outside. Within projection theory, space and time are products of the human being's mind.

It is astonishing that the space–time features, discussed in connection with the projection principle, are also reflected by the ideas of the philosopher Immanuel Kant in the nineteenth century. Let us briefly discuss Kant's ideas.

1.9.1 Kant's arguments

Kant argued that statements about the true reality outside cannot be made. According to him all things we observe are located within space–time and these elements, space and time, are located inside the observer. In Kant's opinion, a human observer can nothing say about the structure of that what we call world outside. In particular, in his opinion a human being is not able to give answers on the following questions: Is there a "one-to-one correspondence" between the structures outside and those in the mind? Does reflect the information, located in the mind in the form of a picture, the "complete" information about physically real (material) world outside? Kant knew nothing about the principles of evolution but this is an essential point in answering such questions (see in particular Section 1.5.1).

According to Immanuel Kant, experiences become only possible through the concepts of space and time. In his opinion, space and time are not empirical concepts, which are determined by abstraction from experience. After that, space and time are not objects, but have to be considered as preconditions for the possibility of all experience.

Although in Kant's opinion space and time are not empirical concepts, they nevertheless have empirical reality. The reason is obvious: All things, which we observe, are located in space and time. In other words, the structures of space and time are reflected in the empirical fact, that is, by the objects. Kant denied the existence of a space and a time independent of brain functions (observations in everyday life, thinking).

1.9.2 Differences to the projection principle

According to Kant, space and time are located inside the observer. Whether space and time are also elements of actual (fundamental) reality outside remains principally an open question within Kant's point of view. There is no doubt that this is a big difference to the projection principle. Within the frame of the projection principle, space and time are not physically real entities and cannot be located in the physically real world outside. This point has been discussed in Section 1.6.2 in connection with Mach's principle.

Kant's perspective is without any doubt important, not only in connection with philosophical questions. But what are the consequences for the theoretical structures of physics?

If we take Kant's view seriously, then the physical laws as, for example, the laws of Newton's mechanics, are merely pictures in the head of the observer and there is principally no way to express them for the reality outside; nothing can be said about the processes in the outside world. If the gravitational law (and all the other physical laws) is merely a picture in the head we get a problem because there can be no gravitational forces in the head of the observer.

1.9.3 Realists and antirealists

Kant's thoughts can lead to considerable problems when we apply them to physics. Barrow remarked: "We can see that Kant's perspective is worrying for the scientific view of the world" [13]. However, Kant's perspective has not taken so seriously in physics and there are other positions. The situation is well analyzed by Barrow:

> There are two poles about the relationship between true reality and perceived reality. At one extreme, we find "realists", who regard the filtering of information about the world by mental categories to be a harmless complication that has no significant effect upon the character of the true reality 'out there'. Even when it makes a big difference, we can often understand enough about the cognitive processes involved to recognize when they are being biased, and make some appropriate correction. At the other extreme, we find "anti-realists", who would deny us any knowledge of that elusive true reality at all. In between these two extremes, you will find a spectrum of compromise positions extensive enough to fill any philosopher's library: each apportions a different weight to the distortion of true reality by our senses [13].

In other words, there are no criteria to decide about the true nature of absolute reality. There are realists and there are antirealists. However, the realists cannot disprove the antirealists and vice versa. The realists more or less assume that there is a one-to-one correspondence between "true reality" and the "perceived reality" (picture). The antirealists maintain that we can say nothing about true reality. Also, the statement that there can be no one-to-one correspondence is against the realist's opinion.

1.9.4 The influence of evolution

When we consider however the basic facts of biological evolution, both viewpoints seem to be not realistic. Evolution is the basic factor here, and evolution teaches us that cognition does not play the important role in nature but the differentiation between "favorable toward survival" and "hostile toward survival," at least at the early phase of evolution [5] (Section 1.6.2). Each picture of reality (perceived reality) is tailor-made to these characteristics. Since the conditions for survival are different for different biological systems, the perceived realities are different for different biological systems. Wolfgang Schleid's experiments (Section 1.8) with a turkey showed that very impressively.

The picture of reality, unconsciously designed by individuals, has to be correct but it may only contain, for economic reasons, information, which is necessary for survival and nothing more. Everything else is unnecessary. The picture of reality does not have to be complete and true. It must not be a precise reproduction of the world outside, but it must be restricted and reliable.

Furthermore, we learned from Schleid's experiments that the conception of the world of human beings and that of turkeys are on the one hand different from each other, but they are on the other hand correct in each case.

This in particular means that neither of these two conceptions of the world outside can be true in the sense that they are a faithful reproduction of nature. In other words, there can be no one-to-one correspondence between the structures in the picture and those in true reality. Objective reality must be different from the images which biological systems construct from it.

The statement that there can be no one-to-one correspondence is, on the one hand, against the realists and, on the other hand, it is simultaneously against the position of antirealists because it is a statement about true reality.

1.10 Summary and final remarks

With nanoscience, we enter a higher level of reality, higher and deeper than that on which traditional physics is based. This step is required when we want to describe the typical features of nanosystems realistically. For the understanding of the effects at the nanolevel, basic conceptions have to be deepened and expanded. In particular, the nature of space and time plays an essential role in this connection. Here, Mach's principle comes into play and has to be fulfilled.

The projection principle is obviously an adequate frame for the realistic description of nanophenomena. In this chapter, some essential items have been summarized and their importance is underlined. The following topics are in the center:

1. In nanoscience, we work at the ultimate level where the properties of matter and biological individuality emerge. Can we transfer the basic laws of traditional physics and technologies to the nanorealm? Here, we have to be careful since new effects at the nanolevel require more.
2. In the quantum-theoretical description of traditional quantum theory, only the classical time τ appears. No other type of time is defined here. Already Erwin Schrödinger, Wolfgang Pauli and Ilya Prigogine strongly argued that we need a

quantum aspect of time. However, a quantum aspect of time cannot be introduced within traditional quantum theory without changing the structure of the physical laws themselves, that is, without changing the structure of Schrödinger's equation itself. Obviously, the most basic features of the principal conception have to be changed.

3. We discussed two fundamental conceptions: The container model and the projection principle. We came to the conclusion that the details of the container model are not realistic and this conception has to be considered at best as an approximation. A more realistic basis for the physical phenomena in the world outside is desired. The quantum aspect of time is only one problem, which the container model is obviously not able to solve. Within the frame of this situation, the so-called projection principle is relevant. Here, the physically real world is not embedded in space and time but it is projected onto space and time. This conception is adapted to the basic features of space and time and is therefore more realistic from the very first.

   In fact, the projection principle exactly delivers what is desired in connection with the "quantum aspect of time." Here, we need an operator for the time, and each system (nanosystem) defines a system-specific time that behaves quantum-mechanically. In other words, the projection principle offers the essential items we need for the understanding of the self-organizing processes.
4. Self-organizing processes belong to heart of nanoscience. It is, however, not possible to describe self-organizing processes adequately within the frame of traditional quantum theory. The basic features, which are essential for the understanding of self-organizing processes, are not given in traditional physics. Here, we need a quantum-theoretical aspect of time, which is not defined in traditional quantum theory. A realistic description of self-organizing processes should however be possible through the projection principle since it contains the characteristic features. Within the frame of the projection principle, a quantum time and an operator for the time are directly being given.
5. It is outlined that space and time are not physically real phenomena. They do not belong to the reality outside. Space and time are the frames on which the physically real objects are projected.
6. The external, metaphysical time is eliminated in projection theory. Here, we have a system-specific time. To each space structure, a system-specific time has to be allocated.
7. The world (picture) before our eyes is observer dependent. This in particular means that the members of other species have possibly other pictures of the world outside before their eyes. That is, the picture of the world outside is not only observer dependent but species dependent as well. This is in fact confirmed through specific studies within behavior research. In this connection, the so-called chick experiment is relevant. There is obviously no similarity between the pictures of reality of a turkey and a human being in the same situation.

1.10.1 Outlook

In Chapters 2 and 3, we will quote traditional methods, which have been used for the characterization of nanosystems. Typical applications will be brought and discussed. Then, in Chapters 4–7, the “reality-observation principle,” important for the understanding of nanosystems, is deepened and the principles for the realistic analytical treatment of nanosystems will be given.

Chapter 2
Directions of nanoscience

In this chapter, we discuss some of the relevant points that could be essential for nanoscience in future. Possible directions in this field are indicated. This is not a complete representation but only a few specific problems have been selected, which appear in the theoretical and computational treatment of nanosystems and are summarized and briefly discussed. In particular, we underline the relevance of the basic physical laws for nanoscience, but also some principal questions in connection with biological systems. Here the principles, which we developed in the first chapter, have been taken into account. Some basic remarks with respect to nonbiological systems are given as well; the determination of atomistic and electronic nanoproperties plays an essential role and will be underlined. What parameters are important here?

2.1 Basic physics and nanoscience

As we already outlined in Chapter 1, the development of the scanning tunneling microscope made it possible to manipulate matter on its ultimate level, where the properties of usual matter emerge and where individual life comes into existence. This situation makes it necessary to apply the physical laws in their basic form. We also remarked in Chapter 1 that in traditional technologies, that is, within the frame of micro- and macrotechniques, scientists do not work on the ultimate level. At these levels, phenomenological descriptions are more or less in the foreground. Each discipline has its own description which, in general, cannot be deduced from the basic physical laws.

In nanoscience and nanotechnology, we have "one" theory for all phenomena and this is given by the basic laws of theoretical physics. This has at least two consequences:

1. Since we are working here on the ultimate level, the properties of matter and functional matter have to be defined in final form because there is no physical level above the ultimate nanolevel and means the following: this is not only the strongest material ever made, but this is the strongest material it will ever be possible to make [1].
2. Working on the ultimate level also means that any change in the basic physical laws will directly influence nanoscience, without any intermediate step. Clearly, this can lead to completely new perspectives with respect to applications. Therefore, to work on nanoscience also means to develop the basic laws further. The necessity for that was demonstrated in Chapter 1 in connection with the quantum aspect of time, and even the basic conception of physics had

https://doi.org/10.1515/9783110525595-002

to be developed further: instead of the container model, the projection principle had to be used [14]. The direct use of the basic theoretical laws in nanoscience particularly means that technology has reached the level of science; nanoscience and nanotechnology are basically undistinguishable.

2.1.1 Why computational and theoretical nanoscience?

Quite generally, we need theoretical and computational nanoscience in order to predict and to describe new nanosystems and, furthermore, to give nanoscience a direction [15, 16].

But there is another point. Self-assembly (self-organizing processes) is a typical nanodiscipline, as we have underlined in Chapter 1. However, self-assembly can be an uncontrolled process. In other words, the final state can be uncertain. This can lead to undesirable developments, just in connection with biological systems. Therefore, the theoretical and computational analysis of such processes is not only desirable but absolutely necessary in order to keep nanoscience under control, just in connection with the self-assembly of artificial biological systems, where the final structure at the beginning of the process in general is not known. These are without any doubt threats, but can be extensively eliminated by the systematic theoretical analyses of the system under investigation using adequate laws from theoretical physics.

Clearly, for a reliable analysis of self-organizing processes we first need a general theoretical framework on which a detailed analysis can be based. As we have outlined in Chapter 1, such a general theoretical framework must contain the quantum aspect of time, which is not defined in traditional physics (traditional quantum theory). Such a quantum aspect of time can obviously only be introduced when we replace the traditional container model through the more realistic projection principle.

2.1.2 Complexity of nanosystems

Nanosystems are without a doubt interesting and they behave quite differently in relevant cases from systems used in micro and macrotechnology. However, many researchers discuss nanosystems on the basis of traditional thinking. Why? They very often study the effects by means of static building blocks as we do in connection with microsystems. Normally, it is assumed that the properties of nanosystems (for example, a nanocluster) are due to the relatively large surface area. The relative number of surface atoms/molecules increases with decreasing size. This effect modifies significantly the inner structure of the building blocks in comparison to that of the bulk [15]. But in most cases, they are theoretically treated as static elements without a changed inner structure, that is, they remain static within this view.

However, this structure aspect is only one typical effect in connection with nanosystems. There is another important point: nanosystems behave strongly anharmonically [2, 15], much more than in its usual solid-state physics and in micro/macrosystems, that is, also the dynamics of the objects is modified, which can lead to unusual effects. The whole nanosystem can transform spontaneously and there can exist many coexisting structural states. Let us briefly discuss the spontaneous transformations of nanoclusters as a typical and relevant example.

Nanoclusters behave in a certain sense like atoms. They can be in an excited state. After a certain time, the excited cluster state transforms spontaneously to the ground state without external influence [2, 15]. The situation can be more complex than in the case of atoms. There can be more than one ground state. It is a matter of chance to what ground state the cluster transforms from the excited state. The various ground states may differ with respect to two features. The internal structure as well as the outer shape are, in general, different for the various ground states. There are grain boundaries, dislocations and other lattice defects. The outer shape of the clusters in the ground state depends on the arrangement of these inner lattice effects.

In other words, a free cluster or a cluster on a surface of a solid behaves for a certain time interval as a static nanosystem. But it transforms spontaneously into another nanocluster with another internal structure and another outer form. In micro- and macrotechniques, we do not observe such effects; here we have for all times one structure and one form. A penny on a table does not transform spontaneously into a 50 cents coin or so; a penny remains a penny forever. However, for such a nanopenny, structural transformations would be possible on nanometer length scales.

We may conclude that systems, relevant in nanoscience (nanotechnology), behave in general complex and they are obviously not comparable with the behavior of micro- and macrosystems. For the description of such nanosystems, the theoretical and computational tools have to be selected very carefully and are partly not developed yet. We discussed this point in Chapter 1 and even the basic conception for the theoretical treatment is under investigation.

In this connection, the interaction laws (potentials) between the objects forming a nanosystem are critical input data, and this is because the structure and the dynamics of such systems are very sensitive to small variations in the potentials. We will discuss this point below, in particular in Chapter 3.

It is generally accepted that nanobiology and nanomedicine will influence our future considerably. Just here, theoretical and computational nanotechnology will play a basic and important role. What mathematical tools and what physical laws do we need here? Are there principal limitations in connection with the understanding of biological systems? The last question directly arises when we consider the human brain as a mathematical object. Is a complete description of the human brain possible on the basis of the laws of theoretical physics? Within the frame of the projection principle, part of the brain is material in character; here the brain is coupled with the mind and this makes the situation complex since the brain must necessarily contain mind states. In Chapters 4 and 5, we will discuss these problems.

2.2 Some principal questions

2.2.1 Nanobiological systems

The impact of nanoscience (nanotechnology) will be tremendous and will influence our future dramatically. The manipulation of atoms and molecules will allow to develop new technological worlds: electronic nanodevices, nanomachines and so on. Nanoscience will particularly develop fundamentally new possibilities in the field of medicine. Just in the case of nanobiotechnology, big changes are expected already in the near future.

A lot of speculations are in discussion. For example, it has been argued that through nanoscience our bodies will be transformed into illness-free, undecaying systems of permanent health. Moreover, it has been prognosticated that it will take approximately 30–50 years to develop nanotechnological means for the creation of superhuman intelligence. That will bring the human era to an end. The following citation is instructive:

> We already have experimental hints of how that might be occur. In September 1999, molecular biologists at Princeton reported adding a gene to a strain of mice, elevating their production of NR2B protein. The improved brains of these "Doogie mice" used this extra NR2B to enhance brain receptors, helping the animals solve puzzles much faster. A kind of genetic turbo-accelerator for mousy intelligence. Human brains, as it happens, use an almost identical protein. It is not far-fetched to suppose that we will learn to tweak or supplement it to increase our own effective intelligence (or that of our children). [17]

Such speculations might become real facts. However, we always have to ask on which basic conception a certain speculation has been made. What is for example the basis of the term "superhuman intelligence"? It is here the container model. Matter, mind and brain are not interwoven within this model, and statements have to be considered with care and are not very much founded. Here we have no order of principle for the brain and the mind to which the matter is coupled. This point will be discussed in more detail in Chapters 4 and 5.

Also here the following question arises: are the theoretical tools developed so far sufficient for the description of such biological phenomena? In the next section we will discuss this point by means of basic principles.

2.2.2 Levels of reality

Since the effects on the nanolevel are principally new, not only improvements and additions of the present theoretical frame are sufficient. We probably have to leave the level of present physical description frames, where the material objects and their interactions are considered and described through the tools of classical mechanics and traditional quantum theory. It is not a speculation when we state that there will be an expansion of physics through nanoscience.

In what direction have we to go? There are some principal questions. For answering them, sophisticated arguments are needed. Can we really develop the theoretical means for the description of superhuman intelligence? In fact, the container model (Chapter 1) is too rudimentary for that. The projection principle is more suitable in such cases, as we have outlined in Chapter 1; this principle is more realistic than the container model. The discussion in the following sections will underline that convincingly.

To explore these new ways experimentally is probably no problem, but is the outcome really what we wanted to produce? The practical way is in most cases simple, but do we really understand the practical steps? In other words, can we fully estimate the consequences for our life and civilization? This seems to be more than doubtful under the present knowledge and conditions. The present theoretical tools and conceptual strategies are probably not sufficient for that. It is questionable whether we can actually understand a “superhuman intelligence” or a “life without death” on the basis of the present human brain and mind structure. Let us briefly discuss why.

2.2.3 On the description of brain functions

To clarify the situation in more detail we have to introduce a useful and convincing conception. This conception is based on what we call reality and its description on certain levels (levels of reality).

The following view has often been discussed in literature, in particular in Chapter 1: Basic reality, that is, reality that exists independently of the observer, is principally not accessible in a direct way. But it is observable or describable by means of pictures (pictures of reality) on different levels, which we called “levels of reality.” What is the order principle for these levels of reality? How are they arranged? The levels can be arranged vertically in accordance with the degree of generality where the level with higher degree of generality is arranged above that with a lower degree of generality (see Figure 2.1).

It is obvious that the description of certain facts, which are located on a certain level of reality, cannot be done on the basis of the information of another level of reality that is located below this level. A more general structure or theory cannot be

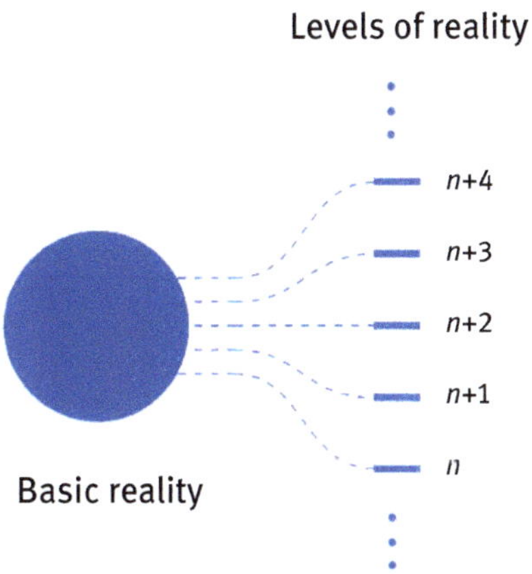

Figure 2.1: Basic reality is not accessible to human beings. They can only perceive the species-specific part of it in the form of a picture. That is, in principle we cannot make any statement about basic reality which is symbolized in the figure by the full circle. But we can observe or describe aspects of it within the framework of levels (levels of reality). These levels are vertically arranged in accordance with the degree of generality (principle of level analysis). The levels up to $n-1$ and those above $n+4$ are not quoted; n is any number.

deduced on the basis of a structure or a theory that is less general. For example, traditional quantum theory cannot be deduced from classical mechanics. In other words, the theoretical structures on a certain level cannot be deduced from the standpoint of a level that is positioned under this level [4].

If we apply these principles on that what we called "superhuman intelligence," we may conclude that such a superhuman intelligence cannot be described on the basis of the present human mind (brain) structures, and this is because the level of the present human intelligence is by definition below the level of superhuman intelligence.

It is obvious that certain experimental brain manipulations can lead to a result that is in principle not predictable. In other words, such manipulations would be risky and, in particular, irresponsible. Moreover, there must be no connection between different mind, brain levels because different pictures of reality (world views) can be incommensurable leading to a completely different assessment of the world outside. Let us briefly discuss an example from the field of behavior research that we have already used in Chapter 1.

Turkey and human being

Working on the basis of various levels is relevant when we analyze different species. This in particular means that the members of different species have possibly different pictures of the world outside before their eyes. We discussed in Chapter 1 that the perception apparatus of a turkey must be quite different from that of a human being. There is obviously no similarity between the pictures of reality of a turkey and that of a human being in the same situation. This can be understood in terms of the principle of level analysis. There is the basic reality (Figure 2.1) and the turkey obtains certain information from it, but also the human being obtains another set of information from the same basic reality. The resulting pictures of reality (that of the turkey and that of the human being) are different from each other because the pulled information contents are different from each other. In other words, turkey and human being are positioned on different levels of reality, which are obviously incommensurable. Clearly, for both levels the projection principle has to be applied.

2.2.4 Living systems described mathematically

There is no doubt that the production of superhuman intelligences is dangerous. Such a step can probably not be done under controlled conditions. Principal questions come into play, which we have quoted earlier. But even when we work only on one level, we get problems. A complete description of the present human mind (brain) structure is obviously hardly possible. Here “Gödels theorem” has to be considered and applied. [18].

Gödel’s theorem

The controlled manipulation of biological systems is a relevant concern in nanomedicine. Here the following principal point has to be considered: to keep the situation under control we have to develop theories for the judge of specific world views on which the biological system under investigation does its actions in its environment. In other words, we must know the world view of the biological system that we would like to manipulate. What are the changes of this world view as a function of specific structure changes in the realm of genes and brain functions? We will at first only give Gödel’s standpoint. In Chapters 4 and 5, we will discuss this point from the point of view of the projection principle.

The developed theory must be able to describe such situations. In accordance with what we have said about levels of reality (Figure 2.1), this theory must be more general than the world view of the biological system itself. The theory is then a world view on a higher level. Is that possible at all? Yes, that is in principle possible but probably not in connection with a biological system that tries to make statements about itself. Here Gödel’s theorem comes into play.

If we try to formulate such a theory mathematically on the basis of theoretical physics, a living system becomes a mathematical system, which obeys certain physical principles. To what extent such kind of mathematical systems can be considered as adequate, remains open in the first place.

Then the question “What is the world view of an individual and how is it described?” is equivalent with the question “Can a mathematical system make statements about itself?” The individual wants to make statements about himself. It is an attempt of a self-related analysis.

In such cases, Gödel’s theorem is relevant. If one wants to make statements with a mathematical system about itself, Kurt Gödel showed [18] that no mathematical system can be complete and consistent simultaneously. That means the following: a consistent system will always contain statements for which one cannot decide whether they are true or not. On the other hand, a system cannot be consistent, if it contains the complete truth. In mathematics consistent systems are relevant.

Let us assume that we have a consistent mathematical system. If the system would contain also wrong statements beside the true ones, the system would be inconsistent. So, in accordance with Gödel's theorem, there will be true statements, which are not included within the system in the case of consistency. The system is missing something, and this is because there are statements that cannot be proven as true or wrong with system-specific axioms and rules. The reason why the system is incomplete is not because certain information is not yet known, but because completeness is principally not achievable: We are confronted with the problems, which appear, if one wants to make statements with a (mathematical) system about the (mathematical) system itself.

In conclusion, the question "Can we develop theories for the judge of world views and the description of their changes as a function of specific gene and/or brain manipulations?" must be answered negatively or at least with care.

A biological system that tries to make statements about itself is immediately confronted with Gödel's theorem, provided this theorem is applicable at all, that is, if the individual can actually be described mathematically sufficiently well on the basis of theoretical physics.

Turkeys, human beings and aliens

The situations in connection with turkeys and human beings suggest the existence of what we denoted earlier by "principle of level analysis". Together with Gödel's theorem, we can estimate certain possibilities for living biological systems and their position relative to others. If, for example, the world view of a human being is more general than that of a turkey, that is, if the level of the human being would be above the level of the turkey (see in particular Figure 2.1), the human being would in principle be able to say something about the structures of the world view of a turkey. On the other hand, a turkey would hardly be able to estimate the world view of human beings, possibly not at all. All that has to be considered when nanotechnological changes of brain functions are planned. We have in particular taken into account that there are mind–brain interrelations, which are necessary within the frame of the projection principle.

In the case of aliens, the situation would be similar, provided Gödel's theorem is applicable here. A realistic assessment of their actions in certain situations can hardly be made, even then when our world view appears to be superior. Aliens could be very different from human beings and they could even not be furnished with senses similar to ours. It is possible that the senses of a human being cannot interact with aliens and their perception as object would not be possible. No doubt, the projection principle is not only more realistic than the container model, but it is obviously more versatile.

2.3 Technological changes

There is a further essential point. We not only leave the level of the usual material description but the developments on this level is essential as well. We observe that the technological change grows more and more like an exponential curve leading to the effect that the future rushes at you.

2.3.1 The trend to drastic technological changes

The more the exponential curve grows,

> the larger is each subsequent bound upward. It takes a long time to double the original value, but the same period again gets you four times farther up the curve, then eight times . . . so that just ten doublings, you've risen a thousand times as far, then two thousand, and on it goes. Note this: the time it takes to go from one to two, and then from two to four, is just the same period needed to take that mighty leap from 1000 to 2000. A short time later we're talking a millionfold increase to a single step, and the very next step after that is two millionfold.
>
> History's slowly rising trajectory of progress over tens of thousands of years, having taken a swift turn upward in recent centuries and decades, quickly roars straight up some time after 2030 and before 2100. That's a Spike. Change in technology and medicine moves off the scale of standard measurements: it goes asymptotic.
>
> So, the curve of technological change is getting closer and closer to the utterly vertical in a shorter and shorter time. At the limit, which is reached quite quickly (disproving Zeno's ancient paradox about the tortoise beating Achilles if it has a head start), the curve tends toward infinity. It rips through the top of the graph and is never seen again.
>
> At the Spike, we can confidently expect that some form of intelligence (human, silicon, or a blend of the two) will emerge at a posthuman level. At that point, all the standard rules and cultural projections go into the waste-paper basket. [17]

This accelerating world of drastic technological change must have consequences. This behavior is the reason why this posthuman phase will be reached relatively soon, and this will possibly have drastic effects. In his book *The Spike*, Damien Broderick formulates that as follows:

> We can expect extraordinary disruptions within the next half century. Many of these changes will probably start to impact well before that. By the end of the twenty-first century, there might well be no humans (as we recognize ourselves) left on the planet – but, paradoxically, nobody alive then will complain about that, any more than we now bewail the loss of Neanderthals. [17]

Such a situation in connection with an accelerating world of drastic technological change is without any doubt dangerous and must inevitably lead to an uncontrolled situation. Each step in relevant technological developments has therefore to be done carefully in order to keep nanoscience (nanotechnology) under control. As we have remarked earlier, the benefits through this new science (technology) can be tremendous but we have to be careful.

2.3.2 Conclusion

These uncontrolled situations, which may appear in connection with the existence of the principle of level analysis (Section 2.2), is superimposed by an accelerating world of drastic technological change (section 2.3), which can lead to uncontrolled situations as well. We have to be careful when we work at the nanolevel. Here the theoretical and computational analysis is important and we must possibly revise our conceptions if necessary. Such theoretical investigations have to be done "before" we develop new nanosystems.

2.4 Brain functions and computational neurogenetics

The impact of nanotechnology will be tremendous. The manipulation of atoms and molecules on the ultimate level of nanoscience will allow to construct new technological worlds and will bring fundamental new possibilities in the field of medicine. Let us give an example. It has been reported in the August 25 (2006) issue of *Science* [19] that scientists at the SUNY Downstate Medical Centre have found a molecular mechanism that maintains memories in the brain, namely by persistent strengthening of synaptic connections between the neurons. It could be demonstrated that by inhibiting the molecule, long-term memories can be erased. Erasing the memory of the brain does not mean that the capability to relearn memory must be lost, that is, this erasing process does not prevent that. The scientists could demonstrate these effects by an enzyme molecule with the name "protein kinase M zeta."

2.4.1 Nanomedicine

Such kind of processes are possibly relevant for certain future directions in medicine (nanomedicine). In other words, the detailed knowledge about such kind of processes could be of fundamental importance, for example, in connection with Alzheimer's disease. In general, we may carefully state that such investigations, such findings, could in future be helpful to treat chronic pain, posttraumatic stress disorder, memory loss and so on.

However, we have to know and to apply reliable models on which such processes are based. The basic conception is relevant here. In the description of such phenomena we have to recourse to the basic laws of theoretical physics. However, it cannot be excluded that the usual laws are not sufficient enough because the basic laws of theoretical physics have not yet been confronted with such kind of phenomena as, for example, the case we have just discussed: a molecular mechanism that maintains memories in the brain.

Self-organizing processes are also possibly relevant and, therefore, what we have discussed in Chapter 1 has to be taken into account. The quantum aspect of time and the projection principle are in the center in the description of self-organizing processes.

2.4.2 Computational neurogenetics

We may formulate the situation quite generally. It is one of the goals of computational and theoretical nanoscience to understand and to describe such molecular mechanisms, which come up in future medicine, in particular, the prognosis is relevant in order to keep nanoscience (nanotechnology) under control but, on the other hand, to profit from nanoscience.

In this connection, it is relevant to note that the relatively new discipline "computational neurogenetics" will be of considerable importance in the future. Here the interaction between brain functions and genes are of specific interest:

> With the recent advancement of genetic research and the successful sequencing of the human and other genomes, more information is becoming available about the interaction between brain functions and genes, about genes related to brain diseases (e.g., epilepsy, mental retardation, etc.) and about gene-based treatment of them. It is well accepted now that brain functions are better understood and treated if information from the molecular and neuronal level is integrated . . . For this purpose, computational models that contain genetic and neuronal information are needed for modelling and prognosis. [20]

Such a scientific program (modeling and prognosis) is necessary in order to eliminate or to reduce the threats that appear in connection with the manipulation of brain functions. Due to the interrelation of the brain with the mind, changes of brain functions can extensively influence the individual behavior of living biological systems, for example, human beings.

Thus, such processes have to be analyzed carefully, and the prediction and development of artificial biological structures are relevant. However, also here we have always to keep in mind that there are possibly principle limits for the complete description of brain functions. Gödel's theorem is important here (Section 2.2.4) and of course the principle of level analysis (Section 2.2.2).

In the following sections, we would like to discuss and analyze phenomena of nanoscience where brain functions and mental states are not involved.

2.5 Theoretical and computational methods

2.5.1 General remarks

The systems used in nanoscience are small but are nevertheless complex. Analytical models for certain classes of nanosystems could not be developed so far. Specific numerical techniques are therefore in the foreground and are used for the understanding and prognosis of such nanosystems. However, the simulation models have to be prepared very carefully in order to be sure that all relevant features and mechanisms of the nanosystem under investigation are considered.

In other words, the input is important in connection with numerical models. Here it is essential to choose the adequate basic conception. Classical mechanics and quantum theory are the basis for such investigations. The formulation of realistic physical equations and also the system-specific boundary conditions are in the focus.

In many cases a lot of simulation techniques and theoretical methods are quoted for the treatment of nanosystems. But almost nothing is said about the nanospecific aspect. Just at the nanolevel the usual theoretical (computational) methods have to be developed further and it is of particular importance to recognize the critical points of these usual methods when they are applied to nanoscience.

In most cases, these methods have been taken over from other disciplines; here solid-state physics and surface science have been applied without considering the nanospecific character of the system under investigation.

Can all theoretical and computational methods be used in this field without applying nanospecific modifications? There is no general rule to answer this question and, therefore, each specific nanoproblem has its own characteristics and has to be analyzed specifically. This has to be done carefully, even when the methods work excellently in the bulk or in the case of microsystems.

2.5.2 The most important techniques

In this section, often applied and successfully used theoretical (computational) methods for the treatment of nanosystems are quoted. All these methods are formulated within the framework of traditional physics (classical mechanics, usual quantum theory), that is, we work exclusively within the "container model." The "time" is given in all these cases as external parameter and is represented in the form of the clock time τ; a system-specific time t is not defined within these treatments and also effects due to the projection principle are not existent. Nevertheless, these common methods work well, but should be considered as recipes.

Some methods of traditional physics

In this part of the chapter we would like to quote some theoretical and computational methods that are important for the description of nanosystems and which have to be proved of value. As already outlined, analytical models for the treatment of the structure and dynamics of nanosystems could not be formulated and also specific simple models have not been introduced with success in this field.

In solid-state physics, the crystalline solid in the harmonic approximation can be considered as a general specific conception. However, the harmonic approximation has no basis in the theoretical and computational treatment of typical nanosystems, and anharmonic effects can only be treated numerically. As we will underline in Section 2.6, a typical nanosystem is rather liquid-like and is not comparable with a solid in the harmonic approximation.

The adequate description of systems with strong anharmonicities requires the application of numerical methods, that is, we have to recourse to simulation techniques. Here the "molecular dynamics method" has been used very successfully and is the foreground. In Chapter 3, this method will be described in detail and typical examples for nanosystems are quoted. In this section, let us bring a brief overview of the molecular dynamics method together with other relevant simulation techniques. In particular, we would like to explain the following techniques that have often been used in the computational treatment of nanosystems:

- molecular dynamics
- quantum molecular dynamics
- nonequilibrium molecular dynamics
- Monte Carlo method
- multiscale modeling

Other important methods used in nanoscience are finite element methods, Brownian dynamics, Langevin dynamics and molecular mechanics, but we do want to quote the details here.

In connection with "electronic properties" the following approaches are relevant: density functional theory, time-dependent functional theory, Hartree–Fock approximation, and the potential morphing method (as the most powerful method for solving Schrödinger's equation). All these methods will not be quoted in this book.

Also within the frame of these methods the "container model" is exclusively applied. Although the electronic properties have to be expressed quantum theoretically, the time here is still a classical parameter, that is, we also have here only one type of time and this is given by the usual clock time τ (Chapter 1).

Molecular dynamics

The molecular dynamics method is important in the computational treatment of nanosystems. Chapter 3 deals exclusively with this method. But let us already quote some essential features of this technique.

The relatively strong anharmonicities in nanosystems are not small perturbations to the harmonic approximation. Even at low temperatures the anharmonics are not negligible. Since the phonon model is not applicable here, we have to describe such nanosystems on the basis of the most general formulation. In the case of classical systems, Newton's mechanics has to be applied in the form of Hamilton's equations.

Here the classical equations of motion are solved by iteration by means of a high-speed computer. As outcome we get for each time τ the coordinates and the momentums for all N objects (atoms, molecules):

$$\tau\colon\ x_i, y_i, z_i, p_{xi}, p_{yi}, p_{zi}, \quad i = 1, 2, \ldots, N \tag{2.1}$$

As an input we need the initial values at time τ_a

$$\tau_a\colon\ x_i^a, y_i^a, z_i^a, p_{xi}^a, p_{yi}^a, p_{zi}^a, \quad i = 1, 2, \ldots, N \tag{2.2}$$

The interaction potential between the atoms (molecules) is needed as well as input, which obeys quantum mechanics; it is the traditional form of quantum theory in this case. The temperature of the nanosystem is determined by the initial values for the momentums [eq. (2.2)]. In Chapter 3, we will discuss the details of the molecular dynamics method.

Equation (2.1) represents the total information of the system in classical form. In particular, the anharmonicities are treated without approximation. No simple models are used here.

Nonequilibrium molecular dynamics

In order to be able to calculate transport coefficients efficiently, nonequilibrium molecular dynamics has been developed. Concerning this method, only a few remarks should be given here. The method has been developed in addition to usual (equilibrium) molecular dynamics and is known since the early 1970s [21–23]. To establish the nonequilibrium situation of interest, an external force is applied to the system. Then, the response of the system to these forces is determined through the simulation process. This method has been used for the calculation of diffusion coefficients, shear and bulk viscosities and thermal conductivities.

The method of nonequilibrium molecular dynamics can be used for the study of non-Hamiltonian systems, for example, dissipative systems, that is, systems that involve friction in one of its various forms. In such cases, particle trajectories are also calculated from the equations of motion. They are however not consistent with any Hamilton function.

Quantum molecular dynamics

The quantum molecular dynamics method directly combines the laws of traditional quantum theory with the classical equations of motion. It has been developed by Car and Parinello [24–26].

The characteristic items of the method are expressed through the following description: The laws of density functional formalism and the classical equations of motion are solved simultaneously. Within quantum molecular dynamics the classical Lagrange equation of the atomic positions and velocities are extended by a fictitious dynamics for the Kohn–Sham wave functions and their time derivations. Also here the classical time τ of traditional physics is used. Calculations of this type are relatively complex and, therefore, the quantum molecular dynamics method is restricted to a few hundred particles.

The Monte Carlo method

In contrast to molecular dynamics, the Monte Carlo is a numerical approach in which specific stochastic elements are used. Whereas the molecular dynamics method enables the study of structural and dynamical properties of a many-particle system, the Monte Carlo method only deals with structural (static) properties, and these are configurational averages and thermodynamic quantities. The results produced by the two methods, molecular dynamics and Monte Carlo, should be in agreement to an order of N^{-1} and, therefore, one can expect agreement of the two methods within the statistical error.

The Monte Carlo method has been used in a diverse number of strategies. The following types have often been used: classical Monte Carlo, quantum Monte Carlo, path-integral Monte Carlo, volumetric Monte Carlo, simulation Monte Carlo, inverse Monte Carlo. Here we do not want to give details about this diverse number of ways.

Multiscale modeling

Statements about large scale behavior can be made with the help of so-called multiscale modeling. Nanoscale properties are reflected at large scales and are treated on the basis of multiscale modeling. In this way we can understand how nanoscale properties give rise to this large-scale behavior. Certain strategies have to be chosen in order to be able to study the properties over a wide range of length and time scales, where the time itself is exclusively defined by the external clock time τ. The methods of multiscale modeling have been developed within the container model. Atomistic and coarse-grained simulations have to be done for multiscale modeling.

In materials science, multiscale modeling is used to predict certain material properties at dimensions ranging from a fraction of a nanometer to meter. For example, the mechanical properties can be explained in this way. Here we have to distinguish between three length scales: atomic scale (nanometer), microscale (micro-meter), and mesoscale (millimeter and above). Multiscale modeling connects these scales. Such investigations are exclusively based on simulations with sufficiently powerful computers.

In connection with multiscale modeling, let us mention an interesting example: the biopolymer translocation through nanopores. This translocation process is a typical multiscale problem. On the one hand, for the understanding of the microscopic details, that is, the structure and dynamics, the polymer-pore system is important and necessary for the understanding. On the other hand, there are processes within this system that take place on a large time scale, for example, the molecular filtering and the protein transport through membranes. In other words, not only the microscopic behavior is relevant but also processes on macroscopic timescales have to be understood. The translocation time is a macroscopic timescale property.

Such investigations are usually done on the basis of a bottom-up approach, starting with atomistic models, where molecular dynamics calculations are relevant. This nanoscale information will be the basis for coarse-grain models.

2.5.3 Simplifying assumptions

Let us summarize the situation: How can we determine the properties of a many-particle system on the basis of a given Hamiltonian, which includes the complete interaction potentials (see in particular Chapter 3).

In the conventional treatment of this problem we need "simple models." In general, simple models can be found by "controlled" approximations from the general case. However, in connection with many-particle systems the general mathematical formulation of the problem is in all known cases very complex, and simplifying approximations have to be chosen.

But, as a rule, such approximations cannot be obtained by controlled simplifying steps from the general formulation, which is, as we remarked, given in terms of the Hamiltonian. It is therefore a fact that simple models are introduced just for convenience. Only in the case of the crystalline solid in the harmonic approximation "simple models" are achievable.

The general description of system properties of classical many-particle systems without the use of "simple models" and other simplifying assumptions can be done by means of molecular dynamic calculations. The general situation is summarized in Figure 2.2.

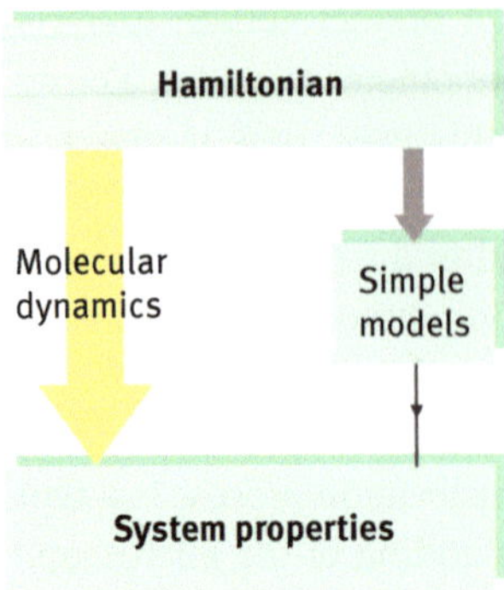

Figure 2.2: The general physical formulation of a classical many-particle system is given in terms of the Hamiltonian and the equations of motion. If a "simple model" is available, the properties of a many-particle system can be determined on the basis of such a specific model; systems of nanometer size are of course included here. In most cases, additional assumptions are necessary. But even then, the description remains rudimentary and is not sufficient for the understanding of nanosystems. The advantage of molecular dynamics is that the description can be done without simple models and other additional assumptions. This is important since in most cases such systems are complex and simplifying models cannot be formulated. This statement is valid for all systems with strong anharmonicities and disordered structure. They have often been chosen merely for convenience and could not always be deduced by controlled simplifying steps from the general case. For nanosystems, a simple model could not be found until now. Therefore, the molecular dynamic methods are of particular importance in the nanorealm.

2.5.4 Some critical remarks

In Section 2.4.2 we have quoted the most relevant theoretical and computational methods for the analysis of atomistic (molecular) properties of nanosystems. Methods for the treatment of electronic properties are not discussed in this monograph.

These methods are not nanospecific; they have been taken over from other disciplines. However, we have to be careful and must critically analyze the specific nanosituations. In general, the usual theoretical and computational tools have to be developed further at the nanolevel, and it is of particular importance to recognize the shortcomings of these usual methods when they are applied to nanoscience. But what are the modifications with respect to nanoscience?

Theoretical preparation of nanosystems

What about the theoretical (computational) preparation under nanoscience conditions? This is always dependent on the system under investigation; nanosystems demand new steps and efforts in their theoretical preparation, at least in some cases. Even the "basic conception" is relevant when we try to make statements about nanosystems and nanoprocesses.

An essential weak point is that all these traditional methods work within the frame of the container model and with only "one" time structure, which is external in character and is defined by the clock time τ.

As we have outlined in Chapter 1, system-specific time structures with quantum aspect are generally needed when we want to understand nanosystems properly, for example, self-organizing processes, which are important in nanoscience.

In conclusion, the experimental preparation (production) of nanosystems on the basis of a realistic basic conception is not only desirable but necessary in order to avoid erroneous developments.

Temperature effects

One of the essential points in connection with nanomaterials is that their properties can respond or switch according to external conditions. The temperature here is an important parameter. This in particular means that the properties of those nanomaterials have to be described as a function of temperature.

Metallic systems are strongly dependent on temperature. However, in literature the electronic properties are exclusively calculated in terms of density functional formalism (DFT) and on time-dependent DFT, but these approaches only deliver zero-temperature facts. The Mermin approach [27] allows to treat the properties of an electron system more realistically, namely for nonzero temperatures. The Mermin-approach should therefore be applied to nanosystems and not the zero-temperature DFT formalisms.

Atomistic (molecular) simulations technique need the precise knowledge of interaction potentials as input, and this is in most cases difficult and their determination has to be performed carefully. This is in fact necessary because nanoproperties are very sensitive to small variations in the interaction potential. However, in many cases the determination of interaction potentials has been done in an uncontrolled manner. In particular, the temperature dependence of potentials in most cases is not considered.

The modeling of potentials has often been done on the basis of relatively drastic and unmanageable simplifications. This loss of precision is often compensated by further inaccuracies in connection with other fit parameters of the system under investigation. In other words, within the frame of such a treatment a certain inaccuracy is compensated by another.

The future challenges in connection with complex nanosystems should not be based on such kind of treatment and, furthermore, the theoretical (computational) methods should not simply be adopted from other disciplines and other systems without checking their applicability for nanoscience. In other words, controlled steps are necessary in the preparation of nanosystems and their interaction laws. The theoretical and computational preparations of nanosystems require care, not only in connection with atomistic simulations but also with respect to electronic properties and their dependence on temperature. We have to be careful with zero-temperature approaches (DFT and time-dependent DFT).

In conclusion, the reliable theoretical analysis of nanosystems inevitably means that the theoretical (computational) methods used in their description have to be precisely under control, that is, with regard to approximations and simplifications. The temperature effects and the fact that nanoproperties are very sensitive to small variation in the interaction potential require that.

2.6 The atomistic standard model

Theoretical materials research at the nanolevel needs statistical mechanics and in particular the theory of liquids. Why? Nanosystems behave strongly anharmonically and their structure is in most cases considerably disturbed even far below the melting temperature of the corresponding bulk state. Therefore, in nanoscience (nanotechnology) one not only needs solid-state physics but the basics of statistical mechanics and the theory of liquids as well. It is very often more appropriate to apply the physics of liquids as the basis for the theoretical description of nanosystems.

2.6.1 Solids and liquids

The standard model of solid-state physics is characterized by an ordered structure and the harmonic approximation for the dynamics, which we want to summarize by the symbol $\omega_p(q)$ for phonon frequencies. This standard model of solid-state physics is not suitable for nanosystems, but the characteristics of liquids are obviously more appropriate.

These liquid characteristics are defined in terms of correlation functions and interaction potentials, which are usually given in terms of $g(r)$ and $v(r)$, where $g(r)$ is the well-known pair correlation function and $v(r)$ is the pair potential. Within this statistical mechanical description time correlations are of course included, for example, the velocity autocorrelation function whose Fourier transform has to be considered as the generalized phonon density of states.

2.6.2 Complex dynamical behavior

Due to the strong anharmonicities, the dynamics of such systems include a broad range of different dynamical states. Here, small local vibrations and complex diffusion processes are typical. This relatively complex situation indicates that the development of a suitable analytical standard model for the theoretical description of nanosystems will hardly be possible; even in the case of liquids in the bulk, a standard model could not be found up to now.

Therefore, for the treatment of nanosystems, which behave essentially anharmonically, the most important tool is the molecular dynamics method since the anharmonics are treated within this method without approximation. In fact, the typical anharmonicities in connection with nanosystems cannot be considered as small perturbations to the harmonic approximation. No other microscopic method in condensed matter physics allows to treat anharmonicities in such a general way. This is only possible in connection with phenomenological models that are however often too rudimentary and detailed statements about the structure and dynamics are hardly possible.

Thus, the molecular dynamics method should be considered as the "standard method" for the theoretical (computational) description of nanosystems. On the basis of these considerations, we come to the following rough classification scheme:

$$\begin{gathered} \omega_p(\mathbf{q}) \\ (\textit{standard model of solids}) \\ \downarrow \\ g(r),\, v(r) \\ (\textit{statistical mechnics}) \\ \downarrow \\ \textit{molecular dynamics} \\ (\textit{standard model for nanosystems}) \end{gathered} \tag{2.3}$$

In nanoscience, we often work on the basis of a few hundred atoms. In general, with decreasing particle number disorder effects and anharmonicities increase. This has an influence on the theoretical treatment of such systems. Here molecular dynamics is the standard model.

2.7 The relevance of basic quantum theory for nanotechnology

2.7.1 General remarks

Quantum effects plays an essential role in nanoscience (nanotechnology). It is not the goal to work under conditions where quantum effects are suppressed and we should not explicitly construct situations where quantum effects are eliminated. In other words, we should profit from the quantum realm in the construction of new nanosystems: electronic nanodevices, the developments in nanomedicine, nanorobotics, drug design, quantum electromechanical systems, quantum computers, and so on. "In the last decade device fabrication and experimental control have progressed to such an extent that one can now see how quantum mechanics will be used to build a new technology" [28].

All these devices and systems operate on certain quantum principles. Milburn [28] gave a list of such principles reflecting the key elements of quantum mechanics relevant for technological tasks: uncertainty principle, superposition, quantization (quantum size effect), tunneling, entanglement and decoherence. In this connection the following remark by Milburn is instructive:

> A number of imperatives will drive the development of quantum technology. To begin, there is the quest for smaller and faster devices taking us to the nanoscale. At this scale quantum principles become manifest at low temperatures. Nanotechnology must take heed of quantum principles at some level. Any technology requires transducers and high precision measurement. Quantum theory has some very important things to say about measurement and its limits. Quantum technology will necessarily lead to new instruments. It is already clear that a number of communication challenges involving bandwidth and energy can be faced within a quantum context, with teleportation being the most surprising protocol. Finally, there is the promise of quantum computing. In each case the quest to harness quantum mechanics for technological ends will bring new science along with new experimental opportunities [28].

This statement refers to the traditional form of quantum theory. However, we have discussed in Chapter 1 that traditional quantum theory, which is based on the "container model," has essential shortcomings when we want to describe nanosystems.

The container model, which is used within traditional quantum theory without exception, is obviously a barrier for further developments in this field of quantum phenomena and other features in the world. The quantum aspect of time is here a typical example.

Again, within the nanorealm self-organizing processes are important and, as we already remarked several times, such kind of processes belong to the heart of nanoscience (nanotechnology). For an adequate description of self-organizing processes, a quantum aspect of time is necessary, but leading physicists were not able to introduce a quantum aspect for the time. The reason for this lack should have its origin in the container model as working hypothesis.

The projection principle allows to describe the time as quantum phenomenon. The transition from the container model to the projection principle does not only mean that there will be specific improvements, but this principle will unlock the view to a more sophisticated world with more realistic features. These new features reflect observable facts which could not be explained so far. Just the connection between matter and mind, mind and brain and other things as evolution can be taken into account and opens the view for a better classification of living biological systems themselves.

The basic laws of theoretical physics are without doubt of particular relevance for the development of nanotechnological devices and for the understanding of living biological systems. In Section 2.1 we have stated the following: working at the ultimate level also means that any change in the basic physical laws will directly influence processes in the nanorealm, that is, without any intermediate step.

This can lead to completely new perspectives in connection with applications. Therefore, to work on nanoscience also means to develop the basic laws of theoretical physics further. This could be important just in connection with quantum theory.

The wave–particle duality is a critical point in traditional quantum theory. New experimental results indicate that we must possibly rethink in this connection. We will discuss this point in more detail in the next section.

2.7.2 Traditional quantum theory: further remarks

The real existence of quantum phenomena destroyed the mechanistic world view. Quantum reality is rationally comprehensible but cannot be visualized like Newton's classical reality. The laws of traditional quantum theory were not easy to recognize.

The most important of them were discovered in the first decades of the last century. An international group of physicists were involved. As is well known, the essential peculiarities of the theoretical formalism have been constructed by Heisenberg and Schrödinger in 1925 and 1926. On this basis of these quantum laws, an enormous number of effects in atomic physics, chemistry, solid-state physics and so on could be predicted and successfully described. However, approximately 100 years after its formulation the interpretation of these quantum laws is

> by far the most controversial problem in current research in the foundations of physics and divides the community of physicists and philosophers into numerous opposing schools of thought [29].

There is an immense diversity of opinion and a huge variety of interpretations. It is typical that the interpretations lead to completely different conceptions of the world. For example, the Copenhagen interpretation of quantum theory (proposed by Bohr in 1927) is quite different from the many-worlds theory (proposed by Everett III in 1957).

The Copenhagen interpretation of traditional quantum theory is presently considered as the most important interpretation frame. It is the most relevant interpretation and most scientists respect it as the standard interpretation of traditional quantum theory. It stands and falls somehow with this interpretation. However, one of the essential points of the Copenhagen interpretation, the principle of complementarity, is obviously not a general feature. New basic experiments indicate that.

2.7.3 Basic new experiments

Again, the Copenhagen interpretation is presently considered as the standard interpretation of usual quantum theory. Nevertheless, it is not really recognized as the final step in the understanding of quantum phenomena. The conception, on which the Copenhagen interpretation is based, seems to be artificial and not convincing. This is in fact the opinion of many scientists and is obviously due to the fact that we do not know the origin of the quantum laws; Schrödinger's equation has been assumed and could not be deduced. The situation is unsatisfactory and can be summarized as follows:

> The Copenhagen Interpretation held away for more than 50 years, from 1930 until well into the 1980s, almost unopposed by the vast majority of physicists. They did not care about the deep philosophical puzzles associated with the Copenhagen Interpretation – indeed, many still do not care – provided that it could be used as a practical tool for predicting the outcome of experiments. But in recent years there has been growing unease about what quantum theory "means", and increasing efforts have been made to find alternative interpretations. [30]

However, it will hardly be possible to formulate a new interpretation without contradictions on the basis of the traditional analytical form of quantum theory, and this is because there are some problems which have not been solved up to now [14]. The unsatisfactory role that time plays within the present form of traditional quantum theory is one of many critical points, but also the particle concept seems to reflect an unclear situation. In the usual form, quantum theory particles are assumed to be local existents. This is an assumption without knowing where it comes from. Another serious problem is the "superposition principle" and requires the collapse of the wave function, and there is no mechanism for that.

All these points indicate that the formalism of traditional quantum theory cannot be considered as a final solution. Therefore, it will probably not enough to reinterpret the Copenhagen interpretation or one of any other proposals made in this field. On the other hand, a changed or extended quantum formalism could lead to completely new perspectives in connection with applications, in particular, within nanoscience.

2.7.4 Simultaneous detection of particle and wave aspects

It is, therefore, of fundamental importance that there are new experimental results which clearly indicate that the "principle of complementarity" cannot be considered as a general fact. These experiments show that the wave–particle duality is obviously not a general feature, that is, Bohr's viewpoint becomes questionable, at least in the original form. If that is really the case, the situation is changed fundamentally. But this is not surprising because it is, together with the lack of a quantum aspect of time, only a further point that points on the limitations of traditional quantum theory.

At the beginning of modern quantum theory, the question was "Is matter (or light) a wave or a particle phenomenon?" In answering this question, Bohr argued as follows: Only one of the two incompatible possibilities is realized, depending on the type of experiment performed on them, and we came to the notion of complementarity.

The situation is clear: either we perform an experiment typical for waves or we choose an experimental arrangement that exclusively marks the particle aspect. This ingenious idea of Niels Bohr made it possible to overcome a serious problem. Only in this way, wave and particle aspects are not incompatible. In other words, not only wave and particle concepts are mutually exclusive but also the corresponding experimental arrangements. However, the new experiments showed that just this point is obviously not the case as demonstrated in the case of light.

The simultaneous detection of particle and wave aspects is possible, despite Bohr's principle of complementarity. There exist optical experimental setups which have been used to prove that experimentally. The experiments have been proposed by Ghose, Home and Agarwal [31] and also by Ghose [32]. In this context, the first experiment has been done by Mizobuchi and Ohtake [33], which was based on the proposal given in Ref. [34]. Although this experiment is without any doubt important, there were experimental limitations. These limitations are mainly due to low statistics, as outlined by Unnikrishnan and Murphy [35].

In order to overcome the uncertainties in connection with the Mizobuchi-Ohtake experiment, Brida, Genovese, Gramegna and Predazzi performed with success a new birefringent experiment (see [36]), and this experiment led to a clear conclusion. A "simultaneous" detection of particle and wave aspects is after all possible.

This is a completely new step in connection with the wave–particle duality. The result is in contrast to Bohr's interpretation in which the principle of complementarity is required, and this principle does not allow the simultaneous detection of particle and wave aspects.

2.7.5 Conclusions

Once again, both experimental groups [33, 36] came to the same conclusion: light is able to reveal particle and wave aspects simultaneously, and this is in contrast to the principle of complementarity which is one of the very basics of traditional quantum theory (Copenhagen interpretation). We have therefore to conclude that the principle of complementarity can obviously not be considered as a general law, and this must inevitably have consequences for the whole picture we presently have of traditional quantum theory.

As we remarked, there are of course other attempts to interpret traditional quantum theory, for example, the so-called many-worlds theory. But the Copenhagen interpretation is, as we already remarked, presently seen as the standard interpretation of traditional quantum theory.

Bohr's law "a phenomenon is always an observed phenomenon" is just based on the principle of complementarity. This principle has been formulated in connection with the problem scientists had in connection with the wave–particle phenomenon. The principle of complementarity is definitely needed within traditional quantum theory in order to avoid logical contradictions.

The principle of complementarity teaches that only "one" of the two incompatible possibilities (wave and particle) can be realized at a certain time. Here the experimental arrangement comes into play. The following is required: the experimental arrangements, which determine those properties, must be similarly mutually exclusive.

Such a concept allows in fact a description of quantum entities (particle and wave) without logical contradiction in ways which are mutually exclusive. However, the new experiments [33, 36] show just the opposite: wave and particle aspects can be observed simultaneously with only "one" experimental setup, that is, the principle of complementarity obviously breaks down and the entire Copenhagen interpretation becomes questionable.

On the other hand, if Bohr's law "a phenomenon is always an observed phenomenon" should no longer be valid, the properties of a system should also be defined between observations and not only in connection with observations. For example, this point would be of interest in connection with the famous double-slit experiment.

Our statements concerning the term "objectivity" must possibly be revised: Within traditional quantum theory, it is meaningless to talk about a phenomenon without observation. Here, the human being's intention is a relevant factor. Since the choice of the experimental arrangement is purely a matter of human intention, we may not consider the properties of quantum systems as objectively real, and this is because the human intention influences the structure of physical reality. This statement is of course based on the Copenhagen interpretation and becomes questionable without the principle of complementarity.

To what extent we have to revise the whole traditional quantum mechanical apparatus has to be investigated carefully. However, through our discussion in Chapter 1, we came to the conclusion that minor changes are not sufficient. The entire basic conception has to be replaced. That is, the container model has to be replaced by the projection principle. This is not only of principle interest but is in particular important for the future developments in nanoscience (nanotechnology). As we have outlined in Section 2.1, the developments in nanoscience are directly influenced by the basic laws of theoretical physics.

2.8 Summary

1. We came to the conclusion that technology has reached the level of science: nanoscience and nanotechnology are basically undistinguishable.
2. Why theoretical and computational nanoscience? Clearly, in order to predict and to describe new nanosystems and to give nanotechnology a direction, but also to keep nanoscience under control, just in connection with the self-assembly of artificial biological systems, where the final structure is in general not known.
3. Nanobiology and nanomedicine will influence our future considerably. Also in the prediction and description of such systems, theoretical and computational nanoscience will play an important role. What mathematical tools do we need here? Are there principal limitations in connection with the understanding of biological systems? The last question directly arises when we consider the human brain as a mathematical object. Is a complete description of the human brain possible on the basis of the laws of theoretical physics? We came to the conclusion that this is probably not possible.
4. Due to the strong anharmonicities, the dynamics of atomic systems is characterized by a broad range of different dynamical states including small local vibrations and complex diffusion processes. This situation indicates that the development of a suitable "analytical standard model" for the theoretical description of nanosystems will hardly be possible; even in the case of liquids in the bulk, a standard model could not be found up to now.
5. Therefore, the most important tool for the investigation of nanosystems is the molecular dynamics method since anharmonicities are treated within this method without approximation, and this is important because the typical anharmonicities in connection with nanosystems cannot be considered as small perturbations to the harmonic approximation.
6. No other microscopic method in condensed matter physics allows to treat anharmonicities without approximation. This is only possible in connection with phenomenological models. Thus, the molecular dynamics method should be considered as the standard method for the computational description of nanosystems.
7. Atomistic simulations (particularly in connection with molecular dynamics) need the precise knowledge of interaction potentials as input, and this is in most cases not simple. But we know that nanoproperties are very sensitive to small variations in the interaction potential. However, in many cases the determination of potentials has been done in an uncontrolled manner. The modeling of potentials has often been done on the basis of relatively drastic simplifications. This loss of precision is often compensated by further inaccuracies in connection with other parameters of the system.

8. It is the goal of nanoscience to profit from quantum effects for the construction of new systems: electronic nanodevices, developments in nanomedicine, nanorobotics, drug design, quantum electromechanical systems, quantum computers and so on.
9. In this connection, the basic laws of quantum theory are of particular relevance. Thus, to work on quantum nanoscience also means to develop these basic laws further. New experimental results with respect to the particle–wave duality indicate that there is possibly a need for that.

Chapter 3
Classical treatment of nanosystems

The quantitative treatment of physical problems started with Newton's classical mechanics more than 300 years ago. Newton based his theory on what he had directly before his eyes. This was the reality before him. Within this view, reality is a "container" in which the material bodies are embedded. The container itself is what we experience as space. The traditional form of quantum theory also works within this container model. Some features of the container model have been discussed in the sections before. Let us give more details in this chapter.

3.1 Container model versus projection principle

The classical description, in the sense of Newton, is not only successful with respect to the motion of planets around the Sun and other macroscopic phenomena, but also the structure and dynamics of atomic (molecular) systems, in particular nanosystems, are described well within this kind of classical frame, if the temperature of the system under investigation is sufficiently large.

We know that this container model of traditional physics does not fulfill relevant items of nature and the so-called projection principle is obviously more realistic and more sophisticated (see in particular Chapter 1). Nevertheless, the classical description of nanosystems within the container model is often used and is still subject of intensive discussion. In other words, such type of investigations should not be ignored, especially they often lead to useful results. The details of the container model are obviously not externalized in the case of such investigations, that is, in classical mechanics.

In nanoscience, the classical treatment of many-particle systems is a relevant task; here, the most important method is the molecular dynamics method (Chapter 2). In this chapter, some details of molecular dynamics will be introduced and some typical applications are given.

Before we discuss nanosystems within classical physics on the basis of the container model, let us briefly underline the main differences between the container model and the projection principle.

The main conceptions used in this monograph are the "container model" and the "projection principle." In this section, we only want to characterize the main differences between these two basic conceptions. A more detailed discussion is given in in Refs. [4, 5] and specific completions are given in Chapters 4–6. The container model is applied within the traditional views of physics (classical mechanics, usual quantum theory).

https://doi.org/10.1515/9783110525595-003

3.1.1 Container model

The situations with respect to the container model and the projection principle are summarized in Figure 3.1 and Figure 3.2. Within the container model, physically real entities are embedded in space and any object at time τ is characterized by a position vector $\mathbf{r} = (x, y, z)$ with the coordinates x, y, z. In Figure 3.1, two $\mathbf{r}$ vectors, $\mathbf{r}_i = (x_i, y_i, z_i)$ and $\mathbf{r}_j = (x_j, y_j, z_j)$, are represented for the objects i and j. They are simultaneously existent at time τ, where τ is again the clock time.

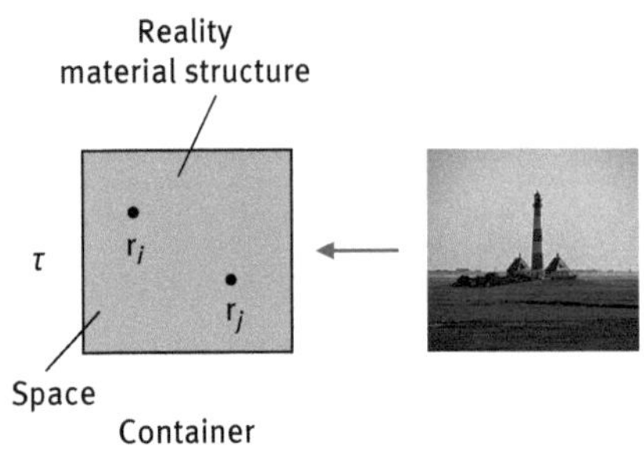

Figure 3.1: Within the container model, physically real entities are embedded in space, Here, two r vectors, $\mathbf{r}_i = (x_i, y_i, z_i)$ and $\mathbf{r}_j = (x_j, y_j, z_j)$, for the objects i and j are indicated and are existent at time τ. Within this model, the lighthouse, shown in the figure, is physically real, that is, it is material within this view. We have at each clock time τ, certain "material structures" in space appear directly before our eyes. The notation "clock time" underlines that it is the time measured with usual clocks of everyday life.

Within this model, the lighthouse, shown in Figure 3.1, is physically real, that is, within this view, the lighthouse is material in character. In fact, we experience it in this way, but it is a rather naive point of view. Within this view, it is assumed that we have at each clock time τ certain "material structures" in space that appear directly before our eyes.

3.1.2 Projection principle

Within the frame of the projection principle, there is the so-called basic reality, which is principally not accessible to human beings and, as we argued in [4, 5], this feature is explained by the phenomenon of evolution. This basic reality is projected onto the space–time and appears at clock time τ as picture in the form of a geometrical structure. Again, within the frame of the container model the space is filled with material entities, in projection theory not.

The space–time itself contains a new type of time. It is the system-specific time t (see Chapter 1), which is an internal entity and belongs to the system under investigation and is therefore system specific. This type of time t has nothing to do with the time τ: τ is external in character and is a reference time shown by our clocks, which we use in everyday life and, therefore, we used for τ the notation "clock time."

In traditional physics, where the container model is applied, the system-specific time t is not known and defined, respectively. Here, the phenomenon "time" only appears in the form of the clock time τ. No doubt, this is a restricted view and has often been criticized, in particular by Ilya Prigogine.

In Figure 3.2, two space–time positions, denoted by $\mathbf{r}_i, t_i = (x_i, y_i, z_i)$, t_i and $\mathbf{r}_j, t_j = (x_j, y_j, z_j)$, t_j for the objects i and j, are indicated that are simultaneously existent at time τ. However, there are no physically real (material) entities positioned in the space–time of projection theory, but exclusively geometrical positions. This space–time is denoted by $(\mathbf{r}, t)$ space.

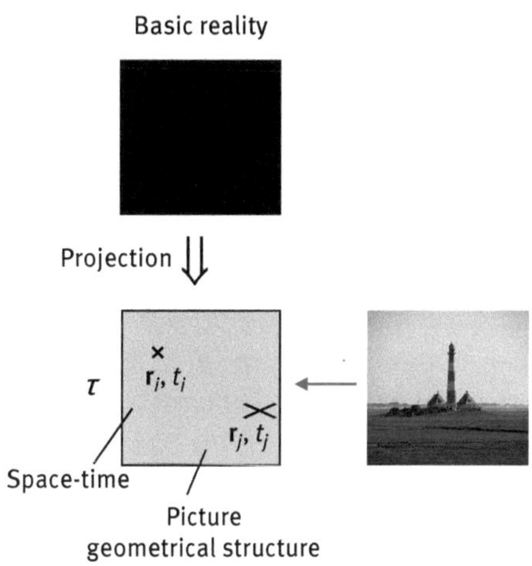

Figure 3.2: Details of the projection principle. Here, we have the so-called basic reality, which reflects the absolute truth but is principally not accessible to human beings; this feature is explained by the phenomenon of evolution. The facts of basic reality are projected onto the space–time and appear at clock time τ as picture in the form of a geometrical structure. Again, within the frame of the container model the space is filled with material entities. In this figure, two space–time positions, denoted by $\mathbf{r}_i, t_i = (x_i, y_i, z_i)$, t_i and $\mathbf{r}_j, t_j = (x_j, y_j, z_j)$, t_j for the objects i and j, are indicated that are simultaneously existent at time τ. However, there are no physically real (material) entities positioned in the space–time of projection theory, but exclusively geometrical positions. In contrast to the container model (Figure 3.1), the lighthouse must be assessed in connection with this figure as strict geometrical structure.

In contrast to the container model (Figure 3.1), the lighthouse must be assessed in connection with Figure 3.2 as strict geometrical structure. Again, the lighthouse is assessed as material object within the container model.

In conclusion, there are two world views: Within the container model, we have physically real (material) objects directly before our eyes. Within the frame of the projection principle, we have geometrical entities before us. As can be argued convincingly [4, 5], both views do not violate the direct observations.

3.1.3 Fictitious realities

Within the frame of the projection principle, the actual physically real (material) world is embedded in basic reality, which cannot be accessed directly by human beings (Chapter 1). Instead, we have to work with fictitious realities, and this kind of reality describes the material part of basic reality.

Fictitious realities are characterized through the momentum $\mathbf{p}$ and the energy E, that is, they are embedded in $(\mathbf{p}, E)$ space. The contents of $(\mathbf{p}, E)$ space is projected onto $(\mathbf{r}, t)$ space; $\mathbf{p}$ and E themselves do not appear in space and time, but form a separate space [$(\mathbf{p}, E)$ space] in analogy to basic reality, which is separate from space and time as well [4].

The situation is summarized in Figure 3.3(b) for two objects, indicated by i and j, having the elements $\mathbf{r}_i, t_i$ and $\mathbf{r}_j, t_j$ in connection with $(\mathbf{r}, t)$ space. The corresponding elements in $(\mathbf{p}, E)$ space are indicated by $\mathbf{p}_i, E_i$ and $\mathbf{p}_j, E_j$, also for the two objects i and j.

That is, within the frame of the projection principle, the quantities $\mathbf{p}$ and E are not embedded in space and time [$(\mathbf{r}, t)$ space], but are located in $(\mathbf{p}, E)$ space. The contents of $(\mathbf{p}, E)$ space is projected onto $(\mathbf{r}, t)$ space and $\mathbf{p}$ and E themselves do not appear in the space–time. Within projection theory, the quantities $\mathbf{p}$ and E are not physically real, but they “describe” physical reality (basic reality) [4, 5]. That is an important point.

The fact that $\mathbf{p}$ and E are used in strict analogy in Newton’s theory, we have to conclude that the picture of reality, located in $(\mathbf{r}, t)$ space, exclusively reflects material states, that is, there are no states of the mind within $(\mathbf{r}, t)$ space which, on the other hand, exists within projection theory [4, 5]. Mind states are discussed in Chapters 4 and 5. Although $\mathbf{p}$ and E are not located in $(\mathbf{r}, t)$ space, the space–time positions $\mathbf{r}, t$ are strictly correlated to the entities $\mathbf{p}$ and E [Figure 3.3(b)].

The situation in connection with the traditional container model are summarized in Figure 3.3(a): Here, the material bodies are positioned in space (without the variable t, which is not defined here). Each of the two bodies, having the position $\mathbf{r}_i, t_i$ and $\mathbf{r}_j, t_j$ at time τ, has a certain momentum $\mathbf{p}$ and a certain energy E at the same time τ: $\mathbf{p}_i, E_i$ and $\mathbf{p}_j, E_j$. In contrast to the projection principle [Figure 3.3 (b)], the quantities $\mathbf{r}_i, t_i$ and $\mathbf{r}_j, t_j$ with $\mathbf{p}_i, E_i$ and $\mathbf{p}_j, E_j$ have to be considered as physically real. In particular, $\mathbf{p}_i, E_i$ and $\mathbf{p}_j, E_j$ are embedded in space, which is in contrast to the projection principle [Figure 3.3(b)].

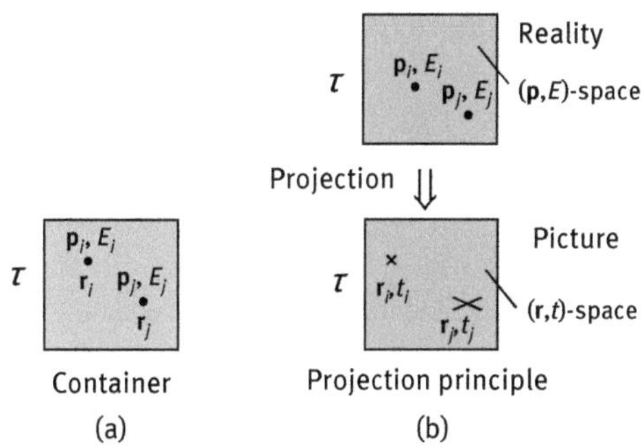

Figure 3.3: (a) Here is the situation in connection with the traditional container model summarized. The material bodies are positioned in space (without the variable t, which is not defined here). Each of the two bodies, having the positions $\mathbf{r}_i$ and $\mathbf{r}_j$ at time τ, has a certain momentum $\mathbf{p}$ and a certain energy E at the same time τ: $\mathbf{p}_i, E_i$ and $\mathbf{p}_j, E_j$. The quantities $\mathbf{r}_i$ and $\mathbf{r}_j$ with $\mathbf{p}_i, E_i$ and $\mathbf{p}_j, E_j$ have to be considered as physically real. In particular, $\mathbf{p}_i, E_i$ and $\mathbf{p}_j, E_j$ are embedded in space. (b) Within the frame of the projection principle, the actual physically real (material) world is embedded in basic reality, which cannot be accessed directly by human beings. Instead, we have to work with fictitious realities, and this kind of reality describes the material part of basic reality. Fictitious realities are characterized through the momentum $\mathbf{p}$ and the energy E, that is, they are embedded in $(\mathbf{p}, E)$ space. The contents of $(\mathbf{p}, E)$ space are projected onto $(\mathbf{r}, t)$ space, and $\mathbf{p}$ and E themselves do not appear in the space–time, but form a separate space [$(\mathbf{p}, E)$ space] in analogy to basic reality, which is separate from the space–time as well. In the figure two objects, again indicated by i and j, have the positions $\mathbf{r}_i, t_i$ and $\mathbf{r}_j, t_j$ in $(\mathbf{r}, t)$ space. The corresponding momentums and energies are indicated by $\mathbf{p}_i, E_i$ and $\mathbf{p}_j, E_j$ in $(\mathbf{p}, E)$ space. Within the frame of the projection principle, the quantities $\mathbf{p}_i, E_i$ and $\mathbf{p}_j, E_j$ are not embedded in the space–time, but the contents (in this case, $\mathbf{p}_i, E_i$ and $\mathbf{p}_j, E_j$) of $(\mathbf{p}, E)$ space are projected onto $(\mathbf{r}, t)$ space and $\mathbf{p}_i, E_i$ and $\mathbf{p}_j, E_j$ themselves do not appear in the space–time. Within projection theory, the momentums and energies are not physically real, but they describe physical reality (basic reality).

3.1.4 The pragmatic view

The container model has been confronted with the projection principle (Sections 3.1.1 and 3.1.2). We have to distinguish between physically real (material) considerations (container model) and the geometrical representation of the world outside; it is the basic reality within the frame of the projection principle. No doubt, the projection principle is more sophisticated than the container model. In particular, the phenomenon, “time” is treated in projection theory more realistically than within the container model. Both views work with the clock time τ, but in the case of the projection principle the system-specific time t appears, which is of fundamental importance for natural science, as we have outlined in Chapter 1.

Within the container model, we have, at clock time τ, the three variables $\mathbf{r}, \mathbf{p}$ and E, that is, in this case, we have at each time τ for each object a position, a momentum and an energy:

$$\tau : \mathbf{r}, \mathbf{p}, E \text{ (container model)} \tag{3.1}$$

In the case of projection theory, we have, in addition, the system-specific time t and we get,

$$\tau : \mathbf{r}, t, \mathbf{p}, E \text{ (projection principle)} \tag{3.2}$$

Although the system-specific time t does not exist within the container model, we also have here time-dependent processes, but only in a restricted form on the basis of the clock time τ. Certain aspects in nanoscience seem to be described well on the basis of τ alone.

The container model is not used because of its ingeniousness, but it simply works well in certain cases without explicitly knowing why. In fact, it has been applied intensively for the analysis of various nanosystems, and it will certainly be used in the future. The container model belongs therefore to the tools of modern nanoscience, even when the future will rather be concentrated on the projection principle.

This is the pragmatic standpoint, but this perspective can be justified because the container model works properly in specific case, just when we deal with classical many-particle systems. But it is merely a pragmatic perspective. There are relevant weaknesses within the container view, which become obvious through the projection principle. The molecular dynamics method, which we will discuss in detail in this section, works well within the container model although it has to be considered as a rough recipe from the epistemological point of view,

Since basic reality, the absolute truth, cannot be accessed directly by human beings, we always have to work with specific recipes, even in the case of the projection principle. This in particular means that we indeed permanently improve our physical views, but we cannot reach the absolute, final truth. This is an important and very principal statement.

In this monograph, we would like to confront the present with the future, the container model with the projection principle. There is an expansion of physics through the projection principle that is initiated by the features of nanosystems. Nanodevices belong to this field but also the possibility to influence brain functions. At the nanolevel, the properties of matter (condensed matter) emerge; it is therefore the ultimate level (Chapter 1).

Remark

The container model is essentially different from the view described within the frame of the projection principle. Although the projection principle is superior against the container principle, the container model works astonishingly well in many cases. What is the reason for that? The interactions used within the container model are strictly adapted to the features of this model. The inadequateness of both, the models for the interactions and the container model, obviously compensate each other. This leads to a useful conception.

3.1.5 Molecular dynamics

In this chapter, we would like to discuss relevant possibilities for the application of the container model for classical considerations where quantum effects are negligible. In this case, eq. (3.1) is responsible, that is, the system-specific time t does not appear within such considerations. Here, N atoms (molecules) are embedded in space and are considered as physically real entities. The N objects interact with each other. For such type of system, Newton's equations of motion are solved by iteration and we get the structure and dynamics for the various nanostates. This method is called "molecular dynamics" (see also Chapter 2). In the following sections, we will bring details and specific applications.

3.2 Atomic nanodesign

The fundamental description of systems in the field of condensed matter, that is, in solid state physics, materials science and nanotechnology, is based on many-particle systems where the particles (objects) are made of atoms or molecules. The objects interact with each other, and this is an essential point because small variations of the interaction potentials influence the properties strongly. The properties of such systems are expressed through the structure and dynamics of the atoms (molecules). Due to the interaction between the objects, the many-particle systems in condensed matter form a unified whole, and the properties of such systems cannot be considered as a superposition of individual atom/molecule properties alone, but there are strong interrelations between them.

3.2.1 What is the relevant input?

In the case of nanosystems, we have the following situation: The number of objects is relevant, that reflects the characteristics of such systems, and this parameter is essential to know and has to be estimated realistically. How many atoms/molecules have to be used in the description? What number of objects is characteristic in the description of specific properties of nanosystems? But the nanofeatures are also reflected by the interaction between the objects. In conclusion, the interaction potentials as well as the number of atoms/molecules of a nanosystem are the essential and critical input information.

Macroscopic systems consist of a large number of atoms, and we know that approximately 10^{23} of these objects form the systems in solid-state physics and materials science. However, the characteristic features and properties are already reflected through a number of particles distinctly smaller than 10^{23}, and the essential characteristics are already reflected within relatively small parts. The particle number that is adequate for the description of a certain material must be investigated carefully.

This is however not necessary when we work within the realm of nanoscience (nanotechnology). Here, no selection of specific parts is possible, but we have to consider the entire system, and this is because the entire system characterizes the properties and not only a part of it. In the nanocase, the real systems normally consist of a few hundred (thousand) atoms or molecules, depending on the complexity of the phenomena, which appear in connection with the system under investigation. Then, the number N of objects determines the characteristics of the system and should be chosen correctly in calculations.

The mechanistic world view

The many-particle systems can be treated by classical mechanics if the temperature of the system under investigation is large enough. Here we work within the frame of the "container model" (see Section 3.1 and Figure 3.4) and the "mechanistic world view" is the background for everything. In other words, not only the container model characterizes this kind of reality but the mechanistic world view as well. Concerning the mechanistic worldview, let us quote the main items here.

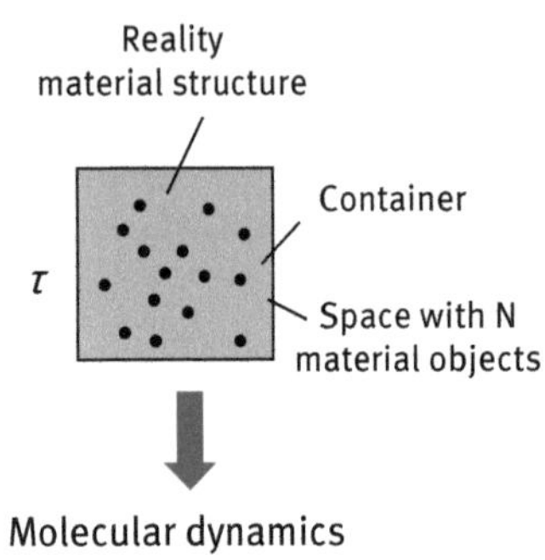

Figure 3.4: The molecular dynamics method is an important tool in nanoscience and this method has often been applied. Molecular dynamics is based on the container model and Newton's equations of motion. Here, *N* atoms (molecules) are embedded in space and are considered as physically real entities. The *N* objects interact with each other at each time τ.

The mechanistic view of nature is based on Newton's mechanical model of the universe. In connection with Newton's theory, two features play an essential role and must be underlined:

Feature 1:

The entities of the world are embedded at each time τ in an absolute space, that is, they move in an absolute space and absolute time. These entities are considered to be small, solid and indestructible objects, always remaining identical in mass and shape.

Feature 2:
Newtonian mechanics reflects a deterministic behavior. Our experiences in everyday life suggest this view. That is, Newtonian mechanics is closely related to a rigorous determinism, that is, the future path of moving objects (e.g., the earth) can be completely predicted and its past can be completely disclosed if its present state is known in all details.

From Newton's point of view, reality is given by the following vision: In the beginning, God had created everything, that is, the material objects, the forces between them and the equations of motion. Then, the entire universe was set in motion and it has continued to run ever since, like a machine, governed by Newton's equations of motion.

In other word, in the Newtonian view, there exists a giant cosmic machine, and this machine behaves completely causally and determinately; a definite cause gives rise to a definite effect, and the future of any mechanical object can be predicted with absolute certainty if its present state and the forces acting upon it are known. Again, this behavior is close to what we have before our eyes.

The tremendous success of Newtonian mechanics in the eighteenth and nineteenth centuries suggested that this mechanical view could be consistently applied to all branches of physics, that all phenomena could be explained in this way, not just in the movements of the planets.

In conclusion, the laws given by Newtonian theory, the laws of mechanics, were seen as the fundamental laws of nature; classical mechanics was considered to be the absolute truth and it was considered for almost three centuries as the ultimate theory of natural phenomena.

It was Lamettrie who required that, in addition to usual objects, living human beings (and also animals and plants) should behave like a Newtonian machine, that is, without any spontaneity. In Lamettrie's opinion, the behavior of human beings was also completely causal and determinate like the dead objects around us. In other words, it was seen as being possible to predict – in principle – with absolute certainty, all the future activities of a man if his present state was known.

Within this view, men are merely automata and not free individuals capable of influencing the course of events by their volitions; the world is completely mechanistic in its ultimate nature, and mind is a production of mechanics. Clearly, all those conclusions about human beings, which were based on religious considerations had no place within such a mechanistic view.

Newton's classical mechanics with all its specific characteristics, just quoted, is applied to the nanorealm as well, at least in some but relevant cases. Also, the method of molecular dynamics, which is discussed in this chapter, belongs to this application field.

3.2.2 The ultimate level

In nanoscience, matter is in the form of many-particle systems described at the smallest level at which functional matter (condensed matter) can exist: It is in fact the "ultimate level," where the properties of usual materials emerge. Here, we deal with the atomic/molecular level, and there is no scientific level above this level. In other words, it is in fact the ultimate level. But most basic biological structures such as DNA, enzymes and proteins also work at this nanoscale, building up, molecule by molecule, macroscopic biological systems we call trees, humans and so on, with their typical intimate features.

For all the systems, the structure and dynamics of the objects is relevant. Since these systems (nanosystems) behave strongly anharmonically, no reliable analytical models, expressed through formulas, exist at the ultimate level. Thus, we have to recourse to numerical methods, which are, in most cases, complex. (In usual solid-state physics, the harmonic approximation works well in most cases and analytical models are expressible.)

Since no analytical models are available, we have to formulate the basic formulas in the most general form and must solve the problem numerically. The structure and dynamics of the atoms/molecules determine the properties essentially and are usually expressed by correlation functions, which can be measured.

3.2.3 Big data sets

As we already outlined in the sections before, the most relevant material properties behave in accordance with classical mechanics if the temperature is sufficiently large, that is, the data can be determined on the basis of Newton's equations of motion. However, as we have outlined previously, not only the object number N is relevant, but also the interaction between the atoms. In the formulation of the interaction, classical mechanics is not sufficient, but quantum effects come into play and cannot be neglected. In the formulation of the interaction laws, the quantum mechanical features of electrons have to be considered carefully.

The procedure is to solve the classical equations of motion numerically, that is, Newton's equations, using the interaction laws as input. These fundamental equations of motion are expressed through coupled differential equations. This procedure is known in literature as "molecular dynamics."

The molecular dynamics method is based on the container model and Newton's mechanistic world view. We know that this conception is not realistic but it works. There are obviously compensating effects with respect to certain assumptions. This point has already been remarked in Section 3.1.4, but we do not want to give more details here.

Within the frame of almost all molecular dynamics calculations, the predictor–corrector technique is used as algorithm for the solution of the equations of motion. Specific predictor–corrector algorithms are necessary and are described extensively in literature [37].

The basic information of 10^2–10^7 particles is very large and we get in such investigations "big data sets." If the many-particle system is treated as classical unit, we obtain, in realistic simulations with 10^2–10^7 particles, for each particle at time τ the position $\mathbf{r}$ and the momentum $\mathbf{p}$, that is, we have

$$\begin{aligned} \tau : \mathbf{r}_i &= (x_i, y_1, z_i), \\ \mathbf{p}_i &= (p_{xi}, p_{yi}, p_{zi}); \; i = 1, 2, \ldots, N \end{aligned} \tag{3.3}$$

where N is again the particle number. This leads to big data sets already in the case of 10^2–10^7 for N.

This big data set can only be used in specific cases but not for the general analysis of the many-particle system [37]. We have to produce measurable quantities, which consist of reduced data sets. However, the calculation of the measurable quantities (reduced data sets) have to be done on the basis of the big data set, that is, the big data set is definitely needed. For the production of the sets and their analysis, sufficiently large and fast computers are necessary.

In this chapter, we will discuss the basic tools for the treatment of nanosystems with many-particle character, which are given by the physical laws. We will talk about analyzing methods that are based on statistical physics and we will quote the main interaction types. Moreover, we will bring examples.

3.3 The physical formulation

3.3.1 The Hamilton function

For the classical description of the many-particle systems Hamilton's equations can be used. The Hamilton function has to be formulated in basic form, which is given in the case of monatomic systems by [15]

$$H = \sum_{i=1}^{N} \frac{p_i^2}{2m} + \frac{1}{2} \sum_{\substack{i,j=1 \\ i \neq j}}^{N} v_{ij} + \frac{1}{6} \sum_{\substack{i,j,k=1 \\ i \neq j \neq k}}^{N} v_{ijk} + \cdots \tag{3.4}$$

where m is the mass of the atoms and p_i is the magnitude of the ith object that can be an atom, molecule or ion.

Under the assumption that the change of the electronic arrangement around each ion, atom or molecule is sufficiently small, the potential energy V may be expanded in terms of many-body interaction potentials and this is reflected through the second and the third term in eq. (3.4). The term with υ_{ij} represents the pair potential contribution and the term with υ_{ijk} describes three-body interactions. Higher order contributions are not explicitly quoted in eq. (3.4) but could be relevant as well, depending on the problem under investigation.

3.3.2 Pair potential approximation

A particular role plays the pair potential approximation. This approximation is realistic if the polarization of the core electrons is negligibly small. In this case, the triplet and higher terms in the interaction (3.4) diminish rapidly in significance when compared with the pair terms. Then, the obvious final step of approximation is to neglect them entirely. This is called the pair potential approximation and is used in almost all molecular dynamics calculation. Furthermore, when we assume that the interaction itself is only dependent on the magnitude of the distances between the objects, we get

$$\sum_{\substack{i,j=1\\ i\neq j}}^{N} \upsilon_{ij} = \sum_{\substack{i,j=1\\ i\neq j}}^{N} \upsilon\left(\left|\mathbf{r}_i - \mathbf{r}_j\right|\right) \tag{3.5}$$

where $\mathbf{r}_1, ..., \mathbf{r}_N$, with $\left|\mathbf{r}_i - \mathbf{r}_j\right| = r_{ij}$, are the positions of the N atoms forming the system under investigation.

Referring to 100 atoms, with eq. (3.5) the problem has been reduced to a 9,900-fold sum of values from one pair potential function υ_{ij} with only one dimension, where i and j refer again to the distance of the two objects. This simplification expands the calculability of the many-particle problem with today's computer power up to millions of particles – at least under certain conditions.

We assumed that the atoms behave classically, which is practically fulfilled in almost all cases. Furthermore, if we assume that only the pair interaction is effective, we obtain the classical Hamilton function in the pair potential approximation, and this is given for monatomic systems by the following expression:

$$H = \sum_{i=1}^{N} \frac{p_i^2}{2m} + \frac{1}{2} \sum_{\substack{i,j=1\\ i\neq j}}^{N} \upsilon\left(r_{ij}\right) \tag{3.6}$$

If three- or higher body forces are effective, instead of the pair potential approximation for the potential energy, the expansion (3.4) has to be used. Pair potentials and many-body forces are discussed in detail in literature (see e.g., Ref. [15]). However, we will also quote some principal points in this monograph, particularly in Section 3.5.

3.3.3 Simple models

How can we calculate and determine, respectively, the properties of a many-particle system on the basis of a given Hamiltonian? In the conventional treatment of this problem, we have to simplify the problem further, and we need "simple models" that have to be reliable enough.

Quite general, simple models can be obtained by controlled approximations from the general case (3.6). In connection with many-particle systems, the general mathematical formulation of the problem in most cases is very complex so that it is unavoidable to choose simplifying models but, in most cases, such simplifying models can only be obtained through uncontrolled steps from the general case (3.6). It is, therefore, a rule to introduce simple models just for convenience. However, a realistic description needs more than that.

The "simple model" of solid-state physics is the crystalline solid in which the dynamics is described within the harmonic approximation. On the basis of this approach, it is, in fact, possible to determine successfully the properties of a lot of solids and materials, respectively. However, there are also a lot of cases where this simple model does not work and it is therefore not applicable in such cases, even when it is extended by specific assumptions. For example, the silver subsystem of the solid electrolyte α-AgI is highly disordered and its dynamics does not behave harmonically [38]. A simple description (model) for α-AgI and similar solids is not available. Also in the case of liquids and gases, "simple models" have not been found.

Even in the case of gases with low densities, we cannot formulate simple models. That is, we cannot simply restrict ourselves to the first terms in the virial expansion in the calculation of the pressure. The reason is obvious: The expansion obviously converges slowly and, therefore, even in the case of low density, one has to consider more than the first two terms. It is well known that the virial coefficients can be expressed in terms of the pair potential $v(r)$, but the corresponding expressions are getting complicated for the higher order virial coefficients and are hardly accessible. In other words, in practical calculations, only the first terms can be calculated. The virial expansion would define a "simple model" if for a broad class of gases a restriction on the first two terms would be realistic, but this is obviously not the case.

In summary, only for a specific class of many-particle systems "simple models" can be introduced: It is the crystalline solid in the harmonic approximation. But in most cases, anharmonicities effects are relevant and cannot be considered as small perturbation to the harmonic approximation.

This is also the case for typical nanosystems. Here, a great fraction of atoms/molecules are more or less close to the surface region and surface particles behave strongly anharmonically. In small nanosystems, as in the case of clusters with a few hundred of objects, even the innermost objects cannot be treated as bulk particles; the cut-off radius for the interaction potential here is larger than the distance to the surface.

The atoms/molecules at surfaces are in general less bonded than in the bulk and, therefore, the mean-square amplitudes of the objects are significantly larger at the surface than in the inner of the nanosystem, leading to relatively strong anharmonicities. This phenomenon can be observed even at low temperatures – low in comparison to the melting temperature of the bulk system. The melting temperature of typical nanoclusters is distinctly smaller than in the bulk state of the same material.

For the precise theoretical description of many-particle systems (nanosystems, gases and other specific solids), simple models in the above sense could not be found. An analytical description for systems with strong anharmonicities has not been introduced so far.

3.4 The numerical treatment: molecular dynamics

Due to the structural and dynamical behavior of nanosystems, we have to choose sophisticated theoretical methods for their treatment. The relatively strong anharmonicities are not negligible even at low temperatures.

The following question is important: How can the structure and dynamics of nanosystems or similar systems be described? What method or basic model is adequate? We already indicated this point previously. Let us give more specific details here.

Only for the harmonic, crystalline solid, a basic model exists. Analytical models could not be found for systems with strong anharmonic behavior. Simplifying principles at the microscopic level for such systems, different from the harmonic, crystalline case, could not be formulated for many-particle systems (see Section 3.3.3 about simple models). Thus, we have to describe such systems on the basis of the most general formulation, and this is given by the molecular dynamics method and that reflects a numerical method.

We will quote here relevant details about the molecular dynamics method. But there are other possibilities for the treatment of many-particle systems, at least in principle. For example, the Monte Carlo approach (Chapter 2) has also been applied. In contrast to molecular dynamics, where we have completely deterministic laws in algebraic form, the Monte Carlo method works with specific stochastic elements, but it is a numerical approach as well. Whereas the molecular dynamics methods enable to calculate structural and dynamic properties, the Monte Carlo method only deals with structural properties, and these are configurational averages and thermodynamic quantities. Quantum molecular dynamics is a suitable method as well, but we will not outline the details here. Relevant methods for the treatment of nanosystems are quoted in Chapter 2.

As we underlined earlier, many-particle systems with strong anharmonicities have to be described on the basis of the most general formulation, and this is given by the molecular dynamics method, which is able to characterize the structure and the dynamics of classical many-particle systems. Molecular dynamics are done in most cases on the basis of eq. (3.6), that is, if the pair potential approximations can be used and the other interaction types in the form of higher order many-body interactions can be assumed to be small.

3.4.1 Basic information

Clearly, only on the basis of the most general formulation the strong anharmonicities, which appear in connection with nanosystems, can be treated without approximation. In other words, the general description of the structure and dynamics of classical many-particle systems can only be done by means of molecular dynamics calculations. Here we do not need "simple models" or other uncontrolled simplifying assumptions (see in particular Figure 2.2).

As already mentioned, such "simple models" and simplifying assumptions, respectively, are not known at the microscopic level and can only be introduced in a phenomenological or empirical way. The harmonic approximation is not valid and the phonon picture is not applicable.

What is here the procedure? Within the framework of molecular dynamics Hamilton's equations

$$\begin{aligned} \dot{p}_{x_i} &= -\frac{dH}{dx_1}, \ldots, \dot{p}_{z_N} = -\frac{dH}{dz_N} \\ \dot{x}_1 &= -\frac{dH}{dp_{x_i}}, \ldots, \dot{z}_N = -\frac{dH}{dp_{z_{Ni}}} \end{aligned} \tag{3.7}$$

are solved by iteration using a sufficiently large and fast computer. We obtain the following information for the N objects:

$$\begin{aligned} & q(\tau_1), p(\tau_1) \\ & q(\tau_2), p(\tau_2) \\ & \vdots \\ & q(\tau_i), p(\tau_i) \\ & \vdots \end{aligned} \tag{3.8}$$

where we have

$$\begin{aligned} q(\tau_i) &= (x_1, y_1, z_1, \ldots, x_N, y_N, z_N) \\ p(\tau_i) &= \left(p_{x_1}, p_{y_1}, p_{z_1}, \ldots, p_{x_N}, p_{y_N}, p_{z_N}\right) \end{aligned} \tag{3.9}$$

In other words, we obtain the coordinates and momentums for all particles of the system as a function of time τ, and this is the information that follows from the solution of the classical equations of motion

The time step in the iteration process has to be estimated and is dependent on the parameters, which characterize the system under investigation. We do not give the details in this chapter. The solution (3.8) represents the most general form for the many-particle system and reflects a large amount of data, that is, it is a "big data set."

3.4.2 Total information

With k iteration steps the information about the system consisting of N particles is given in the time interval

$$\tau_G = k \cdot \Delta\tau \tag{3.10}$$

with

$$\Delta\tau = \tau_{i+1} - \tau_i \tag{3.11}$$

Equation (3.8) contains the total information of the many-particle system, which is given within a certain time interval expressed through eq. (3.10).

On the basis of this information the structure and dynamics of the system, can be determined – at least in principle. In particular, we can express correlation functions, which are the bridge to experimental data. In the following, we will give some remarks about the determination of typical functions on the basis of information (3.8).

3.4.3 Average values

Let us consider a space of $6N$ dimensions and let us represent the system within this space. The system under investigation is given by the points in this space. That is, there are $3N$ coordinates

$$q = (q_1, q_2, \ldots, q_{3N}) \tag{3.12}$$

and there are $3N$ momentums

$$p = (p_1, p_2, \ldots, p_{3N}) \tag{3.13}$$

This space with the coordinates q and p is the so-called phase space, and each point at time τ corresponds to a mechanical state of the system. The evolution of the system as a function of time τ is the complete information and is determined by Newton's equations of motion (Hamilton's equations). Once again, it is a strict classical description.

This collective motion is described by a trajectory; it is a trajectory in phase space [see Figure 3.5(a)]. The trajectory passes through the space element *dqdp* around point q, p of the phase space. We come from the trajectory to an ensemble when we go from Figure 3.5(a) to (b): The points in Figure 3.5(b) indicate how often the elements of the phase space have passed through the trajectory given in Figure 3.5(a), and these points in Figure 3.5(b) define the ensemble.

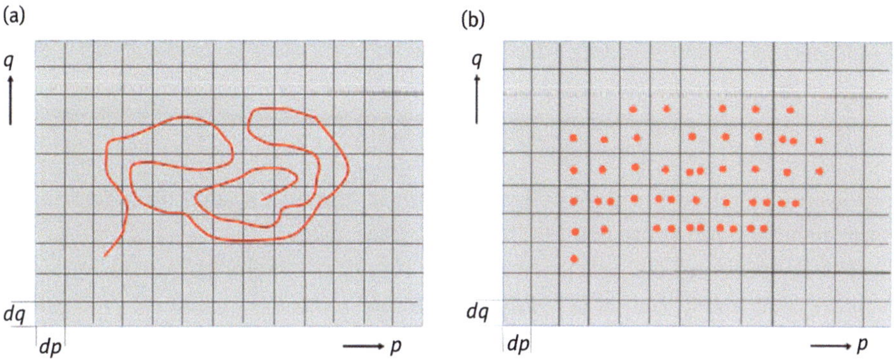

Figure 3.5: Phase space considerations in connection with a classical many-particle system. (a) Trajectory in phase space. At each clock time τ, there is one phase point q, p. (b) Statistical ensemble. For details, see the discussion in the text.

In other words, instead of the trajectory [Figure 3.5(a)] we have now a "cloud" of phase points, and this cloud defines an ensemble. The cloud itself is a big number of systems of the same nature, but differing in the configurations and momentums, which they have at a given time τ.

In summary, instead of considering a single dynamic system [Figure 3.5(a)], we obtain various systems, all corresponding to the same set of equations of motion (Hamilton function). This collection of systems is also called "statistical ensemble." One point of the statistical ensemble is the system under investigation in one of its possible states.

The introduction of the statistical ensemble is very useful regarding the relationship between dynamics and thermodynamics. In particular, it is a tool for the reduction of data sets. This is important because this way also allows the connection to experimental data. In other words, the data reduction performed by this method makes sense.

3.4.4 Density functions in phase space

The statistical ensemble, introduced before, can be expressed through a density in the 6*N* dimensional phase space:

$$\begin{aligned} &\rho(q_1, q_2, \ldots, q_{3N}, p_1, p_2, \ldots, p_{3N}, \tau) \\ &\downarrow \\ &\rho(q, p, \tau) \end{aligned} \tag{3.14}$$

Since the number of points of the statistical ensemble is arbitrary, the function $\rho(q,p,\tau)$ will be normalized as

$$\int \rho(q,p,\tau)dqdp = 1 \tag{3.15}$$

The quantity

$$\rho(q,p,\tau)dqdp$$

can be considered as the probability of finding at time τ a system of the ensemble in the element *dqdp* at the point (q, p) of the phase space.

The definition of the density function (3.14) is meaningful only if this quantity becomes stationary, that is, if $\rho(q,p,\tau)$ takes an asymptotic value for sufficiently large times τ:

$$\lim_{\tau\to\infty} \rho(q,p,\tau) - \rho(q,p) \tag{3.16}$$

When the trajectory given by the elements $q(\tau), p(\tau)$ is able to produce, over a sufficiently long period in time, the τ independent quantity $\rho(q,p)$, the introduction of $\rho(q,p,\tau)$ becomes meaningful. This behavior is known as ergodic hypothesis. In this case, we have equivalence between the averages: The average over the time and the average over the statistical ensemble of a function $f[q(\tau),p(\tau)] = f(\tau)$ are equivalent and we have

$$<f> = \frac{\int f(q,p)\rho(q,P)dqdp}{\int \rho(q,P)dqdp} \tag{3.17}$$

and

$$<f> = \lim_{\tau\to\infty} \frac{1}{\tau}\int_0^\tau f(\tau)d\tau \tag{3.18}$$

Different expressions exist for the density function $\rho(q,p)$, which can be defined in various ways, that is, we obtain different expressions for $\rho(q,p)$. These expressions are dependent on the thermodynamic state of the environment. Without going into detail, let us quote in the following the results for the three most import situations for which the function $\rho(q,p)$ can be expressed.

Microcanonical ensemble

In the case of a microcanonical ensemble, the system is considered to be isolated and the volume V_N, the object number N and the energy E are given. Then, the quantity $\rho(q,p)$ is expressed by

$$\rho(q \cdot p) = \begin{cases} \rho_0 = const \\ 0 \quad otherwise \end{cases} \tag{3.19}$$

for

$$E < H(q,p) < E + \Delta \tag{3.20}$$

where $E \ll \Delta$.

With the formulation (3.17), the average value of a function $f(q,p)$ is given by

$$<f> = \frac{\int\limits_{E<H(q,P)<e+\Delta} f(q,p)dqdp}{\int\limits_{E<H(q,P)<E+\Delta} dqdp} \tag{3.21}$$

The application of the microcanonical ensemble is useful if the thermodynamic state is defined adequately.

Canonical ensemble

Here, the system under investigation is treated as a closed, isothermal system and N, V_N and T are given, where T is the temperature. The density function $\rho(q,p)$ within the canonical ensemble is expressed through

$$\rho(q.p) = \frac{\exp\left\{-\frac{H(q,p)}{k_B T}\right\}}{\int\limits_{-\infty}^{\infty} \exp\left\{-\frac{H(q,p)}{k_B T}\right\} dqdp} \tag{3.22}$$

In other words, if N, V_N and T are given, the density function $\rho(q.p)$ takes the form of expression (3.22).

Grand canonical ensemble

Within the grand canonical ensemble, the system parameters V_N, T and μ are given, where μ is the chemical potential. Then, we obtain for the density function $\rho(q,p)$ the relation that is different from relation (3.22). The expression

$$\exp\left\{-\frac{H(q,p)}{k_B T}\right\} \tag{3.23}$$

in eq. (3.22) is replaced by

$$\exp\left\{-\frac{H(q,p)-\mu N}{k_B T}\right\} \tag{3.24}$$

and instead of eq. (3.22), we get

$$\rho(q.p)=\frac{\exp\left\{-\frac{H(q,p)-\mu N}{k_B T}\right\}}{\int\limits_{-\infty}^{\infty}\exp\left\{-\frac{H(q,p)-\mu N}{k_B T}\right\}dqdp} \tag{3.25}$$

Grand canonical ensembles are used within specific investigations where the object number N varies.

Thermodynamic limit
The thermodynamic limit is defined by the condition

$$\begin{aligned} &N \to \infty \\ &V_N \to \infty \\ &\frac{N}{V_N} = const \end{aligned} \tag{3.26}$$

Within this limit, the microcanonical, the canonical and the grand canonical ensemble are equivalent, that is, the statistical average $<f>$ of $f(q,p)$ is the same for the three methods if the thermodynamic limit is fulfilled, and we obtain, in all three cases, the same value for the statistical average $<f>$ of $f(q,p)$.

3.4.5 Individual and mean velocities

In connection with the statistical ensemble, we may define averages on the basis of what is represented in Figure 3.5(b), but also with respect to the trajectory [Figure 3.5(a)]. Let us repeat. The "ensemble" is defined by all possible states for the system under investigation. In the case of the "trajectory," the system moves through phase space in the course of time τ and occupies the phase space completely, that is, equivalence between the ensemble and the trajectory is given if the trajectory moves through all possible states the system is able to take.

Let us consider a monatomic system with N atoms/molecules and let $v(\tau)$ be the velocity of a single particle at time τ. When we work with classical probabilities, the probability, say dW, to find the velocity of one particle of the ensemble in the interval

$$v, v+dv \tag{3.27}$$

is given by the well-known Maxwell distribution, which is expressed through the following equation:

$$g(v) = \frac{4}{\sqrt{\pi}} \left(\frac{m}{2k_B T} \right)^{3/2} v^2 \exp\left\{ - \frac{mv^2}{2k_B T} \right\} \tag{3.28}$$

and for the probability, we get

$$dW = g(v)dv \tag{3.29}$$

In eq. (3.28), k_B is Boltzmann's constant and m is the mass of the atom/molecule. T is again the temperature.

By averaging of v^2 over all members of the ensemble, that is, over all possible object states [corresponding to Figure 3.5(b)], the mean-square velocity $<v^2>$ of the ensemble is determined. Applying Maxwell's distribution (3.28), we get for the mean-square velocity $<v^2>$ the equation

$$\begin{aligned} <v^2> &= \frac{4}{\sqrt{\pi}} \left(\frac{m}{2k_B T} \right)^{3/2} \int_0^{\infty} v^4 \exp\left\{ - \frac{mv^2}{2k_B T} \right\} dv \\ &= 3\frac{k_B T}{m} \end{aligned} \tag{3.30}$$

We obtain exactly the same value for $<v^2>$ if we average v^2 over the trajectory [corresponding to Figure 3.5(a)], that is, if we average over the object states $v^2 = v^2(\tau)$, which it takes in the course of time τ. If we define a function $v^2(\tau_0)$ by

$$v^2(\tau_0) = \frac{1}{\tau_0} \int_0^{\tau_0} v^2(\tau)\, d\tau \tag{3.31}$$

we must have

$$\lim_{\tau_0 \to \infty} v^2(\tau_0) = <v^2> \tag{3.32}$$

This means that after a sufficiently large time τ_0, the velocity $v(\tau)$ of an atom/molecule of the N-particle system has passed through all the states given by Maxwell's distribution (3.28). In practical calculations, that is, within molecular dynamics models, it turned out that the condition $\tau_0 \to \infty$ in eq. (3.32) is already fulfilled after a relatively short time interval. Let us specify and supplement this point.

How can we calculate average values in connection with molecular dynamics calculations? Here, the following question is of particular relevance: How large is normally the time interval $\Delta\tau$ in order to fulfill eq. (3.32) in a good approximation? Specific calculations showed that the time interval $\Delta\tau$ is relatively small; to fulfill eq. (3.32) one needs not more than 10^3 iteration steps, that is, $\Delta\tau$ is of the order of 10^{-11} s if the time step is 10^{-14} s.

Clearly, within molecular dynamics, calculations averages need not be calculated on the basis of a statistical ensemble using the density function $\rho(q,p)$, but are expressed through the raw data (3.8) that are given by the solutions of the classical equations of motion. The calculated information expressed by eq. (3.8) already represents the data in thermal equilibrium. Therefore, it is straightforward to determine averages on the basis of eq. (3.8). For the mean-square velocity [see eq. (3.30)], the molecular dynamical average is

$$\langle v^2 \rangle = \frac{1}{N_\vartheta}\frac{1}{N}\sum_{j=1}^{N_\vartheta}\sum_{i=i}^{N} v_i^2(\vartheta_j) \tag{3.33}$$

The average is formed over all N particles and various times ϑ_j.

3.4.6 Initial values

The states of a classical many-particle system is described if the coordinates and the momentums (velocities) of all N atoms/molecules are given as a function of time τ. We know that this information is obtained through the solution of Newton's equations of motion. It is the total microscopic information about the system under investigation. However, for the solution of the equations of motion, "initial values" for all the coordinates and all the velocities of the complete ensemble are needed, that is, for all N particles forming the system. Let us quote how these initial values can be obtained.

Initial values for the coordinates

In the case of liquids and gases, the particles can be distributed randomly with appropriate density within a certain volume. The situation is different if we consider crystals (with and without surface). Here, the objects are normally positioned within an array so that the perfect lattice structure, appropriate to the system under investigation, is generated.

This procedure can also be applied for systems in the nanorealm. The system is, of course, not fixed at this structure, but the structure develops in the course of time until a stationary structural state is reached.

Initial values for the velocities

We would like to assume that there are no external forces acting on the system. If that is the case, the directions of the velocities

$$\mathbf{v}_i/|\mathbf{v}_i|, \quad i=1,2,\ldots,N \tag{3.34}$$

at the initial time should be distributed randomly, where $|\mathbf{v}_i|$ is identical with the notation v_i used in Section 3.4.5: $|\mathbf{v}_i| = v_i$ or $|\mathbf{v}| = v$. In the case of a random distribution and without external force, we get for the sum over all velocities a value of zero (or a constant value):

$$\sum_{i=1}^{N} \mathbf{v}_i/|\mathbf{v}_i| = 0 \tag{3.35}$$

This must be valid for all times τ. This condition is necessary because the conservation of momentum must be fulfilled at each time τ during the molecular dynamics calculation.

The magnitudes of the particle velocities are distributed according to Maxwell's distribution if the system is in thermal equilibrium. It is, however, more convenient to choose initially another velocity distribution.

For the initial velocities, we may choose, for example, the same magnitude for all object velocities. However, due to the mutual interactions between the N particles the distribution for the velocities develops in the course of time towards the Maxwell distribution, which is represented through eq. (3.28).

In other words, the probability $dW = g(v)dv$ [see eq. (3.29)] of finding the velocities between $v, v + dv$ is initially expressed by a delta function, and this means that the system is initially not in thermal equilibrium. With the help of the specific function

$$\alpha(\tau) = \frac{\frac{1}{N}\sum_{i=1}^{N}\left[\mathbf{v}(\tau)_i^2\right]^2}{\left[\frac{1}{N}\sum_{i=1}^{N}\mathbf{v}(\tau)_i^2\right]^2} \tag{3.36}$$

we can study the behavior of the velocity distribution. The individual velocities $\mathbf{v}_i(\tau)$, $i = 1, 2, \ldots, N$, in eq. (3.36) are again the velocities obtained from the molecular dynamics calculation. Using this equation, we can study at which point in time Maxwell's distribution is reached. In the case of Maxwell's distribution (3.28), the value for $\alpha(\tau)$ is exactly 5/3 for all times τ. With our initial velocity distribution, we get $\alpha(\tau) = 1$.

It can be shown within realistic molecular dynamics calculations that thermal equilibrium ($\alpha(\tau) = 5/3$) is reached after a few hundred time steps if we start from $\alpha(\tau) = 1$; the time step is of the order of 10^{-14} s. A schematic representation is given in Figure 3.6.

All that reflects realistic situations and would also be observable in nature. Since the number of atoms/molecules is finite ($N \neq \infty$) in such calculations, the system fluctuates around its equilibrium value $\alpha(\tau) = 5/3$. These fluctuations increase with decreasing particle number, that is, in Figure 3.6, we have $N_A < N_B$, where N_A and N_B are the particle numbers of two systems A and B. The variations of the function $\alpha(\tau)$ are object-independent, that is, they are independent of the interaction

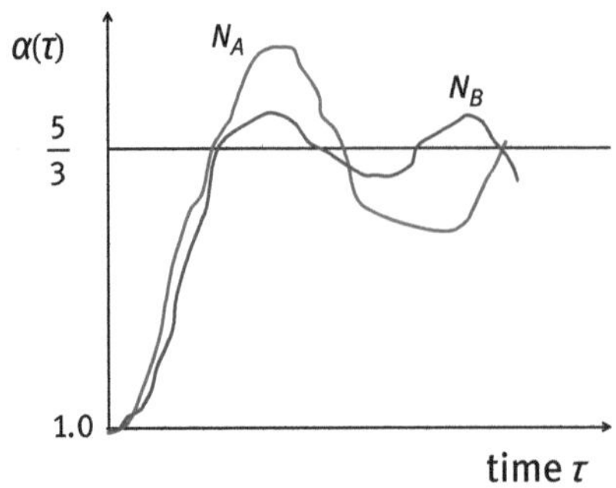

Figure 3.6: Schematic representation of the function α(τ), defined by eq. (3.36). The system under investigation is initially not in thermal equilibrium, which is reflected by $\alpha(\tau) = 1$. After a few hundred time steps, obtained by molecular dynamics, thermal equilibrium (Maxwell's distribution) is reached, and this is expressed through the value $\alpha(\tau) = 5/3$. Since the number of particles is finite ($N \neq \infty$), the system fluctuates around 5/3. These fluctuations increase with decreasing particle number, that is, in the figure, we have $N_A < N_B$, where N_A and N_B are the particle numbers of two systems *A* and *B*. The variations of the function $\alpha(\tau)$ are object independent, that is, they are independent of the interaction potential used in the calculation and they are also independent of the object mass. So, $\alpha(\tau)$ fluctuates around 5/3 for every system.

potential used in the calculation and they are also independent of the object mass. Clearly, in the case of sufficiently large systems, ($N \to \infty$), that is, for macroscopic systems such fluctuations are (almost) zero and $\alpha(\tau)$ is a constant $[\alpha(\tau) = 5/3]$ and is independent of time τ.

Temperature of molecular dynamic systems

How is the temperature expressed in the case of molecular dynamics systems? Also the temperature of such systems is dependent on time τ. It is defined by the well-known relation

$$\frac{1}{2} m < \mathbf{v}(\tau)^2 > \; = \frac{3}{2} k_B T(\tau) \tag{3.37}$$

With

$$< \mathbf{v}(\tau)^2 > \; = \frac{1}{N} \sum_{i=1}^{N} \mathbf{v}_i^2(\tau) \tag{3.38}$$

we get for the temperature

$$T(\tau) = \frac{m}{3Nk_B} \sum_{i=1}^{N} \mathbf{v}_i^2(\tau) \tag{3.39}$$

The temperature fluctuations, represented schematically in Figure 3.7, are of course not artificial but actually appear in the case of small systems like nanoclusters with a finite number of objects. In other words, temperature fluctuations are quite natural. The fluctuations reflect specific material properties like, for example, the specific heat at constant volume. The details are not of interest in connection with this chapter.

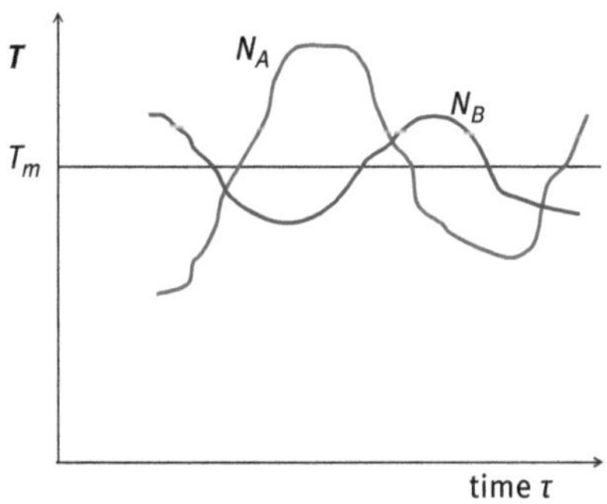

Figure 3.7: The temperature as a function of time τ for two systems, having again the particle numbers N_A and N_B. It is a schematic representation. The mean temperature is here denoted by T_m. Because of the finite number of particles, the temperature fluctuates around T_m. The temperature fluctuations increase with decreasing particle number, that is, we have $N_A < N_B$ in the figure.

3.5 Interaction potentials

For the performance of molecular dynamics calculations, a certain input is necessary. Not only the initial values for the coordinates and momentums (velocities) for all the N atoms/molecules of the system are needed as input (Section 3.4.6), but in particular the interaction laws between the objects on which the system is based. Just the interaction potentials reflect the specific characteristics and features, respectively, of the system under investigation.

The precise knowledge of the interaction potentials is required in order to get a reliable description of material properties. In particular, for the description of surface properties of solids, as well as for the description of nanosystems, reliable potentials are needed.

The system properties are, in general, very sensitive to small variations in the interaction potentials. In fact, small quantitative potential variations can lead to qualitative effects with respect to system properties, just in connection with the structure and dynamics near the surface.

Here we have to distinguish between the repulsive and the attractive part of the potential, and both parts have to be determined carefully. However, the determination of accurate potential functions for systems of interest (e.g., metals and materials with covalent bonding) is rather difficult even for the bulk. The nanoproblem is much harder because a great fraction of atoms/molecules belong to the surface region. The reason is obvious: In the case of nanosystems, the changes regarding the electronic states and other relevant properties near the surfaces are meaningful.

In this subsection, we would like to discuss some relevant points in connection with interaction potentials. In literature, a lot of theoretical studies are given and are also in connection with calculations within the frame of molecular dynamics calculations. However, in most cases, the potentials are used in a rather uncritical way: The interaction laws are often used without testing carefully their reliability, in some cases, even not at all.

Serious research in nanoscience requires just the opposite: Since the properties of nanosystems are particularly sensitive to potential variations, the potentials have to be determined thoroughly for such systems. The performance of a molecular dynamics calculation without reliable interaction potential is hardly relevant and is actually superfluous.

3.5.1 Types of interactions in condensed matter physics

Basically, in condensed matter physics we have to distinguish between various interaction types that differ in their fundamental interaction mechanisms. We have four binding types:
(1) Ionic interactions (4–14 eV)
(2) Metallic bonds (0.7–6 eV)
(3) van der Waals interactions (0.02–0.3 eV)
(4) Covalent bonds (1–10 eV)

Here the numbers denote the typical binding energies. Also, the *hydrogen bridge bond* (0.1–0.5 eV) has to be mentioned and is a characteristic binding type, but will not be discussed in the following.

1. Ionic binding

The ionic interaction is characterized by different atoms (e.g., Na and Cl). In this case, an exchange of electrons between the atoms takes place, so that the many-particle system consists of positive and negative charged ions. The interaction law, as a function of distance r between the ions, is very simple; it is described by Coulomb's law if the distances are larger than $r_+ + r_-$, where r_+ and r_- are the radii of the positively and negatively charged particles. For smaller distances, that is, for $r < r_+ + r_-$ the interaction is repulsive due to the overlap of the electron cores. The critical point is here the repulsive potential parts that have to be determined accurately in the case of large energies.

2. Metallic binding

In the case of metals, we have the following situation: For certain atoms, the electrons of the outer shell are only weakly bonded. If a many-particle system is joined with such types of atoms, the weakly bonded electrons leave the according atoms and move through the whole system. These are the conduction electrons. In other words, in metals, we have the following configuration: There are positively charged ions which move through an ensemble of conduction electrons, often called the sea of conduction electrons.

When we are interested in the structure and dynamics of the ions, we have to perform molecular dynamics calculations for the ions, which behave classically. On the other hand, the conduction electrons form a quantum system. The ion–ion interaction is the input in the molecular dynamic calculations. Thus, in metals the ion–ion interaction is not simply given by the Coulomb potential since the ions are screened by the quantum mechanical conduction electrons and that makes its determination complicated. Such a modeling of the ion–ion potential can be performed, for example, within the frame of pseudopotential theory. However, we do not want to give more details here in this rough overview. It should be remarked that the interaction potentials in metals have, in general, to be considered as temperature dependent. Here the temperature effect is due to the density of the conduction electrons within the many-particle system which is a function of temperature.

3. van der Waals interaction

The electron moves around the atomic nucleus in a nondeterministic way. This means that there is a certain probability that the atoms have, at instant τ, an electrical dipole moment which, however, becomes zero when it is averaged over the time. In a many-particle system, these momentary dipole moments interact with each other leading to an attractive potential between the atoms of the many-particle system.

This is an effect with respect to the long-range part of the interaction. Clearly, at small distances repulsive core effects have to be considered as well in the description of the interaction potential. Typical van der Waals systems are noble gases. Van der Waals interaction has to be considered also in metallic systems and contribute here as well.

4. Covalent bonds

In the case of covalent binding, the interaction mechanism between atoms is explained by the fact that a part of their electrons belongs to several atoms at the same instant, that is, simultaneously. In other words, the probability of finding an electron, which is responsible for the interaction, is relatively large in the space between the atoms, and these electronic states lead to interaction effects between them. In most cases, this process is based on the formation of spin saturated electron pairs, where each atom contributes an electron so that the electron shells take a noble gas configuration. Carbon, amorphous semiconductors, hydrogen molecules and so on are typical examples for substances with covalent bonds.

3.5.2 Remarks concerning phenomenological potentials

In most cases, the interaction is not derived on the basic level, that is, within the frame of Schrödinger's equation and corresponding approaches. Instead, phenomenological potential functions are used very often, which are most often not strictly constructed with respect to the basic interaction types discussed in Section 3.5.1.

As a typical example for a phenomenological potential, let us briefly discuss the potential of Morse [39]. This interaction law has extensively been applied not only in the study of lattice dynamics [40], but also in the investigation of the defect structure in metals [41–49]. The role of the inert gases in metals has been studied on the basis of the phenomenological potential of Morse. [50, 51]. Other applications are the equation of state [52, 53], the study of elastic properties of metals [54], the interaction between crystal surfaces and gas atoms [55]. A lot of other specific problems have been investigated by means of the Morse potential [56–62].

This list indicates that the phenomenological Morse potential has been intensively used in the study of problems in the physics of condensed matter. The Morse potential is, however, merely a typical example for the study on the basis of phenomenological potentials. But what is the physical background of the Morse potential and how is it expressed analytically? Let us briefly give some remarks on the background of this potential.

The basis was not a many-particle system but a single molecule. For the formulation of the interatomic potential of the atoms of the diatomic molecule [63], Morse required some specific conditions in order to be able to describe the spectroscopy of the molecule. Morse chose the following analytic expression for the potential function:

$$v(r) = \alpha_0 \exp[-2\alpha(r - r_0)] - 2\alpha_0 \exp[-\alpha(r - r_0)] \tag{3.40}$$

where α_0 is a constant and r_0 is the intermolecular separation. The solution of the radial part of Schrödinger's equation using this potential form yielded the correct energy-level representation for a diatomic molecule. However, eq. (3.40) has not been derived by means of Schrödinger's equation.

In conclusion, the Morse potential does not explicitly reflect one of the basic interaction types discussed in Section 3.5.1, but it is based on the spectroscopy of diatomic molecules. Nevertheless, the potential obviously works well; it has been applied extensively in the study of various many-particle properties, but there is no physical justification for that.

There are many other phenomenological approaches for the interaction. For example, the so-called Buckingham potential has also often been used and is intensively discussed in literature, but we do not want to discuss this interaction type here and also not other phenomenological approaches. In most cases, there is no or almost no physical background for the introduction of these potentials. They have often been chosen for purely pragmatic reasons and must be considered with care and have to be considered as fitting functions and not more.

As we already remarked at the beginning of Section 3.5, there is a strong potential-sensitivity in connection with the properties of systems with surface and in particular with respect to nanosystems. We have therefore to construct the interaction potentials very carefully. In Ref. [15], we have discussed some relevant theoretical methods for the construction of model potentials. In the next section, let us give some basic remarks concerning the sensitivity of potential functions with respect to system properties.

3.5.3 Potential variations and their influence on system properties

In this subsection, we would like to give some specific remarks on the sensitivity of system properties with respect to variations of the interaction potential. Here, the interaction between the atoms (molecules) is concerned that form a many-particle system as, for example, a nanosystem.

As we remarked several times, the properties of such systems turned out to be very sensitive to small potential variations. This particularly means that in the description of the properties of systems, the interactions have to be modelled carefully in order to get reliable statements.

What does the statement "sensitive to small variations in the interaction potentials" mean? We would like to demonstrate this point with the help a manageable example, that is, we want to do the study on the basis of a relatively simple, but realistic system.

A study for krypton

For this purpose, let us choose the noble gas krypton [64]. In connection with nanosystems, surface properties play a dominant role and are, therefore, of particular interest. Therefore, we want to study the sensitivity of surface properties on the potential function here.

In contrast to metals, the potential functions of noble gas solids, liquids and nanosystems are not dependent on the density and the temperature of the system under investigation. In particular, the interaction laws are the same at the surface and in the bulk of the crystal. This is, in fact, a simple case but, nevertheless, krypton reflects a realistic material.

Such a treatment is, however, not possible in the case of metals where we have ions that are embedded in the sea of conduction electrons. At free metal surfaces, the local electronic background is, in general, different from its average bulk value and produces concomitant changes in the potential functions. In other words, the potential functions in metals are complex. Such kind of effects do not arise at noble gas surfaces and, therefore, in the study of the potential sensitivity, we want to restrict our study to noble gas (krypton) surfaces. We would like to consider here pair potentials as well as three-body interactions.

For krypton, a reliable pair potential is available. It the so-called Barker potential [65]. It is given in analytical form, which is rather complex; we will not outline this complex formula here. The Barker potential is more realistic than the famous Lennard–Jones potential

$$\upsilon(r) = 4\varepsilon\left[\left(\frac{\sigma}{r}\right)^{12} - \left(\frac{\sigma}{r}\right)^{6}\right] \tag{3.41}$$

because it undoubtedly correlates accurately a much wider range of experimental data.

The shape of both potentials, the Lennard–Jones potential and the Barker potential, are similar (see Ref. [15]) and do not differ significantly. Nevertheless, the results for the relaxation are "qualitatively" different from each other. Whereas the Barker potential leads to a contraction of the outer layers, the Lennard–Jones potential leads to a layer expansion, as has been demonstrated by molecular dynamics calculations for the Kr(111) surface and 539 atoms [66]. The layer contraction at the surface is confirmed by LEED data.

As already mentioned, the layer contraction resulting from the model and the experiments is in contradiction to the results of model calculations with the Lennard–Jones potential (leading to an expansion). This result is important for dynamical processes in the surface region; relaxation effects would essentially influence such dynamical processes.

The dynamical processes at the surface for the Lennard–Jones potential (3.41) could be seriously in error. For example, the force constants vary strongly when we go from a layer expansion to a layer contraction.

It should be emphasized here that the Lennard–Jones potential is the most used interaction function in literature. In some cases, it is even used for the description of metallic systems [67], but there is no physical justification for that since van der Waals forces, described by the Lennard–Jones potential or the Barker potential, have nothing to do with the interaction type dominant in metals. Such kind of investigations have to be avoided because they are misleading.

In conclusion, the properties of many-particle systems with a large fraction of atoms/molecules at the surface, in particular nanosystems, are obviously very sensitive to small variations in the interaction between the objects. This follows directly from the study of noble gas systems. The relatively small quantitative differences between the Lennard–Jones potential and the Barker potential [66] give rise to not only quantitative effects but also to qualitative differences. This example shows that the selection of interatomic potentials has be done carefully, especially in the modelling of nanosystems since a great fraction of atoms/molecules are positioned at the surface here.

Three-body interactions

What about many-body interactions? The Lennard–Jones potential as well as the Barker potential are descriptions within the pair potential approximation (Section 3.3.2). However, many-body forces (three-body interactions etc., see eq. (3.4) might also play a role. What can be said, for example, about the influence of three-body interactions, just in connection with our noble gas example? Concerning this point, let us give an instructive statement.

For our model system krypton, a three-body potential exists. It is the so-called Axilrod–Teller potential [68], which is given by the following expression:

$$\upsilon_3(i,j,l) = \nu \frac{1 + 3\cos\theta_1 \cos\theta_2 \cos\theta_3}{r_{ij}^3 r_{il}^3 r_{jl}^3} \tag{3.42}$$

where ν is a constant; the quantities r_{ij}, r_{il}, r_{jl} and $\theta_1, \theta_2, \theta_3$ are the sides and angles of the triangle formed by the objects i, j, l. The three-body potential $\upsilon_3(i,j,l)$ can be repulsive and attractive, and this behavior is dependent on the shape of the triangle formed by the three objects i, j and l.

Are those Axilrod–Teller interactions relevant in connection with the experimental results discussed before? We cannot exclude that but it is not very probable. The three-body potential (3.42) may lead together with the Lennard–Jones potential to give good results for the relaxation at the surface. However, it is shown in Ref. [69] that the influence of $\upsilon_3(i,j,l)$ on structural data is relatively small and, therefore, we cannot expect that the Lennard–Jones potential together with $\upsilon_3(i,j,l)$ gives good results for the relaxation and other quantities of relevance. However, this point has to be investigated in more detail.

On the other hand, the Barker potential is a complex expression with a lot of parameters. These parameters were fitted through a large number of experimental data. This fact can be expressed formally by

$$\upsilon = \upsilon(r, a, b, \ldots) \tag{3.43}$$

where a, b and so on are the fit parameters. In other words, such kinds of potentials should not be considered as pure pair potentials in the sense of eq. (3.4) in Section 3.3, but should rather play the role of effective interaction potentials in which many-body forces are included:

$$\upsilon(r, a, b, \ldots) = \upsilon_{eff}(r, a, b, \ldots) \tag{3.44}$$

Equation (3.43) considers three-body interactions and so on without expressing that explicitly, and this fact expresses a blanket approximation which in particular considers approximately the entire set of possible many-body forces. Therefore, we may express (3.4) in Section 3.3 by

$$V(\mathbf{r}_1, \ldots, \mathbf{r}_N) = \frac{1}{2}\sum_{\substack{i,j\\ i\neq j}}^{N} v_{ij} + \frac{1}{6}\sum_{\substack{i,j,k\\ i\neq j\neq k}}^{N} v_{ijk} + \cdots \cong \frac{1}{2}\sum_{\substack{i,j\\ i\neq j}}^{N} v_{eff}\left(r_{ij}, a, b, \ldots\right) \tag{3.45}$$

That is, the effective potential simulates approximately many-body forces. This is due to the fact that the potential parameters a, b and so on are fitted to experimental data which may be more or less sensitive to many-body forces. The best way is of course not to work with effective potentials, but the use of realistic many-body forces is desirable.

3.6 Structure and correlation functions

The structure of crystalline solids is in general determined through a relatively small number of position vectors. Here, the geometry of the unit cell is essential, and the unit cell is in fact in most cases determined by a few characteristic position vectors for the objects occupying the cell. This is the case for macroscopic solids that can be considered as systems with an infinite extension.

In the case of nanosystems, the situation is different. The structure is already at relatively low temperatures, very often disturbed and disordered, respectively. This is due to the strong anharmonic behavior of nanosystems. In the case of disordered systems, a unit cell cannot be defined and, in general, the structure information cannot be reduced. Then, the information about the relative object positions, that is, the structure, is given by "all" the position vectors

$$\mathbf{r}_i, \quad i = 1, 2, \ldots, N \tag{3.46}$$

of the N particles of the nanosystem under investigation. In other words, the structure information of nanosystems is close to the structure, which is characteristic for liquids.

Although nanosystems are small, this structure information can be very large and the structure of such systems has been characterized by certain averaged quantities. These quantities are the so-called distribution functions, which correlate the positions of the various N atoms, that is, we get correlation functions. Let us briefly discuss this point here.

3.6.1 General remarks

For a classical system (e.g., a nanosystem or a liquid) with N atoms/molecules, which is at temperature T in volume V_N, the probability distribution in phase space is characterized by [70–72]

$$\exp\left(-\frac{H}{k_B T}\right) \tag{3.47}$$

where H is again the classical Hamilton function, expressed here by

$$H = \sum_{i=1}^{N} \frac{\mathbf{p}_i^2}{2m} + U(\mathbf{r}_1, \mathbf{r}_2, \ldots, \mathbf{r}_N) \tag{3.48}$$

Since we are interested in structural configurations, we have to integrate over all N momenta.

The probability

$$P^{(N)}(\mathbf{r}_1, \mathbf{r}_2, \ldots, \mathbf{r}_N)d\mathbf{r}_1 d\mathbf{r}_2 \ldots d\mathbf{r}_N \tag{3.49}$$

that particle 1 will be found in volume element $d\mathbf{r}_1$ around position $\mathbf{r}_1$, particle 2 in $d\mathbf{r}_2$ around $\mathbf{r}_2$, …, particle N in $d\mathbf{r}_N$ around $\mathbf{r}_N$ is expressed by

$$\begin{aligned} &P^{(N)}(\mathbf{r}_1, \mathbf{r}_2, \ldots, \mathbf{r}_N)d\mathbf{r}_1 d\mathbf{r}_2 \ldots d\mathbf{r}_N \\ &= \frac{1}{Z_N} \exp\left(-\frac{U(\mathbf{r}_1, \mathbf{r}_2, \ldots, \mathbf{r}_N)}{k_B T}\right) d\mathbf{r}_1 d\mathbf{r}_2 \ldots d\mathbf{r}_N \end{aligned} \tag{3.50}$$

where $U(\mathbf{r}_1, \mathbf{r}_2, \ldots, \mathbf{r}_N)$ is again the potential energy, which appears in the Hamilton function (3.48). Z_N is the N-body partition function and is in statistical physics given by

$$Z_N = \int \exp\left(-\frac{U(\mathbf{r}_1, \mathbf{r}_2, \ldots, \mathbf{r}_N)}{k_B T}\right) d\mathbf{r}_1 d\mathbf{r}_2 \ldots d\mathbf{r}_N \tag{3.51}$$

Furthermore, we can define the situation for n objects. In this case, the probability that n atoms ($n \le N$) will be in $d\mathbf{r}_1$ around $\mathbf{r}_1, \ldots, d\mathbf{r}_n$ around $\mathbf{r}_n$, regardless of the other $N - n$ objects is obviously expressed by

$$\begin{aligned} &P^{(n)}(\mathbf{r}_1, \mathbf{r}_2, \ldots, \mathbf{r}_n)d\mathbf{r}_1 d\mathbf{r}_2 \ldots d\mathbf{r}_n \\ &= \left\{\frac{1}{Z_N} \int_{-\infty}^{\infty} \exp\left(-\frac{U(\mathbf{r}_1, \mathbf{r}_2, \ldots, \mathbf{r}_N)}{k_B T}\right) d\mathbf{r}_{n+1} \ldots d\mathbf{r}_N\right\} d\mathbf{r}_1 \ldots d\mathbf{r}_n \end{aligned} \tag{3.52}$$

In order to avoid the labeling of the N atoms, the quantity $P^{(N)}$ has to be multiplied by the factor $N!/(N-n)!$ and we get the relation

$$\rho^{(n)}(\mathbf{r}_1, \mathbf{r}_2, \ldots, \mathbf{r}_N) = \frac{N!}{(N-n)!} P^{(n)}(\mathbf{r}_1, \mathbf{r}_2, \ldots, \mathbf{r}_N) \tag{3.53}$$

$P^{(n)}(\mathbf{r}_1, \mathbf{r}_2, \ldots, \mathbf{r}_N)$ is normalized to unity and we get

$$\int_{-\infty}^{\infty} \rho^{(n)}(\mathbf{r}_1, \mathbf{r}_2, \ldots, \mathbf{r}_N)d\mathbf{r}_1 d\mathbf{r}_2 \ldots d\mathbf{r}_N = \frac{N!}{(N-n)!} \tag{3.54}$$

Then, we have the following characteristic: the expression

$$\rho^{(n)}(\mathbf{r}_1, \mathbf{r}_2, \ldots, \mathbf{r}_n)d\mathbf{r}_1 d\mathbf{r}_2 \ldots d\mathbf{r}_n \tag{3.55}$$

is the probability that one atom of the system will be in $d\mathbf{r}_1$ at $\mathbf{r}_1$ another in $d\mathbf{r}_2$ at $\mathbf{r}_2, \ldots$. The function $\rho^{(n)}(\mathbf{r}_1, \mathbf{r}_2, \ldots, \mathbf{r}_n)$ is a relevant quantity in the description of structural properties of many-particle systems and is the so-called n particle distribution function.

Noninteracting objects

In the limit of noninteracting particles, the function $P^{(n)}(\mathbf{r}_1, \mathbf{r}_2, \ldots, \mathbf{r}_n)$ [see eq. (3.52)] takes the following form:

$$P^{(n)}(\mathbf{r}_1, \mathbf{r}_2, \ldots, \mathbf{r}_N) = P^{(n)}(\mathbf{r}_1)P^{(n)}(\mathbf{r}_2)\ldots P^{(n)}(\mathbf{r}_n) \tag{3.56}$$

and with eq. (3.53), we obtain for the distribution function $\rho^{(n)}(\mathbf{r}_1, \mathbf{r}_2, \ldots, \mathbf{r}_n)$ for such a random system the following form:

$$\rho^{(n)}(\mathbf{r}_1, \mathbf{r}_2, \ldots, \mathbf{r}_n) = \frac{N!}{N^n(N-n)!}\rho^{(n)}(\mathbf{r}_1)\rho^{(n)}(\mathbf{r}_2)\ldots\rho^{(n)}(\mathbf{r}_n) \tag{3.57}$$

In the next step in the representation of structural elements of many-particle systems, we would like to introduce "correlation functions." The n-particle correlation function is defined through this relation:

$$\rho^{(n)}(\mathbf{r}_1, \mathbf{r}_2, \ldots, \mathbf{r}_n) = \left\{\rho^{(n)}(\mathbf{r}_1)\rho^{(n)}(\mathbf{r}_2)\cdots\rho^{(n)}(\mathbf{r}_n)\right\}g^{(n)}(\mathbf{r}_1, \mathbf{r}_2, \ldots, \mathbf{r}_n) \tag{3.58}$$

Correlations come into play if the N objects are not randomly distributed. In other words, there are correlations between the N objects of the system (i.e., deviations from a system with randomly distributed N particles) if the function $g^{(n)}(\mathbf{r}_1, \mathbf{r}_2, \ldots, \mathbf{r}_n)$ is different from $N!/(N-n)!N^n$, that is, we have

$$g^{(n)}(\mathbf{r}_1, \mathbf{r}_2, \ldots, \mathbf{r}_n) \neq \frac{N!}{(N-n)!N^n} \tag{3.59}$$

This is a general characterization and needs more specification. In the next section, we will discuss disordered system in more detail.

3.6.2 Disordered systems

A liquid can be considered as a homogeneous and isotropic bulk system. For this case, the single-article distribution function $\rho^{(1)}(\mathbf{r}_1)$ must be a constant, that is, the quantity $\rho^{(1)}(\mathbf{r}_1)$ is independent of $\mathbf{r}_1$. In particular, with eq. (3.54) we immediately get

$$\rho^{(1)}(\mathbf{r}_1) = \rho^{(2)}(\mathbf{r}_2) = \cdots = \rho^{(n)}(\mathbf{r}_n) = \frac{N}{V_N} = \rho \tag{3.60}$$

Within the limit [see also eq. (3.26)]

$$\begin{aligned} &N \to \infty \\ &V_N \to \infty \\ &\rho = \frac{N}{V_N} = const \end{aligned} \tag{3.61}$$

we obtain for a random system (noninteracting many-particle systems) the following relations:

$$\rho^{(n)}(\mathbf{r}_1, \mathbf{r}_2, \ldots, \mathbf{r}_n) - \rho^n \tag{3.62}$$

and

$$g^{(n)}(\mathbf{r}_1, \mathbf{r}_2, \ldots, \mathbf{r}_n) = 1 \tag{3.63}$$

Again, these are the results for noninteracting systems. For interacting objects, the formalism can be extended easily.

Interacting objects

In the case of interacting objects, we get, using the eq. (3.60), instead of eq. (3.58) the following equation:

$$\rho^{(n)}(\mathbf{r}_1, \mathbf{r}_2, \ldots, \mathbf{r}_n) = \rho^n g^{(n)}(\mathbf{r}_1, \mathbf{r}_2, \ldots, \mathbf{r}_n) \tag{3.64}$$

Then, the pair correlation function is given some simple manipulation [using eqs. (3.52), (3.53) and (3.64)]:

$$g^{(2)}(\mathbf{r}_1, \mathbf{r}_2) = \frac{V_N^2}{Z_N} \int_{-\infty}^{\infty} \exp\left(- \frac{U(\mathbf{r}_1, \mathbf{r}_2, \ldots, \mathbf{r}_N)}{k_B T} \right) d\mathbf{r}_3 \ldots d\mathbf{r}_N \tag{3.65}$$

This is a general equation for the pair correlation function and is valid for sufficiently large particle number N. The corresponding expression is given within the pair potential approximation (3.6) by

$$g^{(2)}(\mathbf{r}_1, \mathbf{r}_2) = \frac{V_N^2}{Z_N} \int_{-\infty}^{\infty} \exp\left(- \frac{1}{2k_B T} \sum_{\substack{i,j=1 \\ i \neq j}}^{N} v(\mathbf{r}_i, \mathbf{r}_j) \right) d\mathbf{r}_3 \ldots d\mathbf{r}_N \tag{3.66}$$

In the theory of bulk liquids, the two-particle correlation function $g^{(2)}(\mathbf{r}_1, \mathbf{r}_2)$ is of most interest; it depends only on $|\mathbf{r}_1 - \mathbf{r}_2| = \mathbf{r}_{12} = r$. Then, the pair correlation function takes the form

$$g^{(2)}(\mathbf{r}_1, \mathbf{r}_2) = g(r) \tag{3.67}$$

It represents the probability distribution for the distances r of two particles of the ensemble. Examples for $g(r)$ are given in Figure 3.8. Of all the correlation functions, only $g(r)$ is accessible to experiments.

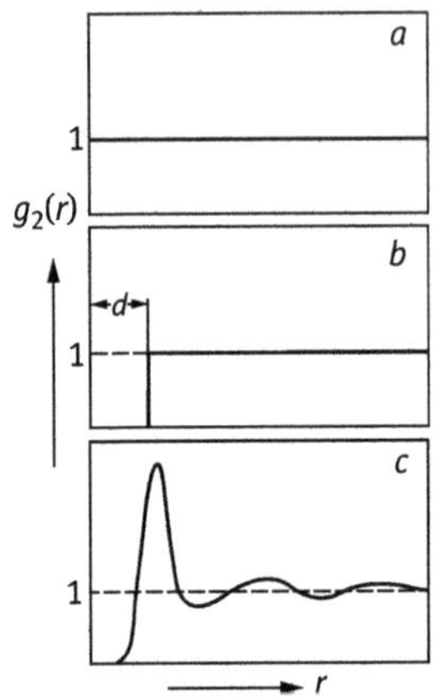

Figure 3.8: The pair correlation function $g_2(r) \equiv g(r)$ for several types of many-particle systems: (a) random system of space points, (b) hard spheres at low density (d is the diameter of the hard spheres) and (c) typical liquid. In the case of nanosystems, the form of the correlation functions is dependent on the distance from surfaces.

The thermodynamic functions

Within the pair potential approximation, the thermodynamic functions are expressed by $g(r)$ alone. The energy, the pressure and so on, of the many-particle system are determined by $g(r)$ and the pair potential $\upsilon(r)$. Although higher order correlation functions are also defined within the pair potential approximation, they do not contribute in this case. For example, in the case of simple liquids, the energy is expressed by

$$E = \frac{3}{2} N k_B T + 2N\pi\rho \int_0^\infty dr \upsilon(r) g(r) r^2 \tag{3.68}$$

where the first term is the well-known mean kinetic energy, and the second term is the mean potential energy:

$$\frac{1}{Z_N} \int_{-\infty}^{\infty} \frac{1}{2} \sum_{\substack{i=1 \\ i \neq j}}^{N} \upsilon(\mathbf{r}_i, \mathbf{r}_j) \exp\left(-\frac{1}{2k_B T} \sum_{\substack{i,j=1 \\ i \neq j}}^{N} \upsilon(\mathbf{r}_i, \mathbf{r}_j) \right) d\mathbf{r}_1 \ldots d\mathbf{r}_N \tag{3.69}$$
$$= 2N\pi\rho \int_0^\infty dr \upsilon(r) g(r) r^2$$

The correlation functions are easily accessible by molecular dynamics calculations, but we do not want to quote examples here.

3.6.3 Theoretical determination of the pair correlation function

Equation (3.66) cannot be used for the determination of $g^{(2)}(\mathbf{r}_1, \mathbf{r}_2)$ from a given pair potential

$$\upsilon(\mathbf{r}_i, \mathbf{r}_j) = \upsilon(r), r = r_{ij} = |\mathbf{r}_i - \mathbf{r}_j| \tag{3.70}$$

and this is because it is impossible to solve the high-dimensional integral, which appears in connection with eq. (3.66). This is not possible even for small systems like nanosystems. However, we can eliminate this problem if we perform molecular dynamics calculations.

On the other hand, the pair potential $\upsilon(\mathbf{r}_i, \mathbf{r}_j)$ could be determined through the pair correlation function $g^{(2)}(\mathbf{r}_i, \mathbf{r}_j)$, but here, we need additional information about the three-body correlation function $g^{(3)}(\mathbf{r}_i, \mathbf{r}_j, \mathbf{r}_k)$. There exists a general connection between the pair potential $\upsilon(\mathbf{r}_i, \mathbf{r}_j)$, the pair correlation function $g^{(2)}(\mathbf{r}_i, \mathbf{r}_j)$ and the three-body correlation function $g^{(3)}(\mathbf{r}_i, \mathbf{r}_j, \mathbf{r}_k)$. In principle, $\upsilon(\mathbf{r}_i, \mathbf{r}_j)$ can be determined if $g^{(2)}(\mathbf{r}_i, \mathbf{r}_j)$ and $g^{(3)}(\mathbf{r}_i, \mathbf{r}_j, \mathbf{r}_k)$ are known:

$$g^{(2)}(\mathbf{r}_i, \mathbf{r}_j),\ g^{(3)}(\mathbf{r}_i, \mathbf{r}_j, \mathbf{r}_k) \rightarrow \upsilon(\mathbf{r}_i, \mathbf{r}_j) \tag{3.71}$$

Then, we have to make approximations for $g^{(3)}(\mathbf{r}_i, \mathbf{r}_j, \mathbf{r}_k)$.

The triplet-correlation function $g^{(3)}(\mathbf{r}_i, \mathbf{r}_j, \mathbf{r}_k)$ is connected to the probability $P^{(3)}(\mathbf{r}_1, \mathbf{r}_2, \mathbf{r}_3)$ (see eq. (3.49)) and the distribution function $\rho^{(3)}(\mathbf{r}_1, \mathbf{r}_2, \mathbf{r}_3)$ through the relations

$$\rho^{(n)}(\mathbf{r}_1, \mathbf{r}_2, \mathbf{r}_3) = \frac{N!}{(N-3)!} P^{(n)}(\mathbf{r}_1, \mathbf{r}_2, \mathbf{r}_3) \tag{3.72}$$

and

$$\rho^{(3)}(\mathbf{r}_1, \mathbf{r}_2, \mathbf{r}_3) = \rho^3 g^{(3)}(\mathbf{r}_1, \mathbf{r}_2, \mathbf{r}_3) \tag{3.73}$$

with

$$P^{(3)}(\mathbf{r}_1, \mathbf{r}_2, \mathbf{r}_3) = \frac{1}{Z_N} \int_{-\infty}^{\infty} \exp\left(- \frac{U(\mathbf{r}_1, \mathbf{r}_2, \ldots, \mathbf{r}_N)}{k_B T} \right) d\mathbf{r}_4 \cdots d\mathbf{r}_N \tag{3.74}$$

where eqs. (3.52), (3.53) and (3.64) have been used. It must, however, be emphasized that all the attempts to work within the frame of eq. (3.71) were not really successful.

The determination of the pair correlation function $g^{(2)}(\mathbf{r}_1, \mathbf{r}_2)$ and all the higher order correlation functions with $\upsilon(r)$ as input can be done most reliably by molecular dynamics (and of course by Monte Carlo simulations). The basic information (3.8) can directly be used for the determination of $g(r) = g^{(2)}(\mathbf{r}_1, \mathbf{r}_2)$. Again, $g(r)$ is a measure of the probability that two objects of the many-particle system have the distance r, and we can simply compute $g(r)$ by

$$g(r) = \frac{1}{4\pi r^2 \Delta r} \frac{n(r, \Delta r)}{\rho} \tag{3.75}$$

where ρ is the macroscopic density and $n(r, \Delta r)$ is the density in the spherical shell around a particle having the radii r and $r + \Delta r$. Equation (3.75) can easily be used in connection with molecular dynamics calculations.

3.6.4 Nanosystems

Systems with surfaces and interfaces are not homogenous and isotropic. In the case of nanosystems or semiinfinite systems, we have, in fact, surfaces and interfaces where a large fraction of atoms/molecules are located. In those cases, the systems cannot be described by scalar functions of the interatomic separation $\mathbf{r} = |\mathbf{r}_i - \mathbf{r}_j|$, but these quantities depend on the vector field $\mathbf{r} = \mathbf{r}_{ij} = \mathbf{r}_i - \mathbf{r}_j$ and on the distances from the surrounded surfaces/interfaces. Such systems are simply not isotropic and homogeneous as in the bulk. Such situations can be complex and are, in most cases, not accessible to an analytical treatment. Numerical molecular dynamics calculations are in the foreground here.

In the case of a semiinfinite system, which is less complex, only the vector field $\mathbf{r} = \mathbf{r}_{ij} = \mathbf{r}_i - \mathbf{r}_j$ and the distance z, as measured along the axis perpendicular to the surface/interface, are involved.

In connection with surfaces and interfaces, the single-particle distribution function $\rho^{(1)}(\mathbf{r})$ cannot be treated as a simple constant as in the case of bulk states [see eq. (3.60)], but it is dependent on the coordinate z, but not on the coordinates x and y if we deal with semiinfinite systems. Thus, instead of a constant, we have now the relation

$$\rho^{(1)}(\mathbf{r}) = \rho^{(1)}(z) \tag{3.76}$$

The corresponding correlation function can be defined and is, in analogy to eq. (3.64), expressed by

$$\rho^{(1)}(z) = \rho g^{(1)}(z) \tag{3.77}$$

Also, the function $g^{(1)}(z)$ is interpreted in terms of probabilities: $g^{(1)}(z)$ is the probability that an object of the system is between the vertical positions z and $z + dz$. This function can be used for density considerations, that is, it reflects the density profile within the surface/interface zone. In principle, $g^{(1)}(z)$ is accessible to experiments.

Furthermore, the two-particle distribution function for systems, which are not isotropic and homogeneous, must be expressed in the form of

$$\rho^{(2)}(\mathbf{r}_i - \mathbf{r}_j) = \rho^{(2)}(z_i, z_j, |\mathbf{r}_i - \mathbf{r}_j|) \tag{3.78}$$

that is, this function cannot be treated as a quantity that is only dependent on the distance $\mathbf{r} = |\mathbf{r}_i - \mathbf{r}_j|$ as in the case of an isotropic and homogeneous bulk system, but is dependent on the distances z_i and z_j as well, where z_i and z_j are the distances of the objects i and j from the surface/interface.

That is, many-particle systems with surface/interface, in particular nanosystems, are dependent on their relative distance $|r_i - r_j|$ and their distances z_i and z_j from the surface/interface. This is expressed by eq. (3.78).

Clearly, the inhomogeneity in the surface/interface zone leads to the effect that the single-particle distribution function $\rho^{(1)}(r)$ is not a constant as in the case of disordered bulk systems.

Semiinfinite systems are relatively simple. In nanoscience, the systems are usually more complex. Nanoclusters are not semiinfinite in character. As we have remarked several times, the most reliable method to analyze the structure of such systems is molecular dynamics. The resulting information (3.8) can directly be used for the calculation of correlation functions like $g^{(1)}(z)$.

An example for a simple nanocluster is given in Figure 3.9 [15]. In Figure 3.9, the single-particle correlation function for nanoclusters is shown. They consist of $N = 500$ krypton atoms; for krypton a realistic pair potential exists (Section 3.5.3), which has been used in the molecular dynamics calculations presented here. Figure 3.9 shows the results for two temperatures: $T = 20$ K and $T = 85$ K. The structural effects are large,

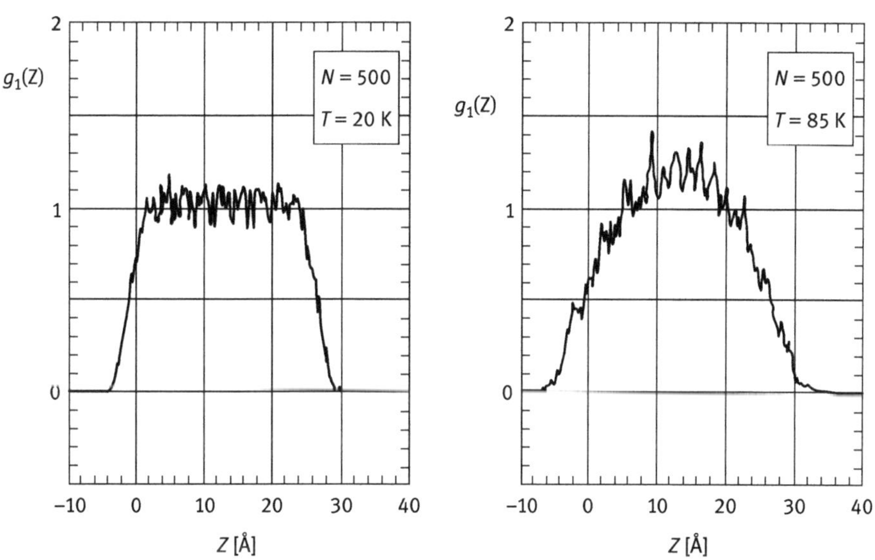

Figure 3.9: Single-particle correlation function $g^{(1)}(z)$ (z is the coordinate perpendicular to the surfaces) for simple nanoclusters consisting of $N = 500$ krypton atoms at $T = 20$ K and $T = 85$ K. The structural changes within the clusters are large and they are strongly dependent on temperature, although the melting temperature for the bulk is 116 K (© 2011, American Scientific Publishers).

although the temperatures are below the melting temperature of krypton in the bulk, which is 116 K.

In the calculation for $T = 20$ K, we recognize that $g^{(1)}(z)$ mainly takes values of one. This structure of $g^{(1)}(z) = 1$ means that the local density is (almost) identical with the macroscopic density ρ of the bulk. In the surface region, the thermal expansion becomes relevant. Here, thermal expansion is relatively large leading to a relatively smooth transition from $g^{(1)}(z) = 1$ to $g^{(1)}(z) = 0$. Clearly, the large thermal expansion reflects the fact that there are strong anharmonicities. The reason for that is obvious: The mean-square amplitudes of the atoms/molecules increase strongly in the surface region, that is, the mean-square amplitudes are significantly larger at the surface than in the bulk of the system. This makes it clear that anharmonic effects are effective; in such cases, the harmonic approximation is limited to (very) low temperatures.

In the case of 85 K, the local density inside the cluster is significantly larger than in the case of the bulk state. On the other hand, the surface region, where the transition from $g^{(1)}(z) = 1$ to $g^{(1)}(z) = 0$ takes place, is more extended than for the lower temperature ($T = 20$ K). In other words, there is a relatively large region in which the cluster density is distinctly smaller than the bulk density ρ, and this situation leads to the effect that the cluster is highly disordered in the surface region and has no similarity with a crystalline solid, although the temperature is far below the melting temperature ($T = 116$ K).

Furthermore, there are diffusion processes. The analysis of the molecular dynamics date shows that the diffusion constant within the surface region is relatively large, larger than in the case of the liquid in the bulk and also larger than for semi-infinite surfaces in a state which shows the effect of premelting.

All that indicates that the study of single-particle correlation functions is useful just for the structural analysis of nanosystems. Numerical studies on the basis of molecular dynamics models are here in the foreground.

3.6.5 The dynamics of nanosystems

In solid-state physics, the phonon concept, often denoted as standard model of solid-state physics, is of particular interest. The dynamics of an ordered crystal is completely described by noninteracting phonons. This conception is, however, strictly restricted to the harmonic approximation.

3.6.6 The dynamical matrix

Let us consider, for example, a monatomic many-particle system. The equation of motion for the kth atom (molecule) having the mass m is expressed in the harmonic approximation by

$$\ddot{u}_i^k + \frac{1}{m}\sum_{l,j} \phi_{i\,j}^{kl} u_j^l = 0\ ;\ \ i = x, y, z \tag{3.79}$$

where $\phi_{i\,j}^{kl}$ are the force constants and u_i^k is the ith component of the time-dependent displacement from the mean position r_0^k.

The solutions of eq. (3.79) have the form of three-dimensional Bloch functions if the system is infinite, that is, if we work with a system without surface. The Bloch functions are relatively simple and are expressed through

$$u_i^k \propto \exp\{i(\mathbf{q}\cdot\mathbf{r}_0^k - \omega\tau)\} \tag{3.80}$$

The phonon density of states is relevant here. It is straightforward to extract the phonon density of states $g(\omega)$ from the dispersion curves $\omega = \omega(\mathbf{q})$, which are accessible to measurements.

In the case of systems with free surfaces (semiinfinite crystal), there is only two-dimensional invariance, and this has consequences for the force constants. There is only a two-dimensional invariance of the force constants and this implies that the normal mode solutions, eq. (3.80), have the form of two-dimensional Bloch functions.

3.6.7 The anharmonic case

As we have already outlined, the harmonic approximation does not work for a lot of systems. In particular, to those with surface like nanosystems. Here, the harmonic approximation often breaks down already far below the melting temperature, and the phonon picture cannot be applied. Phonons are even not the starting point for the description of the particle dynamics and we have to choose new approaches without the phonon conception as background.

Liquids in the bulk, semiinfinite systems with one surface (for example, nanofilms) as well as small nanosystems show in general strong anharmonic behavior even at low temperatures. Nanoclusters (Figure 3.9) are typical examples. In all these cases, the anharmonicities cannot be considered as small perturbations. We have to apply statistical mechanics.

Concerning the dynamics, “time correlation functions” are in the center. The distributions in space have been treated also through correlation functions, and the pair correlation function is here of particular relevance.

3.6.8 Time correlation functions

Let us start with some definitions. In the analysis of many-particle systems, time correlation functions of type

$$\langle a(\tau')b(\tau'')\rangle \tag{3.81}$$

are of interest, where

$$\begin{aligned} a(\tau') &= a(q(\tau'),p(\tau')), \\ b(\tau'') &= b(q(\tau''),p(\tau'')) \end{aligned} \tag{3.82}$$

Here, τ' and τ'' are time values, which are in general different from each other, that is, we have

$$\tau' \neq \tau'' \tag{3.83}$$

The angular brackets $\langle \cdots \rangle$ in eq. (3.81) denote statistical averaging within the frame of thermodynamics on the basis of ensembles. This point has been outlined in Section 3.4 in connection with molecular dynamics calculations; various ensembles have been introduced in Section 3.4.

The time evolution of a many-particle system is given in statistical mechanics by the operator $\hat{S}(\tau)$ and is defined as follows:

$$a(q(\tau'),p(\tau')) = \hat{S}(\tau' - \tau'')a(q(\tau''),p(\tau'')) \tag{3.84}$$

where $\hat{S}(\tau)$ can be written in terms of the well-known Liouville operator $\hat{L}$ as follows

$$\hat{S}(\tau) = \exp\left(i\hat{L}\tau\right) \tag{3.85}$$

with

$$\hat{L} = i\sum_{i=1}^{3N}\left[\frac{\partial H}{\partial q_i}\frac{\partial}{\partial p_i} - \frac{\partial H}{\partial p_i}\frac{\partial}{\partial q_i}\right] \tag{3.86}$$

where q_i and p_i denote again all the $3N$ coordinates and the $3N$ momentums of the system. The quantities q_i and p_i have been introduced in Section 3.4 and represent the most basic and complete information about a classical many-particle system. When we have q_i and p_i, we have everything about a many-particle system under investigation.

In general, it is difficult to determine the time evolution of nanosystems and other systems on the basis of law (3.85). However, with information (3.8) that follows from molecular dynamics information, it is relatively easy to calculate correlation functions of type (3.81). However, in this monograph, we will not deepen this point, but we refer to the literature.

In connection with time correlation functions (3.81), it is important to mention that this kind of functions are not dependent on the time origin, that is, in eq. (3.81) not τ' and τ'' are relevant but $\tau'' - \tau'$:

$$\langle a(\tau')b(\tau'')\rangle = \langle a(0)b(\tau''-\tau')\rangle \tag{3.87}$$

The fact that time correlation functions are not dependent on the time origin is an important point.

3.6.9 Generalized phonon density of states

Let us discuss an important example: With $a(0)=\mathbf{v}(0)$ and $b(\tau)=\mathbf{v}(\tau)$, we get, with eq. (3.81), the velocity autocorrelation function

$$\varphi(\tau) = \langle \mathbf{v}(0)\cdot\mathbf{v}(\tau)\rangle \tag{3.88}$$

which is given in normalized form by

$$\phi(\tau) = \frac{\langle \mathbf{v}(0)\cdot\mathbf{v}(\tau)\rangle}{\langle \mathbf{v}(0)^2\rangle} \tag{3.89}$$

where $\mathbf{v}(\tau)$ is the velocity at time τ for one atom of the many-particle system (ensemble). In connection with the velocity autocorrelation function, the Fourier transform is relevant. With the property $\phi(\tau)=\phi(-\tau)$, the Fourier transform of $\phi(\tau)$ is given by

$$f(\omega) = \frac{2}{\pi}\int_0^\infty \phi(\tau)\cos\omega\tau\, d\tau \tag{3.90}$$

where $f(\omega)$ is normalized to unity:

$$\int_0^\infty f(\omega)d\omega = 1 \tag{3.91}$$

The Fourier transform $f(\omega)$ has a definite meaning. In the case of the harmonic solid, the frequency spectrum $f(\omega)$ is just the spectrum of the normal modes, that is, the phonons. The function $f(\omega)$ is often called “generalized phonon density of states.”

The frequency spectrum $f(\omega)$, defined by eq. (3.90), is a quite general quantity and is, in particular, applicable to systems with strong anharmonicities like nanosystems.

The generalized phonon density of states $f(\omega)$ describes the complete dynamics of the many-particle system under investigation and is not restricted to the harmonic case, that is, all kinds of dynamical excitations are included and even diffusion processes are involved. The diffusion constant is directly expressed by $f(\omega)$:

$$D = \frac{k_B T\pi}{2m} f(\omega=0) \tag{3.92}$$

Using $f(\omega)$ in the description of the dynamics, no problems appear in connection with systems with surfaces where the dynamical matrix solutions for the calculation of the modes are obviously not complete.

Molecular dynamics results

Molecular dynamics results for the generalized phonon density of states are shown in Figure 3.10. These are the curves for krypton clusters consisting of 500 atoms at $T = 20$ K and $T = 85$ K. The nanoclusters have an atomic structure that are shown in Figure 3.9.

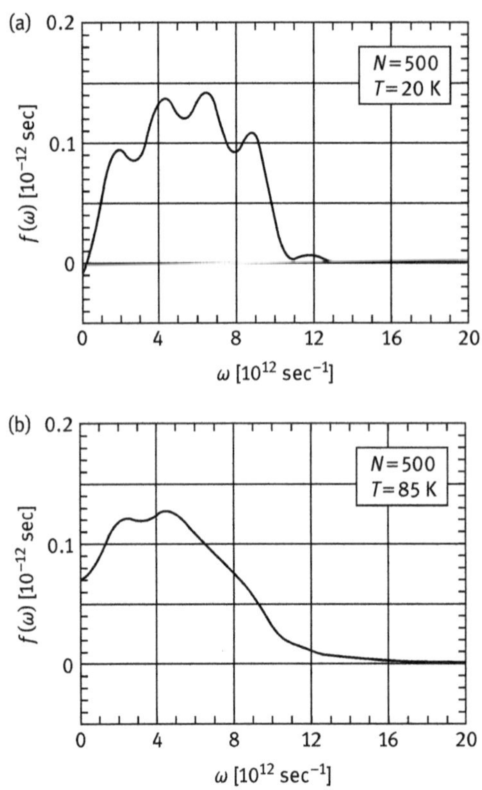

Figure 3.10: Generalized phonon density of states. Molecular dynamics results for krypton nanoclusters consisting of $N = 500$ atoms particles. The calculations have been done for $T = 20$ K (figure a) and $T = 85$ K (figure b). The corresponding structures for these systems are given in Figure 3.9 (© 2011, American Scientific Publishers).

The result for $T = 20$ K shows that the velocity correlations are long-living compared to the correlations at $T = 85$ K. Therefore, in the calculation of $f(\omega)$ by eq. (3.90), the cut-off effects in the calculation of the integral for the $T = 20$ K case are larger than for $T = 85$ K. For example, at $T = 20$ K the value $f(\omega = 0)$ becomes negative, although $f(\omega)$ must be a positive definite quantity.

With increasing temperature, the generalized phonon density of states is shifted to low frequencies. This is particularly reflected in the value for the diffusion coefficient, which has been determined by eq. (3.92). The diffusion coefficient D takes at $T = 85$ K the value of $D = 0.92 \times 10^{-5}$ cm^2/s. This value reflects an average over all $N = 500$ particles. However, the inner particles of the cluster are less mobile than those at the surface. The reason is obvious and is explained through the relatively large particle density in the inner of the cluster (see Figure 3.9).

The generalized phonon density of states for the bulk is shown in Figure 3.11 for two temperatures. The results are based on molecular dynamics calculations. Also, here the calculations have been performed for 500 krypton atoms. To avoid surface effects, periodic boundary conditions were imposed on the systems. Figure 3.11 shows the results for $T = 7$ K (a) and $T = 102$ K (b). The melting temperature is 116 K. The arrows in Figure 3.11 ($T = 7$ K) indicate the positions of the peaks in $f(\omega)$, which have been obtained on the basis of a three-nearest-neighbor fit to experimental dispersion curves. The positions of the peaks obtained by molecular dynamics are the same, indicating that the calculations are realistic.

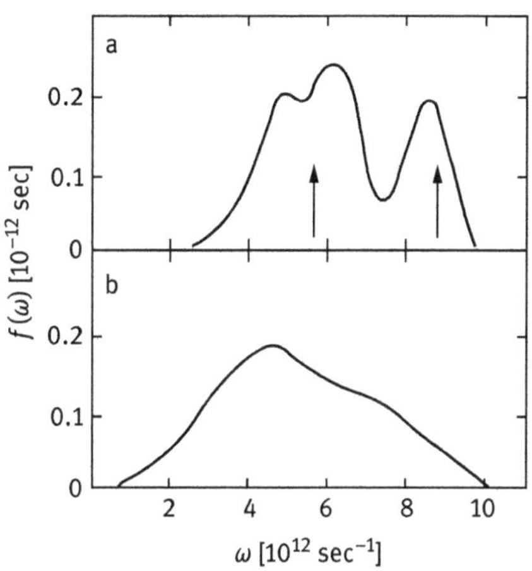

Figure 3.11: Generalized phonon density of states. Molecular dynamics results for krypton in the bulk having 500 atoms. The calculations have been done for $T = 7$ K (a) and $T = 102$ K (b). The melting temperature is $T_S = 116$ K (© 2011, American Scientific Publishers).

3.6.10 On the modeling of the generalized phonon density of states

As we remarked several times, molecular dynamics calculations are important for systems with strong anharmonicities since the anharmonicities are treated without approximations. Therefore, the most reliable results for the velocity autocorrelation function $\phi(\tau)$ [eq. (3.89)] could be obtained on the basis on molecular dynamics calculations.

On the other hand, in the theoretical modelling of $\phi(\tau)$ the so-called memory function $\Gamma(\tau)$ could be of interest for nanosystems. So far, the memory function $\Gamma(\tau)$ has been used for the description of $\phi(\tau)$ in the liquid state. $\Gamma(\tau)$ is related to $\phi(\tau)$ by the well-known integrodifferential equation

$$\frac{d\phi(\tau)}{dt} + \int_0^\tau \Gamma(\tau - \tau')\phi(\tau')d\tau' = 0 \tag{3.93}$$

and $\Gamma(\tau)$ is given by

$$\Gamma(t) = \frac{< \dot{\mathbf{v}} \exp\left\{ it(1-\hat{P})\,\hat{L}^N \right\} \dot{\mathbf{v}} >}{< \mathbf{v}(0)^2 >} \tag{3.94}$$

where $\hat{P}$ is a projection operator, and is defined through

$$\hat{P}\mathbf{G}(q,p) = \mathbf{v}\frac{1}{Z_N}\exp\left\{ -\frac{H}{k_B T}\right\} \int_{-\infty}^{\infty} dq' dp' \mathbf{v} \cdot \mathbf{G}(q',p') \tag{3.95}$$

$\mathbf{G}(q,p)$ is a phase function and $\hat{L}^N$ is again the classical Liouville operator given by eq. (3.86).

Concerning the memory function $\Gamma(\tau)$, we do not want discuss the details here, but let us nevertheless give few remarks: Langevin diffusion is described by a delta function:

$$\Gamma(\tau) = C_L\,\delta(\tau) \tag{3.96}$$

where C_L is a constant characterizing of the diffusion process.

On the other hand, in the case of the well-known Einstein oscillator, the memory function takes a constant value:

$$\Gamma(\tau) = const \tag{3.97}$$

The realistic case for nanosystems should be just between these two cases, that is, between a liquid and a solid. In this way, we can construct models for the generalized phonon density states with the help of the memory function $\Gamma(\tau)$, and vibrational aspects can be simply combined with diffusion effects, independent of the structure of the system under investigation. But it must be underlined that the most realistic way to determine $f(\omega)$ is by molecular dynamics calculations.

3.7 Nanoengineering

So far, the general procedures for designing molecular dynamics models have been discussed without any specific condition regarding the shape of the nanosystem. In many cases, single and simple-shaped objects are of interest; cubical nanoclusters is an example. In such cases, the setup of an appropriate molecular dynamics model is relatively easy.

3.7.1 Complex nanomodels

However, in the field of nanoengineering, such simple objects are only of minor interest. Within nanoengineering, structures with complex shapes or the interactions of various nanosubsystems play an important role. Already, cluster–cluster or surface–cluster interactions represent complex situations. Just in these and similar cases, the molecular dynamics method is an indispensable tool. Analytical models can be excluded in such complex cases. For the study of nanomachines, the molecular dynamics models have to contain moving parts in addition to parts of miscellaneous materials.

In contrast to common mechanical engineering, the preparation of nanomodels is done under specific procedures. For usual nanomodels, the initial atomic positions are chosen in accordance with the perfect lattice structure that corresponds to a temperature of 0 K. The next step is to accelerate the atoms until an equilibrium is reached according to a prescribed specific model temperature. Such an equilibration with respect to a desired temperature is performed during a sufficient number of molecular dynamics calculation steps by rescaling the velocities of all particles:

$$\mathbf{v}_i(t) \leftarrow \sqrt{\frac{T_d}{T}}\mathbf{v}_i(t), \quad i=1, \ldots, N \tag{3.98}$$

In eq. (3.98), T_d is the desired temperature, and T and $\mathbf{v}_i$ are the current values for the temperature and N is the number of velocity vectors.

3.7.2 Nanoengineering and common mechanical engineering

The initial molecular dynamics model, according to the perfect lattice structure, is not a stable, stationary configuration but, due to the temperature and the necessary equilibration process, the outer shape normally takes a disturbed geometry. Figure 3.12 illustrates the typical design steps in the case of a metallic nanostructure. First, a computer-aided design (CAD) model is established, which defines the rough layout in accordance with common mechanical engineering. After that, the molecular dynamics model is created by filling the CAD model with atoms. Then, the atoms are accelerated until the desired

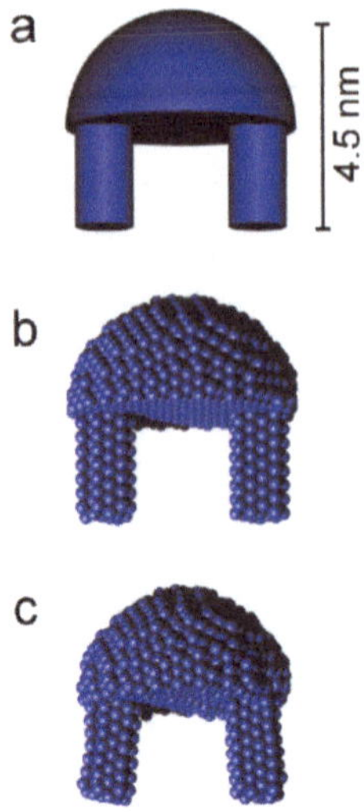

Figure 3.12: Design of realistic nanosystems consisting of aluminum atoms (ions). (a) There are three steps within this process: first, a CAD model has to be produced. (b) After that, the molecular dynamics model is created by filling the CAD structure with 1958 aluminum atoms. (c) Then the atoms are accelerated until the desired temperature is reached, in this case 300 K (© 2006, American Scientific Publishers).

temperature is reached. Finally, the realistic nanostructure results from the equilibration process which applies a prescribed temperature to the molecular dynamics model.

Figure 3.12 shows that the shape of the realistic atomic model with a certain temperature differs significantly from the CAD model, which simulates the macroscopic case.

The atomic positions of the molecular dynamics model have been chosen with respect to the perfect lattice structure. Such a structure is a stable configuration in the macroscopic realms but not for nanometer-sized systems. The atoms of the molecular dynamics model rearrange their positions during the equilibration phase until a stable configuration is reached (Figure 3.12(c)). This leads most often to quite different shapes in comparison to the initial input shape.

In conclusion, in comparison to common mechanical engineering, nanodesigning strongly depends on the outer shape of the considered system and, therefore, nanodesign is different from the usual construction conception of macroscopic technology. No doubt, this is relevant. In the next section, we would like to deepen this point.

3.7.3 Constructions at various levels

Macroscopic systems

Let us consider a macroscopic system as, for example, a car which represents a macroscopic machine. If we work below the melting temperature, the shape of the various parts of the car do not vary with temperature. Clearly, we have certain thermal expansions but this effect can be considered in good approximation as a uniform effect and does not affect the shapes of the workpieces.

Furthermore, the shape of a certain workpiece of the car is hardly or not influenced by the other car workpieces around it. This is a relevant point in connection with the production of macroscopic machines, in this case the car.

In other words, the shape of the isolated workpiece is exactly the same as in the composed state of the machine. So, the production of a certain part of a macroscopic machine may be done independently from the other parts forming the macroscopic machine, that is, in our example the car.

In conclusion, the parts of a macroscopic machine can be treated in all cases as continuous blocks without the consideration of the atomic structure. In particular, their shapes are practically not dependent on temperature and are also not influenced by the other parts of the machine. All that appears to be trivial and is seen by human beings as a matter of course. Our constructions of everyday life are based on this feature.

The bulk states are equivalent to what we have discussed with respect to macroscopic machines. In the case of the bulk states, the number of surface atoms is relatively small, and we have the situation as in the case of macroscopic machines, although we deal here with relatively small, microscopic systems with atomic structure.

Let us underline once again that the shapes of the parts of a macroscopic machine do not vary with temperature. The effect of thermal expansion is not relevant here because it produces a uniform effect which does not affect the shapes of the workpieces. Also, the bulk states of a nanosystem are influenced by temperature, but also here only the thermal expansion is relevant.

As we have already outlined, in the case of macroscopic machines the shape of a certain workpiece is not influenced by the other workpieces around it. This must have relevant consequences. All parts of a macroscopic machine can be produced separately, that is, the other workpieces around it have not to be considered when a specific workpiece is under production.

The situation is really simple: If the macroscopic machine consists of n different parts, we produce each of the n parts independently from the others. After this process we can put these n parts together in order to get the macroscopic machine.

However, this procedure does not work in the case of nanosystems as, for example, a nanomachine. Here the situation is quite different. Let us briefly discuss this point.

Within the nanomachine, the atomic structure of a certain part, "surrounded" by other parts of the nanomachine, might have for example the shape given in Figure 3.12(b), which is however not identical with the shape of an "isolated" atomic structure, given for example in Figure 3.12(c).

In other words, within the nanorealm, the surroundings have an influence on the shape. All parts are assumed to have the same temperature T_a. Though the other parts of the nanomachine, the single parts are more or less in bulk-states.

Isolated parts and surrounded parts

In conclusion, the structure in Figure 3.12(b), which is thought to be embedded within the nanomachine, develops differently from the case without surroundings. In other words, the outer shape of a certain part of a nanomachine is dependent on the surroundings, that is, whether it develops together with the other parts of the machine or without them. Again, the temperatures of the structures in Figure 3.12(b) and (c) are assumed to be identical.

In other words, the isolated part is different from that, which is surrounded by the other parts of the nanomachine. That is, such a system cannot be used for the composition of a nanomachine, because it is not identical with that inside the nanomachine.

A certain part of the nanosystem, which is surrounded by the other parts of the nanomachine, might have the shape given in Figure 3.12(b). However, if the same system is isolated, we obtain another shape for this part of the nanosystem (see Figure 3.12(c)). Both pieces are assumed to have the same temperature T_a.

Let us repeat:

The isolated part is different from that, which is surrounded by the other parts of the nanomachine. In other words, this system cannot be used for the composition of the nanomachine, because it is not identical with that inside the nanomachine.

This must have consequences. Parts of macroscopic machines can be exchanged. If the engine of a car is defect, it can be exchanged without any problem by a new engine. However, this is not possible at the nanolevel. Such a procedure does not work here. The reason for that has been just discussed. Because the shape of a part inside a nanomachine is different from that outside the machine, which is not surrounded by other parts, an exchange is not possible. As is demonstrated in Figure 3.12, this fact is due to the unavoidable temperature. The nanoreality is in fact different from the macroscopic world

3.7.4 Nanomachines

So far, the production of real nanomachines is not yet a topic of industry and has even in science not yet been performed. But it is a mostly important to perform molecular dynamics calculations for specific nanomachines. How and under what conditions do nanomachines work? The answering of this and further questions is of scientific and technological relevance, regardless of the question of how to assemble nanomachines in reality.

Nanoturbine

An example of such a study is presented in Figure 3.13. This is a nanomodel for a simple turbine consisting of a paddle wheel with an axle and two bearings resting on a substrate. The propulsion of the turbine could be established by a particle or laser beam [2, 15].

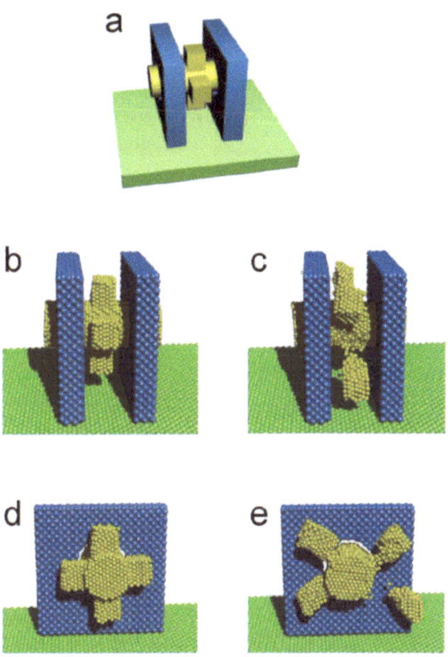

Figure 3.13: A simple nanoengineered turbine. The nanoturbine consists of a paddle wheel with an axle and two bearings. (a) Mechanically engineered model built with a common CAD software. (b) Perspective view and cross section of the molecular dynamics model with its atomic structure. The paddle wheel rotates with 5×10^{10} revolutions per second. (c) The number of revolutions is increased to 10^{11} per second and the paddle wheel ruptures. The side views of both situations are shown in (d) and (e): For (b) in (d) and for (c) in (e). The bearings (blue) consist of 4,592 silicon atoms, each with ideal shaped surfaces. The bearings have been treated as rigid objects within the molecular dynamics calculations, and this is possible because of the strong covalent bonds of silicon. The axle with the attached paddle wheel (light green) consists of 5,226 aluminum atoms and for the substrate (green) krypton atoms have been chosen (© 2006, American Scientific Publishers).

When comparing Figure 3.13(a) with Figure 3.13(b), the differences between common mechanical engineering and the nanodesign are recognized immediately. While mechanical engineering usually works with sharp edges and smooth surfaces (Figure 3.13(a)), nanodesign and nanoengineering, respectively, are dominated by single atoms, which exhibit the typical grainy appearance of nanostructures. In particular, tolerances may be kept very small within mechanical engineering but, due to the atomic structure, nanodesign allows to measure only in certain steps, dictated by the atomic overall structure, that is, the steps depend on the according material. Therefore, the fit of axle and bearing is rather loose, as can be recognized by the cross sections of Figure 3.13. However, the molecular dynamics studies have shown that such nanoturbines could be working stable for revolutions of up to about 5×10^{10} per second. This is in fact a lot.

Electrical nanogenerator

Another computational molecular dynamics model in the field of nanoengineering is represented in Figure 3.14. It is an electric power generator [2, 15] with a simple structure: It consists of one winding, which is stabilized by an isolating kernel. The kernel rotates between two bearings. It might be driven in various ways: Either by a nanoturbine similar to that of Figure 3.13 or by attaching paddles to the rotor kernel for propulsion with laser or particle beams. Such a nanogenerator would be able to perform a large number of revolutions. Stable operations of about 5×10^9 revolutions per second would be possible, which is based on the model calculations. With the magnetic field of the earth alone, an electrical voltage of about 10^{-11} V between the axles could be predicted. It is demonstrated in Figure 3.15 that the designed generator is very small.

The straight line in Figure 3.15 is the hair. The nanogenerator could be positioned on a very small area of the hair. The hair has a diameter of 80000 nm. On the other hand, the nanogenerator has an extension of 20 nm. In other words, the diameter of the hair is 4000 times larger than the size of the nanogenerator This situation is almost unimaginable. Another example is instructive as well: approximately 1 million of such electrical nanogenerators could be arranged side by side within a space interval of 1 cm.

Also, the number of revolutions per second is tremendous. The number for the revolutions per second is 10^9. If the wheels of a car could rotate with such a large number of revolutions, the car would circle the earth more than 80 times in a second. This is unimaginable.

3.7.5 Wear at the nanolevel

Within the frame of molecular dynamics, we deal with Hamiltonian systems. Here friction in the macroscopic sense is not defined. At the microscopic, molecular dynamics level, the forces are formulated in terms of structural configurations (particle positions), but they are not dependent on the particle velocities. Therefore, a force which is proportional to the velocity cannot be introduced at the nanolevel and, therefore, a friction constant in the macroscopic sense is not definable here. This is valid at the microscopic level of modern materials research. In the case of non-Hamiltonian systems,

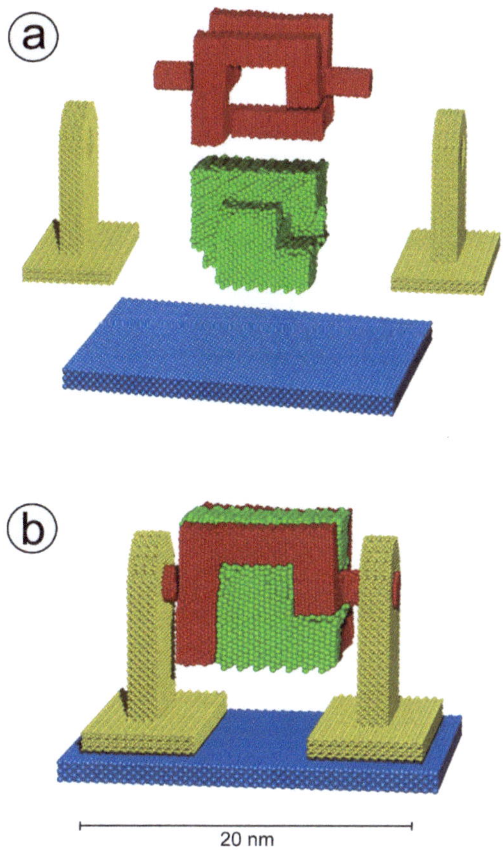

Figure 3.14: Nanogenerator for electric power: (a) The single parts of the machine consist of a winding with axles, an isolating kernel, and the substrate. (b) The assembled nanogenerator is able to rotate with 5×10^9 revolutions per second. The winding consists of 25.433 aluminum atoms and the rotor kernel has been assembled of 11.341 krypton atoms. The bearings are made of silicon and are treated in the calculations as rigid objects (© 2006, American Scientific Publishers).

friction can be studied by nonequilibrium molecular dynamics; this topic will not be discussed in in this monograph, but is indicated in Section 2.5.

At the nanolevel, wear can be understood in terms of specific complex processes. Let us underline this point by means of an example: Figure 3.16 shows a molecular dynamics model for a wheel which is pressed on a surface with atomic structure. The wheel rotates with 10^{12} revolutions per second and it has a temperature of 300 K. The wheel has a diameter of approximately 10 nm. Wear effects emerge when the wheel is pressed on the surface. According to the magnitude of the force vertically applied, the wheel can even be destroyed, as can be recognized in Figure 3.16.

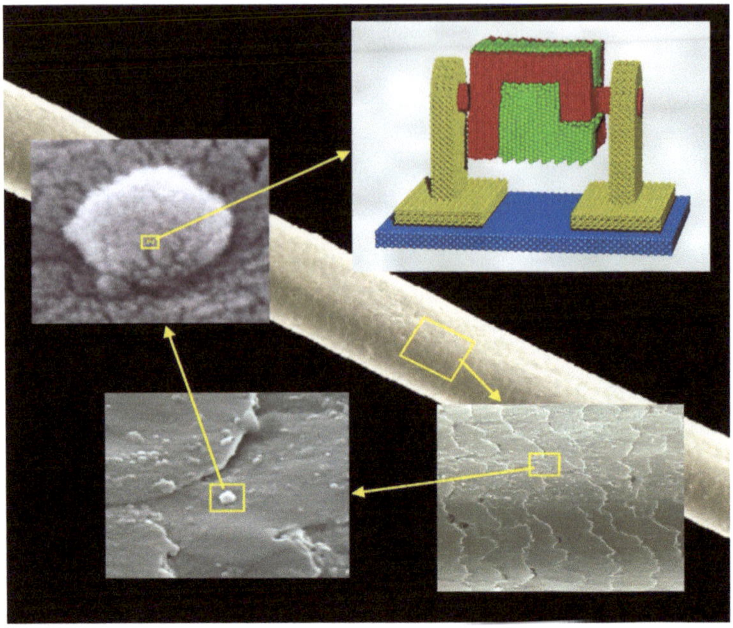

Figure 3.15: The figure shows the electric nanogenerator in comparison to a hair. The size of the generator is approximately 20 nm; the hair shown here has a diameter of 80 μm, that is, the diameter of the hair is 4,000 times larger than the size of the nanogenerator (© 2006, American Scientific Publishers).

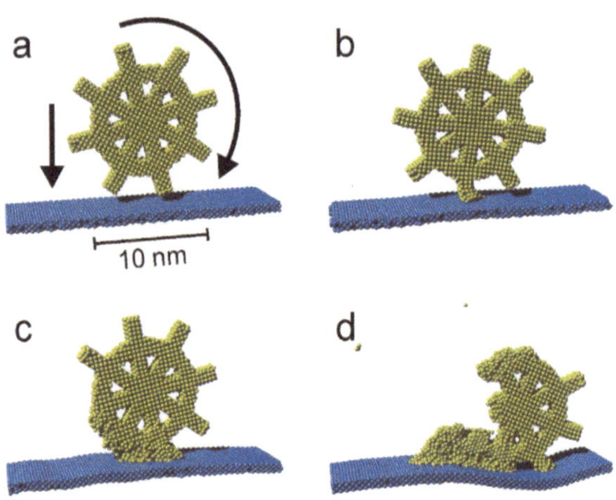

Figure 3.16: A molecular dynamics model for a wheel consisting of aluminum atoms. It rotates with 10^{12} rev/s and its diameter is approximately 10 nm. The wheel is pressed on a surface and wear effects emerge. According to the magnitude of the force vertically applied, the wheel can even be destroyed, as can be seen in the figures. The forces increase when we proceed from (a) to (d) (© 2004, American Scientific Publishers).

In summary, friction (wear) at the microscopic level is a complex process and cannot be characterized by only one constant (velocity). This process has nothing to do with what we call friction in the macroscopic realm. In our example (Figure 3.16), wear is dependent on the specific atomic structure of the surface and also on the wheel design.

3.8 Summary and final remarks

In nanoscience, we work at the "ultimate" level at which the properties of matter emerge. What kind of reality do we have here? This question has to be answered, especially when we consider effects at the ultimate level in nanoscience.

The starting point of the quantitative treatment of physical problems was Newton's classical mechanics more than 300 years ago. On what kind of reality is Newton's theory based? For Newton, the reality was what he had directly before his eyes. Within this view, reality is a "container" in which the material world is embedded. The container itself is just what we experience as space. The traditional form of quantum theory works also within this container model. But do we really have a world before us that is filled with material entities?

3.8.1 Classical descriptions

We know that this container model of traditional physics does not fulfill relevant items of nature [15]. The so-called projection principle is obviously more realistic and more sophisticated.

Nevertheless, the classical container model works well in most cases. We, therefore, discussed in this section about nanosystems within classical physics on the basis of the container model. Let us briefly underline the main differences between the container model and the projection principle.

The following features are relevant: Within the container model, we have physically real (material) objects directly before our eyes. Within the frame of the projection principle, we have geometrical entities before us. As can be argued convincingly [15], both views do not violate the direct observations.

The notion "time" is a relevant factor in nanoscience, just at the quantum level. However, certain aspects in nanoscience seem to be described well on the basis of the external time (clock time) τ alone, and the container model becomes relevant. This model is not used in applications because of its ingeniousness, but it simply works well in certain cases without explicitly to know why. In fact, it has been applied intensively for the analysis of various nanosystems, and it will certainly be used in the future. The container model belongs therefore to the tools of modern nanoscience, even when the future will rather be concentrated on the projection principle. This is the pragmatic standpoint, but this perspective can be justified because the container model works properly in specific cases, just then when we deal with classical many-particle systems; here the molecular dynamics method is relevant. In this chapter the essentials of molecular dynamics have been discussed, the principles and applications.

3.8.2 Molecular dynamics

The molecular dynamics method has often been used in the description of nanosystems. Here, Newton's equations of motion are solved by iteration. As input, we need the interaction potentials between the members of the many-particle system. We distinguished between four basic interaction types: Ionic interactions, metallic bonds, van der Waals interactions and covalent bonds. Phenomenological potentials (e.g., the Morse potential) have often been used in molecular dynamics calculations.

The properties of such systems are very sensitive to small potential variations. Therefore, for the description of system properties the interaction laws have to be modelled very carefully.

The molecular dynamics method is based on the container model and Newton's mechanistic world view. The many-particle systems (nanosystems) can be treated by classical mechanics if the temperature of the system under investigation is large enough, that is, the properties of the systems can be often determined on the basis of Newton's equations of motion. We know that this conception is not realistic but it works. There are obviously compensating effects with respect to certain assumptions.

3.8.3 The ultimate level

In nanoscience (nanotechnology), matter in the form of many-particle systems is described at the smallest level at which functional matter (condensed matter) can exist: It is in fact the "ultimate level" where the properties of materials emerge. Here we deal with the atomic/molecular level, and there is no scientific level above this level that would be relevant for nanoscience.

Most basic biological structures such as DNA, enzymes and proteins also work at this nanoscale, building up, molecule by molecule, macroscopic biological systems we call trees, humans and so on, with their typical intimate features.

3.8.4 Structure and dynamics at the nanolevel

Nanosystems are in general not crystalline. A great fraction of atoms/molecules are located near surfaces (interfaces) and, due to the strong anharmonicities at surfaces, the structure of nanosystems is disturbed and we need correlation functions for the structure description. The pair correlation function is important here but the single particle correlation function as well.

Also, the particle dynamics cannot be described by phonons but we have to use the so-called generalized phonon density of states, which is defined through the velocity autocorrelation function.

Specific material properties of nanosystems can be essentially different from the corresponding properties of macroscopic systems (the solid in the bulk). This has been demonstrated in connection with the thermal stability and the melting temperature of specific nanostructures. An essential reason for this tendency is the fact that a great fraction of the particles (atoms, molecules) of such small systems belong to the surface region and the surface particles are less bonded than the particles in the bulk, leading to relatively strong anharmonicities even at low temperatures, that is, far below the melting temperature.

Even the melting process takes place far below the bulk melting temperature. The thermal behavior of such systems is a complex function of the particle number and the outer shape of the systems. This must have consequences for the theoretical description of the material properties for systems of nanometer size.

We investigated, as an example, an electrical nanogenerator, which is relatively complex, but it is nevertheless very small. The size of this nanogenerator is approximately 4,000 times smaller than a hair (see Figure 3.15). The nanogenerator with all its parts is represented in Figure 3.14.

Nevertheless, the classical treatment of nanosystem is in many cases not sufficient, that is, nanosystems have to be considered, in general, as quantum systems.

In this connection, it is important to remark already here that the usual form of quantum theory has to be expanded in order to obtain reliable descriptions. This is particularly obvious in the case of "self-organizing processes." Self-organizing processes belong to the heart of nanoscience and we cannot resign on them in this field.

Chapter 4
Mind, brain, reality and nanoscience

For human beings and other living biological systems, the "observation" of the world outside is a relevant concern. Observations are the basis for "activities" in the environment of the living system in order to improve life and to avoid biodestruction. An observation and an activity as well can occur spontaneously, but they are in most cases done on the footing of conscious decisions. In other words, the material environment is changed in connection with mind states.

Matter and mind are somehow interrelated. But how? No doubt, this is an important question and has to be analyzed critically and carefully as well. In particular, the role of the brain needs to be elucidated, which is, as we know, a relevant connector in this case.

It is planned to manipulate brain functions at the nanolevel and, as we remarked in Chapter 1, the nanolevel reflects the ultimate level where the properties of condensed matter emerge. In other words, basic changes are planned and we have to know what we do and, therefore, we must find out what the notions "mind" and "brain" in connection with matter constitute. This is essential because these entities are of particular relevance for observations and activities. As we remarked, observations and activities are of vital interest for human beings and for other biological systems as well.

We know from behavior research that the observation activity realm is species dependent (Chapter 1): a turkey assesses the world outside differently than a human being, obviously quite differently. The results of Wolfgang Schleidt's experiments surprised the scientific community, and they are a mystery up to the present day. All that demonstrates that the traditional conceptions concerning the mind–brain–matter complex is still incomplete and not usable, respectively. We therefore need more sophisticated scientific conceptions for the solution of the mind–brain–matter problem.

The following questions arise: What is an observation? How is an activity explained within the frame of the complex consisting of matter, mind and brain? How are space and time inserted?

We definitely know that the entities matter, mind and brain exist and we know as well that space and time are of particular relevance. But how is this complex ordered? How are these entities arranged relative to each other? What are the connections between them? These questions have to be answered seriously before we try to manipulate brain functions at the nanolevel.

In this chapter, we discuss these and similar questions. We treat these points in connection with traditional views and new aspects, which are quoted in recent literature. We do not primarily consider the matter–mind–brain realm as a philosophical problem but rather as a question of practical relevance. For nanoscience, the

https://doi.org/10.1515/9783110525595-004

"interrelations" between matter, mind and brain are of interest. We treat the problem with respect to this aspect and find in this way the features, which connect matter, mind and brain as conditions so to speak. Such kind of information (conditions) is relevant for the nanorealm. What matter, mind and brain actually constitute in detail is not in the foreground here and must not be deduced within this frame.

4.1 Relationships

The impact of nanoscience (nanotechnology) will be tremendous and will influence our future dramatically. The manipulation of atoms and molecules will allow the development of completely new technological worlds: electronic nanodevices, nanomachines and changing brain functions. Nanotechnology will particularly develop fundamentally new possibilities in the field of medicine. Just in the case of nanobiotechnology, big changes are expected already in the near future. In particular, the changing of brain functions will be a challenge and needs sophisticated conceptions.

But do we really know what we try to do? Does the traditional physics really deliver adequate tools for such far-reaching interventions? Just the planned changes of "brain functions" throw up fundamental questions. What is the brain? What is its relation to that what we call "mind"? How are the brain and mind connected to physical (material) reality? We would like to answer these questions within the framework of the "projection principle." Remarks with respect to the traditional view (container model) are included.

4.1.1 Three facts

Nanoscience works at the ultimate level. As we have outlined in Chapter 1, this fact requires a reliable "basic conception" for the description of nanoproperties. The traditional view, based on the container model, does not allow the comprehensive treatment of the facts at the nanolevel; this model is simply too rudimentary. The lack of a quantum time is an essential example (Chapter 1). This point indicates that we need quite generally a more sophisticated view, in particular for the theoretical description at the nanolevel. The nanotechnological treatment of brain functions belongs to this category and is without any doubt a big challenge. An extension of the traditional view is required.

There are three characteristic facts that are equally experienced by each human being in everyday life. Let us first quote them, and after that we will analyze them adequately.

1. At time (clock time) τ we have objects directly before our eyes that are located in space. When we touch an object, for example a stone, we "feel" it, and we

are convinced that this object (stone) reflects a "material" state. The stone is considered as "matter." Theoretical physics is not involved yet.
2. Independently from these matter experiences, which are defined as x, y, z, τ structures before us and reflect unconscious phenomena, the effect of "thinking" is a real phenomenon. Thinking exists independently from the matter impression that we have at clock time τ before our eyes. Thinking takes place in the mind, and the creations of mind do not appear as picture before us, but appear as phantasy.
3. Whether the mind is made of matter or not is not touched through these direct experiences. Also the role of the brain, in particular its relation to the mind, is not explicitly considered within this direct view.
4. Ideas are mind activities and can therefore be considered as entities of the mind. Due to an "idea" the body, which consists of matter, can come in action; arms, legs and so on can be moved in order to perform specific changes in the environment. That is, the actions reflect the ideas of the mind and there is a "coupling" between mind and matter. This is obvious in this case. But how is such kind of situation explained?

4.1.2 Analysis

So far, no specific "world view" has been applied for the understanding of the experiences just quoted and which are a common feature for all human observers. In the following, we will analyze these three facts within the frame of traditional physics and in connection with the projection principle. For the understanding and the adequate description of nanophenomena, we have to start in fact at the very beginning of physics.

Container model

As we outlined several times, in traditional physics we work within the container model. At clock time τ, the objects before us are considered as physically real (material) entities, that is, the objects before our eyes refer directly to "matter." Within the container model, we do not touch geometrical objects with our fingers but these objects are made of matter, which are embedded in space and time. Within this view, there is nothing outside of space and time.

What the "mind" actually is remains an open question within the container model. Therefore, also the coupling between matter and mind (point 3) cannot be explained but defines a separate field, open for hypotheses and speculations. How the impressions before our eyes come into existence cannot be said within the frame of the container model. The fault of this view is obvious: physically real entities cannot be embedded within a space–time, which itself is not physically real (see also Chapter 1).

Projection principle

The projection principle is undoubtedly more sophisticated than the container model. The four points quoted earlier can be explained consistently within the framework of "one" conception. It particularly explains how the impressions in space, which appear at clock time τ directly before our eyes, come into existence. The projection principle is also able to give statements about the influence of changes at the ultimate nanolevel on the behavior of human beings; the manipulation of brain functions belongs to this realm. In particular, such artificial changes at the nanolevel can initiate changes with respect to the entire impression before the eyes of human beings, that is, there can be an impact of what we experience in everyday life.

4.1.3 The conception

Why the projection principle turned out to be more sophisticated than the container model? There are obviously no inner contradictions within this conception. But why the projection principle is without inner contradictions? The answer is straightforward: the most fundamental entities in natural science, that is, space and time, are treated more realistically in projection theory. Here the following point is essential: the elements of space and time, that is, the coordinates x, y, z and the time τ are not observable and a procedure (measuring process) for their detection is even not thinkable. Only distances of objects and time intervals in connection with processes can be registered (more details are given in Chapter 1).

Clearly, the objects and the processes have to be specified within the frame of the projection principle. Here we have to consider the specific features of projection theory. The essential characteristics of the projection principle and its relation to the often discussed statements in literature are briefly summarized in the following section.

Mach's principle

Once again, entities that are in principle not observable (the coordinates x, y, z and the time τ) cannot be considered as physically real and are therefore not be able to create physically real processes. We cannot embed physically real entities within a frame, which is not physically real. This is an extended form of Mach´s principle (Chapter 1). Ernst Mach required that space and time cannot be the source of physically real effects. Mach's principle is not grantable in traditional physics. Within the frame of the projection principle, Mach's principle is however accomplishable in detail.

Thus, the space–time structure before us, which emerges at the time τ, is not represented by objects made of matter. What we experience at the clock time τ are actually "geometrical" positions. The "material" entities (matter) must be located outside space and time.

The statement by C. G. Jung (Chapter 1) reflects this situation explicitly. Also the philosopher Immanuel Kant argued in this direction. The basic ideas by Kant are outlined in Chapter 1 as well.

Basic reality and pictures

From our discussion in the last section, we have to conclude that the physically real world must be outside space and time. We denoted this world without space and time as "basic reality." In other words, the physically real (material) objects are embedded in basic reality and the space–time structure before our eyes, which we experience at the clock time τ, is therefore a "picture," consisting of geometrical positions. Matter is projected onto space and time and we get a picture of matter. This defines the basic conception of projection theory and is summarized within the following scheme:

$$\begin{array}{c} \text{Matter} \\ \text{in} \\ \text{basic reality} \\ \downarrow \\ \text{Picture of matter} \\ \text{in} \\ \text{space and time} \end{array} \tag{4.1}$$

How do we come from basic reality to its picture? What steps are necessary? What items have to be considered? One of the essential items is here the phenomenon of evolution. It is particularly essential to know what information is actually depicted in the "picture of reality," which a human being experiences in everyday life.

4.1.4 Evolutionary processes

The effect of evolution belongs to the realm of physically real processes and takes place within the material world. That is, the evolution of material entities takes place in basic reality. However, we have to emphasize that evolutionary processes must not be restricted to matter.

Evolution here means the development of individual entities in basic reality, where the term "individual entity" refers to dead matter as well as to living biological systems (animals, human beings, etc.). Although basic reality does not contain space and time, we may order the evolutionary processes from the point of view of an external observer, who orders the things notionally in connection with the clock time τ when we identify the external observer with a human being. The situation is summarized by the following scheme:

$$\text{Matter in basic reality} \xrightarrow{\tau} \text{individual material entities in basic reality} \tag{4.2}$$

Scheme (4.2) is a postulate and is not further explainable. The reason is obvious: we cannot describe the evolutionary processes (4.2) quantitatively because human beings are caught in space and time, and they have no direct access to basic reality, that is, its structure remains hidden for human beings, but it definitely exists within the frame of this conception.

The principle of usefulness

Relation (4.2) represents the first step with respect to evolution. There are further evolutionary steps, which consider the existence of "species." Here the mind of an individual comes into play as well. This case is treated in the following sections and is a necessary feature of the physically real world.

A certain species is organized through the "principle of usefulness." As we already remarked in Chapter 1, it is not cognition that plays the important role in nature but the differentiation between "favorable toward survival" and "hostile toward survival." This is in fact the case at least at the early phase of evolution and everything, which has been developed evolutionarily, is based on the features developed earlier.

Not the "absolute truth" is essential for the members of a species, but the principle of usefulness. Each picture of reality, which is actually perceived by the members of a certain species, is tailor made to the principle of usefulness. Since the conditions for survival are different for different species (biological systems), the perceived realities are also different for the members of different species. The chick experiment, discussed in Chapter 1, demonstrates that such discrimination is necessary. Such species-specific situations require a second evolutionary step. As we will outline later, also this second evolutionary step is performed in basic reality, and this is because physically real processes are involved here.

The individual material entities are located in basic reality. They are developed through evolutionary first-step and second-step processes. These individuals, prepared evolutionarily, are pulled by species-dependent tools. That is, each species selects just this kind of information, which is "useful" for it. For this purpose, the evolution uses just the species-dependent tools, which we would like to call "tools of evolution."

Where does nature transfer this optimal species-dependent information, which is pulled from basic reality by the species-dependent tools of evolution? There is only one possibility for that: since the optimal information is individual in character, it can only be transferred to the "mind" of an individual member of the species.

The strategy of nature

How does nature organize such information transfer? The perception of the complete basic reality would mean that with growing fine structures, increasing information of the outside world is needed. Then, the evolution would have furnished the tools of evolution with the property to transmit as much information from basic reality as possible. But the opposite is correct: It is the strategy of nature to take up as little information from the world outside (basic reality) as possible, just that what is needed (principle of usefulness). As we remarked earlier, the world outside is not assessed through "complete" and "incomplete" but by "favorable toward life" and "hostile toward life." Concerning this point, Hoimar von Ditfurth stated the following [73]:

> No doubt, the rule "As little outside world as possible", only as much as is absolutely necessary is apparent in evolution. It is valid for all descendants of the primeval cell and therefore for ourselves. Without doubt, the horizon of the properties of the tangible environment has been extended more and more in the course of time. But in principle only those qualities of the outside world are accessible to our perception apparatus which, in the meantime, we need as living organisms in our stage of development. Also our brain has evolved not as an organ to understand the world but an organ to survive.

In conclusion, the perception of reality by biological systems is essentially influenced by the principle "as little outside world as possible" and this fact is species dependent. This situation characterizes that what is pulled from basic reality by the tools of evolution. Again, within the first evolutionary step [scheme (4.2)] the individuals are created, and the second evolutionary step underlines that there are biological blocks (species) existing as physically real groups, also located in basic reality.

Remark

The tools of evolution pull only those facts from the world outside (basic reality), which are useful for the observer; the pulled information is, in particular, species dependent. If a human being observes, for example, a certain object or a specific process, not the characteristics of the object or the process themselves are reduced, but all that what is superfluous (needless) around them. This means the term "as little outside world as possible."

4.1.5 Information transfer

In the case of a certain physically real object, say γ, which is embedded in basic reality, the individual tools of evolution pull an optimal part of γ and transfers it to the mind of the individual. Let us denote the species-dependent optimal part by the Greek letter α; α is a raw information and is in this step not given through a formula. The reason is obvious. The tools of evolution pull information, independent of an observer, and only an observer vests this information by a mathematical expression, that is, by a formula.

The effect through the tools of evolution leads to the following scheme, which reflects a projection:

$$\begin{array}{c} \gamma\,(\text{basic reality}) \\ \big\downarrow \text{tools of evolution} \\ \alpha\,(\text{mind}) \end{array} \tag{4.3}$$

The selected information α varies from species to species. Let us assume here that α is the raw information of human beings, and α occupies the mind of a human observer. The picture of reality is given in space and time; the information in the picture contains exactly the raw information α. In fact, it must be exactly the same information because changes with respect to α are only thinkable in connection with physically real processes, and such processes are by definition not possible in the human being's mind but exclusively in basic reality. There must therefore be a one-to-one correspondence between α and the structure in the picture of reality.

The raw information α is not directly observable, but must be representable in space and time and, as we remarked several times, this is because human beings are caught in space and time. The transition from α to the space–time structure is processes through the mind. There is in fact no other possibility.

In other words, in order to get at clock time τ the x, y, z structure, which appears before the eyes of human beings and which has exactly the same information as α, the quantity α has to be processed through the mind. The mind must have the ability to process selected information, which has been pulled by the tools of evolution from basic reality. For this purpose, the mind must contain a certain processing characteristic that we want to denote by the Greek letter η. In other words, η processes α and creates the picture of reality that appears at clock time τ as x, y, z structure before the eyes. Let us denote here the picture of α by the Greek letter θ. The picture before the eyes human beings comes into existence in an unconscious way.

This picture θ before a human being is not somewhere outside, but it occupies, like α, the mind of the human observer as well. It is merely an impression that the picture is outside us, but it is actually within the observer's mind. This point is particularly discussed in Chapter 1; we will deepen the situation as we go. The picture before a human being reflects a geometrical structure and has nothing to do with a physically real structure as in the case of the container model of traditional physics. The physically real objects are exclusively located in basic reality.

At this point, it must be underlined that the processing characteristic η creates the picture together with the coordinates x, y, z and the time (clock time) τ. This is an important point and shows that the space–time entities x, y, z, τ are not physically real. Again, the entities in the mind are not physically real. Only objects in basic reality are physically real. The notion “matter” will be defined and analyzed in Chapter 5.

The process in mind, from the information α to θ (the x, y, z, τ structure), is represented and summarized in scheme (4.4):

$$\begin{array}{c} \alpha\,(\text{mind}) \\ \downarrow \eta \\ \theta\,(x, y, z, \tau \text{ structure in mind}) \end{array} \tag{4.4}$$

The way from basic reality to the picture of reality is summarized in scheme (4.5):

$$\begin{array}{c} \gamma\,(\text{basic reality}) \\ \downarrow \text{tools of evolution} \\ \alpha\,(\text{mind}) \\ \downarrow \eta \\ \theta\,(x, y, z, \tau \text{ structure in mind}) \end{array} \tag{4.5}$$

The tools of evolution pull the species-specific part α from the physically real object γ, and α is transferred to the mind. The raw information α will be processed by the mind on the basis of the processing characteristic η and we get a picture θ of the physically real entity γ in the form of an x, y, z structure that is observed at clock time τ and appears unconsciously before the eyes of human observers. This is the situation within the projection principle where space and time are treated realistically and where evolutionary processes are taken into account in accordance with what we actually observe.

Observation means to register at clock time τ the physically real object γ as a picture in the form of a species-specific information within an x, y, z structure. Only this information should appear in the picture, not the activities of the mind, which creates the picture.

4.1.6 What is in the picture?

Once again, the picture before a human being reflects a geometrical structure and cannot be identified with physically real entities. The geometrical structures in the picture are not physically real objects but they represent them. Thoughts and ideas cannot be depicted in space and time. The experience teaches us that thoughts have no geometrical form. No doubt, certain thoughts and ideas, respectively, can be translated into a space–time impression in order to formulate thoughts and so in, in terms of familiar structures.

The possibility for the presentation of physically real objects in space and time belongs to physics and requires in particular that only the object itself appear in space and time and not other features, for example features of the observer's mind, which creates the picture of the physically real object as space–time structure. In fact, we do not observe thoughts in space and time; thoughts cannot be depicted as geometrical structures. This fact is in accordance with the projection principle. Thinking does not mean to create geometrical forms in space and time. Thinking is a process, which takes place at a level above the usual space–time activities.

4.1.7 Observer-dependent Worlds

The entity θ is the representation of the material object γ in space and time. A human being is caught in space and time and is not able to observe entities that are located outside space and time. The object γ is located in basic reality and is therefore not accessible to human beings.

The object γ is observer independent. This is however not the case for the picture θ, which is represented in space and time, and θ is "observer dependent." The reason is obvious: the picture θ comes into existence through the mind of the human observer [see scheme (4.5)]. All pictures come into existence through the mind of any human being and are therefore in any case observer dependent. Since we are caught in space and time, a human being is only able to register observer-dependent worlds.

Furthermore, the pictures are "species dependent." This is due to the fact that the tools of evolution are effective [scheme (4.5)]. All members of a certain species work with the same tools of evolution, but these tools vary from species to species.

4.2 The barrier

The processing characteristic η and the tools of evolution are not described through space and time, that is, they are not represented in the form of x, y, z, τ structures. The physically real world outside, which we called "basic reality," does not contain space and time as well. That is, the physically real entities are embedded in basic reality without the existence of space and time, but are projected onto space and time as pictures.

4.2.1 Only selected information is describable

The evolution teaches us that we cannot know the complete information of an entity, say γ, but merely a selected part α of it. Only the information α is accessible to human beings and is in everyday life given in processed form just in the form of a x, y, z, τ structure before the eyes of human beings. This refers also to human beings themselves, at least with respect to their physically real part. There is a principal barrier (see also Figure 4.1).

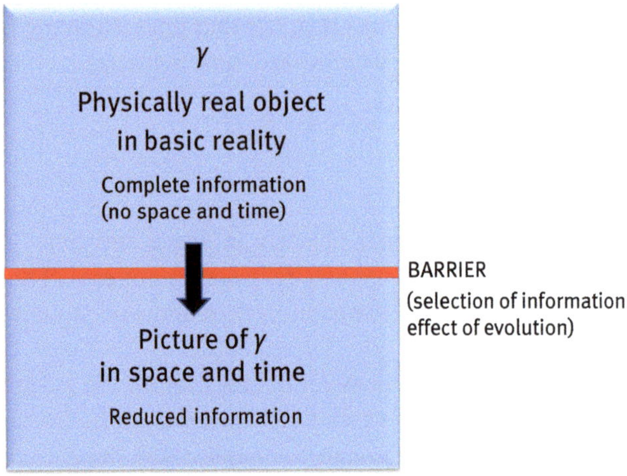

Figure 4.1: The world outside (basic reality) contains the physically real objects and, therefore, space and time, which cannot be considered as physically real entities (Chapter 1), do not belong to basic reality. A certain object, say γ, is embedded in basic reality and represents the complete information. However, this complete information of γ is, due to the effect of evolution, not accessible to human beings. There exists a principal "barrier," which human beings cannot exceed. Also human beings are captured through the biological evolution. A human being can only recognize a reduced information of himself that is represented in space and time. At the barrier, a selection takes place. Only that information about the world outside is recognizable for human beings, which is "useful" for him/her in daily life (Chapter 1). The absolute truth is not relevant in this connection. This statement also refers to human beings themselves, at least for its physically real part.

4.2.2 No one-to-one correspondence

Because of the lack of the total information, we cannot describe the transition from γ to α and also not the processing from α to the x, y, z, τ structure. The function of the entire complex, consisting of basic reality, mind, brain and space and time, (in connection with the tools of evolution and the processing characteristic η) can only describe the situation in a "qualitatively" way. A "quantitative" description is principally not possible, unless we work with unprovable statements. This is a principal point and reflects the limits of physical science.

The structures in basic reality and those of the pictures must be different from each other. We are caught in space and time and theses entities do not appear in basic reality. Thus, there can be no one-to-one correspondence between the structures in the picture and those in basic reality. This is the only what we can say in connection with basic reality.

4.2.3 Conscious and unconscious actions

At time τ we touch with our fingers an object, for example a stone, and feel it immediately. This situation takes place in space and time: The stone is positioned at time τ in space. However, also the fingers are represented as geometrical objects in space, that is, the geometrical position of both, the stone and the fingers, touch each other, but we "feel" the stone simultaneously. However, the touch of the geometrical positions cannot initiate this "feeling" of the stone that the human being has at clock time τ. The feeling can exclusively come into effect through a physically real process and must therefore take place in basic reality. The "touch" in space at time τ and the "feeling" in basic reality take place at once, that is, also the feeling takes place at clock time τ. In other words, there is a strict correlation between the "touch" and the "feeling," although there is no space–time in basic reality. The situation cannot be analyzed further since basic reality is not accessible to human beings.

4.3 Representation and observation

Let us consider a human being and let us denote him/her by the letter A. Scheme (4.5) does not mean that object γ is actually registered by A or by any other observer, but it merely indicates how physically real objects are represented as picture in space and time through the mind of any human being, for example, human being A. But the object γ cannot directly be registered by A, but only the picture of the object γ. In order to get a picture of γ the observer A has to interact with γ, and this is a physically real process, which takes place in basic reality.

4.3.1 Observation procedure

We want to express human being A in basic reality symbolically by γ_A. In order to be able to observe a certain object, say γ, the observer A has to interact with γ. For simplicity, let us assume that γ is a stone denoted here by γ_{STONE}; γ_{STONE} is the physically real form of the stone in basic reality. The interaction process itself in basic reality is symbolized by

$$\gamma_A \leftrightarrow \gamma_{\mathrm{STONE}} \tag{4.6}$$

We would like to assume that the interaction process (4.6) is assessed by another human being, denoted here by B. Observer B pulls this information through his/her tools of evolution and is, in reduced form, transferred to the mind of B. In analogy to scheme (4.5) we get

$$\alpha_A \leftrightarrow \alpha_{\mathrm{STONE}} \tag{4.7}$$

The processing in the mind of human being B is performed by his/her processing characteristic η_B. We obtain, also here in analogy to scheme (4.5), for the corresponding situation in space and time the relation

$$\theta_A \leftrightarrow \theta_{\text{STONE}} \tag{4.8}$$

Then, the case of observation scheme (4.5) has to be replaced by a similar procedure, which is given by the following block:

$$\begin{aligned} &\gamma_A \leftrightarrow \gamma_{\text{STONE}}\ (\text{basic reality}) \\ &\downarrow \text{tools of evolution } B \\ &\alpha_A \leftrightarrow \alpha_{\text{STONE}}\ (\text{mind of } B) \\ &\downarrow \eta_B \\ &\theta_A \leftrightarrow \theta_{\text{STONE}}\ (x, y, z, \tau \text{ structure in mind of } B) \end{aligned} \tag{4.9}$$

The tools of evolution and also the processing characteristic refer of course to human being B. More details will not be given here.

4.3.2 Levels of observation

In principle, the entity $\theta_A \leftrightarrow \theta_{\text{STONE}}$ is open for various observation levels. In other words, $\theta_A \leftrightarrow \theta_{\text{STONE}}$ contains various facets. The observation processes themselves take place of course in basic reality, but we may discuss them as phenomena in space and time, that is, in the form of observer-dependent x, y, z structures at clock times τ. The realization of a certain facet depends on the applied observation procedure. In the following we will discuss this point in more detail.

The macroscopic level

At the macroscopic level, where the everyday life experiences are located, the complete information of $\theta_A \leftrightarrow \theta_{\text{STONE}}$ is recorded on the basis of the five senses, which are expressed experimentally by sight, touch, taste, hearing and smell processes. A stone, its picture is given by θ_{STONE}, is essentially characterized through seeing and touching.

The atomic level

On the other hand, the application of suitable measuring instruments allows the investigation of the atomic structure and dynamics of the object under investigation, which is given in our case by the stone.

This atomic level is above the macroscopic level because at this level the information is more sophisticated and detailed, respectively. The complete information of $\theta_A \leftrightarrow \theta_{\text{STONE}}$ at this atomic level is here expressed by the structure and dynamics of the system. The levels are indicated in Figure 4.2 by $[\theta_A \leftrightarrow \theta_{\text{STONE}}]_{\text{macroscopic}}$ and $[\theta_A \leftrightarrow \theta_{\text{STONE}}]_{\text{atomic}}$.

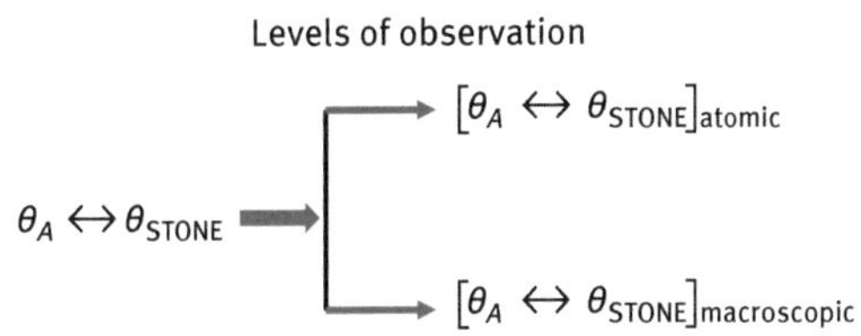

Figure 4.2: In principle, the entity $\theta_A \leftrightarrow \theta_{\text{STONE}}$ is open for various observation levels. Each level is characterized by specific experimental methods, which are different from each other. In everyday life, we deal with the macroscopic level. At this level, the five senses are relevant and the observations lead to a complete information at this macroscopic level. The application of suitable measuring instruments allows to investigate the atomic structure and dynamics of the object under investigation, for example, a stone. This atomic level is above the macroscopic level.

Representations

Since there is no space and time in basic reality, the structure of the complex $\gamma_A \leftrightarrow \gamma_{\text{STONE}}$ must be different from the picture $\theta_A \leftrightarrow \theta_{\text{STONE}}$ of $\gamma_A \leftrightarrow \gamma_{\text{STONE}}$, which is a space–time representation. We have

$$\gamma_A \leftrightarrow \gamma_{\text{STONE}} \neq \theta_A \leftrightarrow \theta_{\text{STONE}} \tag{4.10}$$

Again, the interaction between human being A and the stone takes place in basic reality. The symbols are γ_A and γ_{STONE}, and the symbol for the interaction process in basic reality is $\gamma_A \leftrightarrow \gamma_{\text{STONE}}$. Since we have principally no access to basic reality, we cannot say anything about the interaction process. We only know that it takes place, but we do not know how it works.

4.3.3 Observations

Let us first study the observation of the stone through human being A and we want to assume that it is an everyday-life observation. An observation in everyday life means that γ_A interacts with γ_{STONE} at the macroscopic level where the five senses are effective. In this case, the interaction with light is relevant. In other words, the usual sight process is in the focus, here investigated on the basis of the projection principle.

Human being *A* is the observer and wants to study (observe) the stone. The entire situation is represented in Figure 4.3, which reflects a scheme in analogy to (4.9).

The stone appears spontaneously before the eyes of human being *A*. Clearly, parts of human being *A* himself/herself are observable as well in this way and reflect a certain kind of self-observation, which is however not indicated in the picture (x, y, z, τ structure of Figure 4.3).

$\gamma_A \leftrightarrow \gamma_{\mathrm{STONE}}$ (basic reality)

$\downarrow$ tools of evolution *A*

α_{STONE} (mind of *A*)

$\downarrow \eta_A$

θ_{STONE}

(x, y, z, τ structure in the mind of *A*) $\longrightarrow$

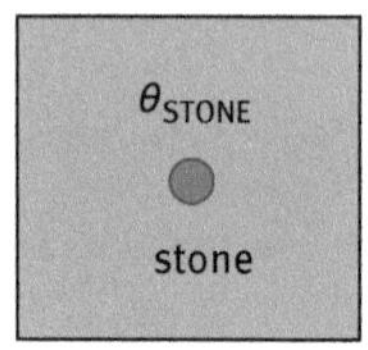

x, y, z, τ

(human being *A*)

Figure 4.3: Human being *A* observes the stone; both entities are again given in basic reality by γ_A and γ_{STONE}. It is an everyday life observation, which takes place at the macroscopic level where the five senses are effective. In this case, the interaction with light is relevant; it is a sight process. The tools of evolution of observer *A* pull α_{STONE} from γ_{STONE}, and α_{STONE} is processed by the processing characteristic η_A of the mind of *A*. In this way, the picture of the stone θ_{STONE} is obtained and appears as x, y, z, τ structure in the picture.

If γ_A and γ_{STONE} do not interact, we have merely representations of *A* and the stone, that is, in this case human being *A* does not observe the stone and we obtain Figure 4.4. Then, human being *A* knows nothing about the existence of the stone. However, human being *A* and the stone can be observed both from the point of view of another observer, here denoted by *B*. Then, human being *B* interacts macroscopically with human being *A* as well as with the stone.

4.4 Interactions at the macroscopic level

Let us emphasize once again that we have principally no access to basic reality and, therefore, we know nothing about the details of the interaction process between physically real entities. We only know that it takes place, but we do not know how it works. A human observer is caught in space and time and, therefore, we have to judge certain interaction types from the point of view of space and time.

$\gamma_B \leftrightarrow \{\gamma_A, \gamma_{STONE}\}$ (basic reality)

$\downarrow$ tools of evolution B

α_A, α_{STONE} (mind of B)

$\downarrow \eta_B$

θ_A, θ_{STONE}

(x, y, z, τ structure in the mind of B)

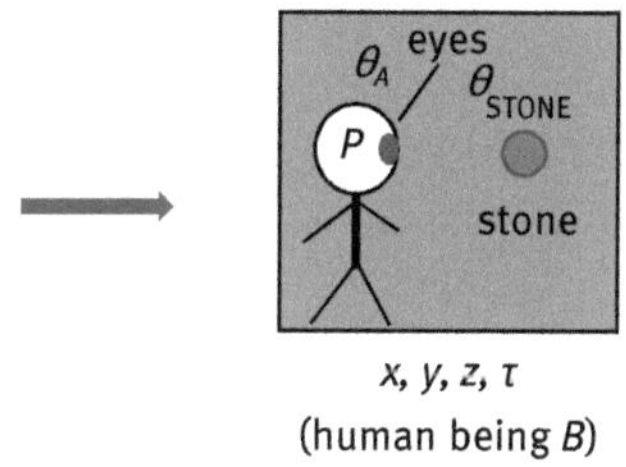

Figure 4.4: There is no interaction between human being A and the stone. The physically real objects γ_A and γ_{STONE} exist in basic reality side by side; human being A knows nothing about the existence of the stone. For the representation of both, γ_A and γ_{STONE}, the tools of evolution of any other human being, say B, pull α_A and α_{STONE} from basic reality. The mind of human being B processes α_A and α_{STONE} leading to the pictures θ_A and θ_{STONE} in space and time (x, y, z structures at time τ). The brain of A is indicated by the letter P. The situation is assessed from the point of view of human being B.

We would like to assume tentatively that this procedure is possible. In Section 4.6 we will discuss this question in detail.

Then, we can assess from the point of view of space and time to which interaction type the corresponding process in basic reality has to be related. This is a general rule in projection theory and, as we will still justify in Section 4.6, this point may in particular be applied in connection with representations of geometrical configurations at the macroscopic level.

In Figure 4.4, human being B observes the stone and human being A. Human being A and the stone do not interact, that is, human being A does not observe the stone. If the space–time structure of an experimental setup leads to the x, y, z, τ structures given in Figure 4.4, we may conclude on the footing of our daily experiences that a sight process is concerned; touch, taste, hearing and smell processes have no influence here. Then, due to the rule introduced earlier, the interaction in basic reality corresponds to what we usually call sight process.

Other geometrical constructions at the macroscopic level have to be assessed accordingly. For example, touch effects judged from the point of view of space and time have to be assessed in basic reality also as touch effect, but only qualitatively.

4.4.1 Analysis

In Figure 4.4, the stone and human being A do not interact with each other, but there is human being B who interacts with both, that is, with A and the stone. This has been symbolized in Figure 4.4 by $\gamma_B \leftrightarrow \{\gamma_A, \gamma_{STONE}\}$. If human being A interacts with the stone, we get instead of $\{\gamma_A, \gamma_{STONE}\}$ the term $\{\gamma_A \leftrightarrow \gamma_{STONE}\}$, and this interacting complex is again observed by human being B. Also in this case observation provides a physically real interaction process in basic reality, that is, γ_B interacts with $\{\gamma_A \leftrightarrow \gamma_{STONE}\}$ and we obtain $\gamma_B \leftrightarrow \{\gamma_A \leftrightarrow \gamma_{STONE}\}$. The observation itself is exclusively done on the basis of the space–time picture. In the following, we will analyze the situation in detail.

Pictures in the mind of A and B

Let us analyze the situation in more detail: The stone is represented in the "mind of B" as x, y, z, τ structure, which is indicated by θ_{STONE} (see Figure 4.4). On the other hand, the "observation" of the stone through human being A must lead to a picture as well but in this case in the "mind of A," and this picture comes into existence via specific brain functions of human being A. What does this procedure mean?

The brain functions are physically real in character. Observation provides an interaction process between human being A and the stone (usually called sight process) and is here judged from the point of view of space and time, that is, on the basis of the picture. Or more precisely, there must be an interaction between the brain of A and the stone. In this way, human being A observes the stone and this must lead to a picture of the stone in the mind of A. How does it work?

As in the case of Figure 4.4, we would like to denote the brain of A by the letter P. Due to the interaction between the brain P and the stone θ_{STONE} (Figure 4.4), the brain itself is changed and converts into an excited state here and in the following indicated by P^*. That is, we have the transition

$$\begin{array}{c} P^* \\ \uparrow \\ P \end{array} \tag{4.11}$$

Again, we may read (analyze) the situation in this way, that is, as interaction in space and time. Clearly the physically real (actual) interaction process takes place in basic reality and not in space and time.

Excited brain states

What does the term "excited brain state" mean? It reflects the brain activities in the physically real realm and characterizes characteristic brain currents. These brain activities are physically real in character and are initiated by the interaction process (sight process) in basic reality or we may say between P and θ_{STONE} in space and time.

There is a coupling between the mind of A and its brain state P^*. The information within the brain is transferred to the mind. The mind processes this information by its processing characteristic η_A and a space–time picture of the stone is obtained. That is, the coupling between the mind and the brain leads to the picture X of the stone as x, y, z, τ structure, where the picture X refers to human being A. That is, we get the scheme

$$\begin{array}{c} \text{mind of } A \leftarrow P^{*} \\ \downarrow \\ X \end{array} \tag{4.12}$$

The picture of the stone is projected onto space and time, but we have the impression that the stone is directly before us.

Note that the picture θ_{STONE} refers to human being B, but the picture X refers to human being A. Without the coupling between mind and brain, we would not be able to observe the physically real world outside and also conscious actions would not be possible; the facts are not outlined here.

Clearly, a coupling mechanism remains an open question at first. At this stage we are only interested in the main steps in the cooperation between mind, brain, matter together with space and time.

4.4.2 Illustration

Space–time structures with respect to *B*

Figure 4.5 shows the principles of what we have discussed in connection with scheme (4.9), that is, the transition from basic reality to the picture. We have the following situation:

The interaction process $\gamma_A \leftrightarrow \gamma_{\text{STONE}}$ in basic reality is observed by human being B, that is, human being B, symbolized in basic reality by γ_B, interacts with the complex $\gamma_A \leftrightarrow \gamma_{\text{STONE}}$ and we have $\gamma_B \leftrightarrow \{\gamma_A \leftrightarrow \gamma_{\text{STONE}}\}$.

Clearly, parts of human being B himself/herself are observable as well in this way and reflect a certain kind of self-observation, which is however not indicated in the picture (x, y, z, τ structure of Figure 4.5 and also not in Figure 4.4). Human being A wants to learn something about the stone and this is the reason why he/she observes it. Here, the stone reflects an "objective" attribute, and observer A plays here the role of a "subjective" item. Subjective facts are superfluous in the picture and are therefore not quoted in Figure 4.5. The same argument holds of course for Figure 4.4.

The interaction process $\gamma_A \leftrightarrow \gamma_{\text{STONE}}$ is pulled through the tools of evolution of human being B and we get the entities $\alpha_A, \alpha_{\text{STONE}}$, which reflects $\gamma_A \leftrightarrow \gamma_{\text{STONE}}$ in reduced form. This information is transferred to the mind of B. Through processing in the mind of human being B by the processing characteristic η_B, we obtain the pictures

$\gamma_B \leftrightarrow \{\gamma_A \leftrightarrow \gamma_{STONE}\}$ *(basic reality)*

$\downarrow$ *tools of evolution B*

α_A, α_{STONE} *(mind of B)*

$\downarrow \eta_B$

θ_A, θ_{STONE}

(x, y, z, τ *structure in the mind of B*)

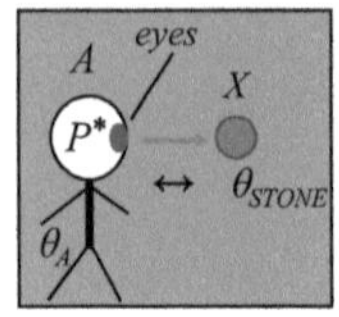

x, y, z, τ
(*human being B*)

Figure 4.5: The symbols $\gamma_A \leftrightarrow \gamma_{STONE}$, α_A, α_{STONE} and θ_A, θ_{STONE} are explained in connection with Section 4.3.2 and Figure 4.4. The x, y, z, τ structure contains the representations θ_{STONE} and θ_A for the stone and for human being *A* with eyes. The entire situation is again assessed from the point of view of human being *B*. The stone θ_{STONE} is represented in the mind of *B* as x, y, z, τ structure (see also Figure 4.4). The "observation" of the stone through human being *A* must lead to a picture in the mind of *A* and is here denoted by the letter *X*. It is an observation at the macroscopic level, where the five senses are effective. This is indicated in the figure by the double arrow. Other entities of the environment (photons, trees, etc.) are not indicated here. The position of *X* has to be identical with the position of θ_{STONE}, which is indicated in the figure by an arrow going from *A* to θ_{STONE}. The pictures θ_{STONE} and *X* are projections at each time τ on the two spaces located in the mind of *B* and in the mind of *A*. The conditions are formulated by eq. (4.13).

θ_A and θ_{STONE} resulting from the interaction process $\gamma_A \leftrightarrow \gamma_{STONE}$. The pictures θ_A and θ_{STONE} reflect the x, y, z, τ structures in the space–time of human being *B*.

Space–time structures with respect to *A*

We would like to discuss the picture of the stone in the mind of human being *A* on the basis of space and time of human being *B*. In this case, we work within the picture represented by Figure 4.5. The procedure in the picture corresponds to what actually takes place in basic reality. The interaction process can therefore be judged from the point of view of space–time pictures, in this case from the space–time picture of human being *B*, represented in Figure 4.5.

As in the case of Figure 4.4, we would like to denote the brain of *A* by the letter *P*. Due to the interaction between the brain *P* and the stone θ_{STONE} (Figure 4.5) via light (not indicated in the figure), the brain of human being *A* is changed and converts into an excited state, which is indicated in Figure 4.5 by the letter P^* [see eq. (4.11)].

Since there is a coupling between the mind of *A* and its brain state P^* [see scheme (4.12) and Section 4.4.1], the information within the brain of *A* is transferred to the mind of *A*. The mind of *A* processes this information by its processing characteristic η_A and creates a space–time picture of the stone, which we denoted by the letter X. In other words, the picture X is created through the mind of A together with space and time. It is a projection onto the space–time frame of human being A.

However, as we know, "mind features" are not representable in space and time, and this is because the unit "mind" does not belong to the class of physically real entities. So, the features of mind A cannot appear as x, y, z, τ structure within the space–time frame of human being A. The stone is created by the mind of A together with space and time, and its x, y, z, τ structure, that is, the picture X, reflects a projection, but it remains in the mind of A.

That is, the "features" of the mind of human being A cannot appear within space–time frames, but the picture of the stone appears as projection onto the space–times of A and B, as picture X and as picture θ_{STONE} (see Figure 4.5). The x, y, z, τ structure in Figure 4.5 is located in the mind of observer B and it is just that what observer B has directly before his/her eyes. The other features of the mind have no imprint in space and time.

Conditions

The space–time picture X of the stone appears as projection onto the space–time structure $\{x, y, z, \tau\}_A$ of mind A. On the other hand, the space–time picture θ_{STONE} of the stone appears as projection onto the space–time structure $\{x, y, z, \tau\}_B$ of mind B. In other words, the picture of the stone appears twice, as space–time structure in the mind A and also as space–time structure in the mind of observer B.

The x, y, z, τ structure reflects a projection of what actually takes place in basic reality as physically real interaction process. There are no processes in space and time, that is, the x, y, z, τ structure given in Figure 4.5 is a projection from basic reality and contains everything in frazzled form.

We have to consider the following items: since nature does not create the same fact twice, we have inevitably to conclude that the space–time position of picture X, indicated by $[X]_{\text{position in}\,A}$, is identical with the space–time position $[\theta_{\text{STONE}}]_{\text{position in}\,B}$ of picture θ_{STONE}.

Furthermore, the space–time frames $\{x, y, z, \tau\}_A$ and $\{x, y, z, \tau\}_B$ of human being A and human being B must be congruent and have to be identical. Then, we have the following conditions:

$$
\begin{aligned}
&[X]_{\text{position in}\,A} \equiv [\theta_{\text{STONE}}]_{\text{position in}\,B} \\
&\{x, y, z, \tau\}_A \equiv \{x, y, z, \tau\}_B
\end{aligned}
\tag{4.13}
$$

Two pictures of the same thing would make no sense and would be a superfluous information. Both observers are members of the same species.

4.4.3 Brain and body functions

Once again, the picture of the stone is projected onto space and time and we have the impression that the stone is directly before us. This situation is in fact reflected by Figure 4.5: human being A has the stone θ_{STONE}, X [eq. (4.13) is valid] before his/her eyes, which is indicated in Figure 4.5 through an arrow going from A to θ_{STONE}, X. The results are in accordance with our daily experiences. However, we have to underline that nothing moves through space in the course of time τ. The entire picture is in the mind and appears as projection in space and time.

We may state quite generally that the x, y, z, τ structures are not produced by the brain. In the picture, for example in Figure 4.5, only the physically real part of the brain appears, which is of course located in the head and is denoted by P^* in Figure 4.5. But the brain is more than that. It connects mind states with body functions. The brain has the ability to interact with the mind and with parts of the body as well. All that can appear consciously but also unconsciously.

To construct an image of the stone θ_{STONE} within the brain of human being A using the space time frame $\{x, y, z, \tau\}_B$ of human being B is not possible, and this is mainly due to fact that the mind states of A do not appear in space and time. Such a construction would lead to a "picture of the picture," which makes however no sense within the frame of the projection principle.

4.4.4 The sight process within projection theory

What does sight process mean within projection theory? It reflects an observation of physically real objects, and let us explain the details by means of the stone. For this purpose, the study of Figure 4.5 is suitable. In Figure 4.5, human being B assesses an observation process with respect to human being A who observes the stone θ_{STONE} in the space–time picture of human being B; the space–time frame is again indicated by $\{x, y, z, \tau\}_B$. Clearly, also human being B plays the role of an observer, but here we are only interested in the observation procedure in connection with human being A.

Human beings are caught in space and time and they can only make statements about the physically real world within the frame of space–time pictures. Here human being A perceives the stone through "correlations." There can be no physically real "interactions" in space and time but exclusively in basic reality. Let us explain the situation.

Correlations

"Observation" means "interaction" and, as we have remarked several times, the physically real interaction processes take place in basic reality. That is, human being A

observes the stone in basic reality through the interaction process $\gamma_A \leftrightarrow \gamma_{\mathrm{STONE}}$. This process is observed by human being B and is projected onto the space–time of human being B, that is, onto the frame $\{x, y, z, \tau\}_B$. As picture, we get the x, y, z, τ structures θ_{STONE} and θ_A (Figure 4.5).

The physically real process $\gamma_A \leftrightarrow \gamma_{\mathrm{STONE}}$ must also be identifiable in the space–time picture, that is, the process has its counterpart in space and time. We know that there are no physically real "interactions" in space and time but "correlations." In other words, there are interactions between γ_A and γ_{STONE} in basic reality and correlations between the x, y, z, τ structures θ_{STONE} and θ_A. Figure 4.6 summarizes the facts.

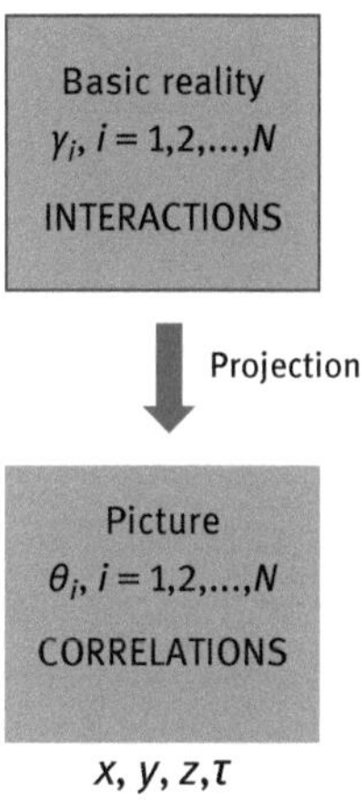

Figure 4.6: The physically real objects in basic reality (world outside) are denoted again by γ_i, $i = 1, 2, \ldots, N$, where N is the total number of objects in basic reality. In general, these objects interact with each other. For example, object k with object m: $\gamma_k \leftrightarrow \gamma_m$. This interacting complex is projected onto space and time and we obtain a picture. The projected physically real objects become geometrical entities in the picture, denoted by θ_i, $i = 1, 2, \ldots, N$. In particular, we have the geometrical entities θ_k and θ_m in the form of x, y, z, τ structures. The physically real "interactions" in basic reality effectuate that there are "correlations" in the picture.

We may call these correlations in space and time "interactions." For example, we may say that θ_{STONE} interacts with θ_A. We used this formulation above, but these are not interactions in the physically real sense. Within this notation, the situation becomes possibly more conceivable.

How have we to assess the sight process in Figure 4.5? The complex, consisting of "light, stone, eyes and brain," is a strict correlated ensemble. The correlation sequence light $\rightarrow$ stone θ_{STONE} $\rightarrow$ eyes $\rightarrow$ brain P^* is compatible with our daily experiences. The components of the sequence are recognizable in the space–time picture and are therefore physically real entities in basic reality.

The brain of human being A, denoted by P^* in Figure 4.5, is coupled to the mind of A. However, as we already remarked in Section 4.4.2, the mind itself is principally not representable in space–time pictures, also not within the frame $\{x, y, z, \tau\}_B$ of human being B. This coupling between the mind of A and his/her brain effectuates that the information of the stone located in the brain of A and which is transferred to the mind of A will be processed by the processing characteristic η_A within the mind of human being A. This procedure leads to a space–time picture X of the stone, represented within the frame $\{x, y, z, \tau\}_A$. We do not know the details of this procedure but this is the general line. It is underlined earlier that the space–time structures $\{x, y, z, \tau\}_A$ and $\{x, y, z, \tau\}_B$ must be congruent [eq. (4.13)].

In summary, the perception of physically real entities comes into existence through "correlation effects." There are no physical real interactions and entities in space and time.

System-specific time

This correlation sequence is a global effect and is effective at each clock time τ. Observation (human being A observes the stone θ_{STONE}) takes place if such a global situation is given. The globality reflects nonlocal situations and has to be described quantum theoretically.

This can hardly be done within the frame of the typical quasiclassical considerations of traditional quantum physics, but a more sophisticated theoretical frame is necessary in which the projection principle is applied. In fact, traditional quantum theory is quasiclassical in character, and this is because it works still with the external classical time τ. As we outlined in Chapter 1, a quantum aspect of time is not known in traditional quantum theory, but is required. Usual quantum theory is not sufficient for the complete description of quantum phenomena [7].

The projection principle allows the introduction of the required quantum aspect of time; the projection principle even demands it. This quantum time is expressed by the so-called system-specific time, denoted by the letter t (Chapter 6). The clock time τ remains an essential element in projection theory, but here it is merely a reference time. Moreover, within projection theory an operator for the time comes into play (see Chapter 1).

Let us already state here that we have x, y, z, t quantum structures within projection theory, but only the part $x, y, z, t = \tau$ is experienced at clock time τ. The discussion is continued in Chapter 6.

4.5 The atomic level and the atomic facet

When we want to explore the stone at the "atomic level," which reflects a facet above the macroscopic level, we leave the realm of the five senses. The atomic level is characterized through the application of specific measuring instruments. That is, instead of the five senses, measuring devices interact now with the object under investigation, that is, the stone.

The measuring instruments themselves have to be invented by the observer, which is in our case human being A. Their construction provides a conclusive theoretical conception. This theoretical conception has also to be used for the evaluation of the measured events.

In conclusion, the atomic level is accessed through "conscious" activities. In everyday life (macroscopic level), the world before us is experienced spontaneously, that is, it is observed in an "unconscious" way. Let us give more details concerning the exploration of the atomic level and the atomic facets of physically real objects.

4.5.1 Measurements at the atomic level

The atomic level is identical with the ultimate nanolevel (Chapter 1). Human being *A* wants to measure the atomic structure and the atomic dynamics of physically real objects; it is in our case a stone. The structure and dynamics are the adequate information at the atomic level.

Human being *A* can do the measurements by means of scattering experiments. For this purpose, he/she needs a source, which emits particles, for example, electrons, neutrons or photons. Let us assume here that observer *A* works with electrons.

The emitted electrons are scattered by the stone and are registered by a detector after the scattering process and leads to a characteristic spectrum, which contains the information about the atomic structure and dynamics of the stone. In other words, the experiments reveal the relevant information about the stone at the atomic level (Figure 4.2). It is a facet of the stone, which is different from the facet at the level of everyday life and reflects the situation at the "macroscopic level" (Figure 4.2) as obtained on the basis of the five senses; it is a sight process in the specific case of Figure 4.5.

For the investigation of the "atomic level," sophisticated experimental setups are needed. Figure 4.7 shows such a setup: the electron source *S*, the detector *D*, the stone θ_{STONE} and human being θ_A. Human being *A* observes the entities *S*, *D* and the stone θ_{STONE}, and all that is judged from the point of view of human being *B*.

Human being *A* makes his/her observation at the macroscopic level; as in the case of Figure 4.5, the pictures in the mind of *A* are identical with the entities with respect to the pictures in the mind of human being *B*. Also here the space–time frames of *A* and *B* are congruent, that is, we cannot distinguish between the pictures of the two space–time structures.

In other words, the situation in Figure 4.7 is identical to that in Figure 4.5 except the number of observed objects is different: human being *A* does not observe only "one" object (the stone) but "three," which are given by *S*, *D* and the stone. It is again an observation at the macroscopic level, but it reflects a setup for the investigation of the atomic facet of the stone.

As in the case of Figure 4.5, the entire complex is observed by human being *B*, indicated in basic reality by γ_B. That is, *B* interacts with the entire complex consisting of *A*, *S*, *D* and the stone, all that is projected onto the space–time frame $\{x, y, z, \tau\}_B$ of human being *B*.

$$\gamma_B \leftrightarrow \left\{ \begin{matrix} \gamma_A \leftrightarrow \gamma_{\text{STONE}} \\ \gamma_A \leftrightarrow \gamma_S \\ \gamma_A \leftrightarrow \gamma_D \end{matrix} \right\} \quad \text{(basic reality)}$$

$\downarrow$ Tools of evolution B

$\alpha_A, \alpha_{\text{STONE}}, \alpha_S, \alpha_D$ (mind of B)

$\downarrow \eta_B$

$\theta_A, \theta_{\text{STONE}}, \theta_S, \theta_D$

(x, y, z, τ structure in the mind of B)

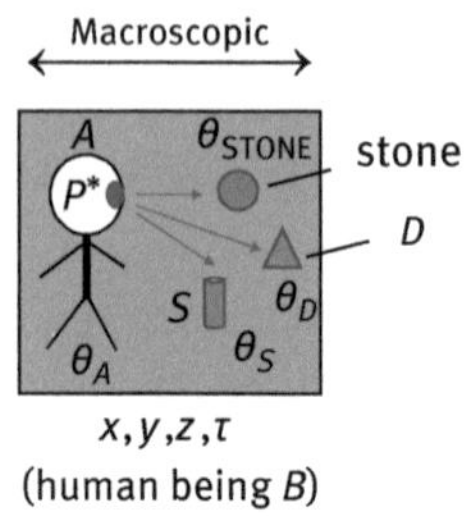

Figure 4.7: The electron source, the detector and the stone are observed by human being A and are also observed by human being B, indicated in basic reality by γ_B. The details: The physically real entities,γ_{STONE}, γ_S and γ_D interact macroscopically with γ_A, which is again human being A. The interactions take place in basic reality and are represented in scheme (4.14). All that [given in scheme (4.14)] is pulled through the tools of evolution of observer A and those of human being B and are transferred to the mind of A and also to the mind of B. The entities α_{STONE}, α_S and α_D are located in the mind of A and in the mind of B. The contents of the mind of A and that of B are then processed by the processing characteristic of A and B, and the complex with θ_A, θ_{STONE}, θ_S and θ_D appear in the pictures in the form as indicated by scheme (4.16). But these pictures, represented by scheme (4.16), do not reflect physically real interaction processes but "correlations." Correlation here means "perceivable." That is, human being A perceives the stone, the detector and the source. Within the mind of human being A, there are the pictures θ_{STONE}, θ_S and θ_D. On the other hand, within the mind of human being B there are the pictures θ_A, θ_{STONE}, θ_S and θ_D.

4.5.2 Composition of interactions

What are processes in Figure 4.7? Once again, the material objects γ_{STONE}, γ_S and γ_D interact at the macroscopic level with human being A that is symbolized in basic reality by γ_A. This block is observed by human being B, indicated in basic reality by γ_B. That is, B interacts with A, S, D and the stone. All that is projected onto the space–time frame $\{x, y, z, \tau\}_B$ of human being B via the tools of evolution of B and the processing characteristic η_B.

The interaction processes take place in basic reality and are indicated through the following interaction scheme:

$$\gamma_B \leftrightarrow \left\{ \begin{matrix} \gamma_A \xleftrightarrow{\text{macroscpic}} \gamma_{\text{STONE}} \\ \gamma_A \xleftrightarrow{\text{macroscpic}} \gamma_S \\ \gamma_A \xleftrightarrow{\text{macroscpic}} \gamma_D \end{matrix} \right\} \tag{4.14}$$

All that, which is given in scheme (4.14), is pulled through the tools of evolution of human being A and those of human being B. This information is transferred to the mind of A but to the mind of B as well. The entities α_{STONE}, α_S and α_D are located in the mind of A and in the mind of B in the form of

$$\alpha_A, \alpha_{\text{STONE}}, \alpha_S, \alpha_D \tag{4.15}$$

The contents of the mind of A and B are then processed by the processing characteristics η_A of A and η_B of B. This procedures effectuate and the complex with θ_A, θ_{STONE}, θ_S and θ_D appear in the picture of human being B, and the complex with θ_{STONE}, θ_S and θ_D in the picture of human being A. Concerning B, we have

$$\theta_A, \theta_{\text{STONE}}, \theta_S, \theta_D \tag{4.16}$$

But these pictures, represented by scheme (4.16), do not reflect physically real interaction processes but "correlations" (see the statements in Section 4.4.4). Correlation here means "perceivable." There are correlations between the x, y, z, τ structures of θ_A and θ_{STONE}, θ_A and θ_S, θ_A and θ_D. That is, human being A perceives the stone, the detector and the source.

Within the mind of human being A, there are the pictures of θ_{STONE}, θ_S and θ_D. On the other hand, within the mind of human being B there are the pictures θ_A, θ_{STONE}, θ_S and θ_D.

The source S, the detector D and the stone are perceived through human being A at the macroscopic level, that is, the atomic level is not touched yet. However, the features at the macroscopic level are needed in the introduction of the atomic level.

Remark

The situation described in Figure 4.7 reflects a judgment from the point of view of human being B, but the observer is human being A; A observes the electron source, the detector and the stone. It is an information at the macroscopic level. Human being A needs this macroscopic information for the determination of the properties of the atomic facet of the stone.

4.5.3 Transition to the atomic level

In the next step of our analysis, we will discuss the atomic level, where another facet of the stone is expressed. That is, we will construct space time pictures at the atomic level for physically real objects. Clearly, the atomic level is identical with the ultimate nanolevel.

The transition from the macroscopic level to the atomic (nano) level is not trivial. This topic will be discussed in Section 4.7. In Section 4.6, we will investigate how the sight process has to be assessed when we work exclusively in space and time without recourse to basic reality. In this way, comparisons can be made with respect to the usual views of traditional physics.

4.6 The sight process within projection theory

We have identified the situations in Figures 4.5 and 4.7 with sight processes. Therefore, also the corresponding process in basic reality must be in analogy to this sight process, although we cannot actually state what happens in basic reality. As we remarked several times, this is a basic feature (Chapter 1).

The picture of the stone is not welded in the physically real structure of the brain, but the brain contains merely the "information" of the stone picture. The picture itself is created, as we know, in the following way: the excited brain, denoted by P^* in the Figures 4.5 and 4.7, is coupled to the observer's mind and the mind takes this information on and processes it in order to obtain the stone picture in space and time. That is, we get a x, y, z, τ structure of the physically real part of the brain.

Only the mind creates the space–time elements x, y, z, τ and not the brain. The physically real brain structures are representations in space and time, that is, we need the space–time elements x, y, z, τ to "characterize" the physically real brain structures and, therefore, the brain does not create the elements x, y, z, τ, which are needed to depict the brain. No doubt, this is a relevant statement and follows directly from the elementary features of the projection principle.

4.6.1 Details concerning the interrelations

The details of the sight process are given by a specific construction. The construction itself describes the sight process "qualitatively," and this is because the physically real process with respect to this sight process (and other physically real processes as well) does of course not take place in the picture with its x, y, z, τ elements. Again, the physically real objects are embedded in basic reality and also the physically real processes take place in basic reality and nowhere else. Since human beings have no direct access to basic reality, a "quantitative" description of the objects and processes is not possible. This situation indicates that we have reached the fundamental view.

What happens? Light comes from outside and interacts with the stone and the eyes, and this information is transferred to the brain of human being A. This light information with respect to the stone is recorded by the brain, which we denoted by the letter P and, due to the light impact, P converts into an excited state P^* in analogy to eq. (4.11).

The excited brain P^* contains now the information about the stone outside, which is represented in Figures 4.5 and 4.7 within the x, y, z, τ frame of human being B. It is observed by human being A and appears before the eyes of A. Both pictures, θ_{STONE} and X, are congruent [condition (4.13) is valid].

Since the brain is generally coupled to the mind (Section 4.4), there is a coupling between the excited brain P^* and the mind of human being A, which can be expressed through

$$\begin{array}{c}\text{mind}\\ \uparrow\\ P^*\end{array} \tag{4.17}$$

Again, we want to construct a space–time picture of the stone (Figure 4.5) or the space–time picture of all: the stone, the electron source S and the detector D in Figure 4.7. For this purpose, x, y, z, τ frames are needed, which are not definable by the brain as we already mentioned.

Thus, the information given through P^* has to be processed by the processing characteristic η of the mind (see Section 4.1.5), and the mind creates the x, y, z, τ frame that is needed for the representation of the physically real objects like S, D and the stone. That is, in analogy to eq. (4.12), the following scheme is valid for Figure 4.7:

$$\begin{array}{c}\text{mind of } A \leftarrow P^*\\ \downarrow\\ \text{picture of}\\ \text{the stone, source } S \text{ and detector } D\\ (x, y, z, \tau \text{ structure})\end{array} \tag{4.18}$$

Exactly the same scheme holds for Figure 4.5; in Figure 4.5, only the picture of the stone is represented within the x, y, z, τ frame.

As we have outlined in Section 4.4, nature does not create the same fact twice. Therefore, we have inevitably to conclude that the space–time positions of S, D and the stone within the x, y, z, τ frame of observer A has to be identical with the space–time positions of S, D and the stone within the x, y, z, τ frame of human being B. In analogy to eq. (4.13), we obtain therefore the following relation:

$$[\text{stone}, S, D]_{\text{position in } A} \equiv [\text{stone}, S, D]_{\text{position in } B} \tag{4.19}$$

Two pictures of the same thing would make no sense and would be a superfluous information. Also here the space–time frames of human being A and human being B are congruent. Both observers are assumed to be members of the same biological species.

In other words, eq. (4.19) is valid. Otherwise, we would get an extra picture, a "picture of the picture" so to speak. However, this feature, that is, the term "picture of picture," does not belong to projection theory. Strictly speaking, this term is forbidden here because it makes no sense to work with it within this conception. Thus, the picture of observer A has to be identical with the picture of human being B.

The x, y, z, τ frames have to be congruent; this is a “construction instruction.” Specific boundary conditions in connection with self-observations require that too. Once again, two simultaneously existing pictures of the same thing would be superfluous.

4.6.2 Assessments of processes in the picture

As we have remarked several times, the assessment of physically real processes in space–time pictures should principally be possible, but has to be done with care. The reason is obvious: physically real processes take place in basic reality and not in the picture of reality. As we know (Chapter 1), the processes in basic reality are not accessible to human beings. But we may use the picture, that is, its space–time structure, to judge “qualitatively” what actually takes place in basic reality. Let us choose an example: the stone in space and time.

What can we say about the situation in space and time? Can we judge the problem analogous to what we know from traditional physics? We cannot. There is no complete construction of physically real processes in space and time possible. We need more. Let us explain why.

No picture without mind

The sight process in connection with the complex consisting of stone, light, eyes and brain, as is indicated in Figure 4.5, can be used for the assessment of what actually appears in basic reality, but only “qualitatively.” Light, stone, eyes and brain do not interact in space and time, but their geometrical structures are “correlated” in space and time. Again, the interaction takes place in basic reality but we may “read” this process from the space–time picture (e.g., see Figure 4.5). Light interacts with the stone. This information flows via the eyes to the brain. The brain is coupled with the mind and transfers the information about the stone to the mind. Then, the mind processes this information and we obtain a picture of the stone in space and time. The mind projects the picture of the stone onto space and time.

The mind itself is not representable in space and time because it has no physically real basis, but it is its task to process physically real information (Section 4.4). Also the transfer from the brain to the mind cannot be represented in space and time. In other words, the extremity of the process description is given by the brain; the activity of the mind cannot be “read” from Figure 4.5.

The picture of the physically real object (stone) comes only into existence (as x, y, z, τ structure) if the mind is active, that is, the mind is essentially involved and not only the light, the eyes and the brain. However, as we mentioned earlier, the mind itself is not presentable as x, y, z, τ structure. The role of the mind has to be classified through other facts.

That is, the picture of the object (stone) is not given alone by the entities shown in the space–time picture but the mind is involved as well, which is however located outside space and time and does not appear in the picture. The mind itself does not appear as space–time information but, as we already know, it creates space and time. The mind projects the physically real information about the stone onto space and time and, simultaneously, it creates space and time. That is the situation from the point of view of projection theory.

How can we characterize the brain? The physically real part of the brain is located in the head of the human being (see Figure 4.5), but the brain is more than that because it is able to couple to the mind and must therefore have the ability to give information to the mind (eq. (4.18)). Other parts of the body (arms, legs, etc.) do not have this property.

The physically real processes in basic reality, which are relevant for the representation of objects, can be qualitatively constructed in the space–time picture. The relevant entities of the sight process (the light, the eyes and the brain) are correlated in space and time. Therefore, a simulation of the physically real process, taking place in basic reality, becomes possible. A correlated sequence between the four entities light, stone, eyes and brain is intuitively recognizable. In this respect, the simulation procedure delivers a complete reproduction as far as the interaction processes in basic reality are concerned. But this is not sufficient yet when we want a representation of the physically real object (stone) in the space–time picture. The mind is essential as well.

No influence of the transition elements on the picture

The structure of the world outside, represented in the picture, is not created by the tools of evolution and also not by the processing characteristic η. Thus, there is a direct evidence for what actually happens in basic reality (corresponding to a physically real process in basic reality). It is justified to "read" from the picture what actually takes place in basic reality but, as we already remarked, the activity in basic reality is not sufficient for the creation of a picture of the world outside. The mind is essentially involved.

Qualitative descriptions

The transition from basic reality to the picture is performed in umpteen steps. Here the tools of evolution, the mind as well as the processing characteristic η play an important role. But we may assume that these various steps do not produce the systematic correlations that can be "read" (recognize) from the structures in the picture. Therefore, the systematic correlations reflect the physically real interaction processes taking place in basic reality. In other words, we may use the structure in the picture, that is, the x, y, z structure at clock time τ, for the qualitative description of the physically real processes in basic reality. But we have to mention once again that the entities in the picture (the light, the eyes and the brain in the case of the sight process) are not sufficient for the understanding of the representation of the geometrical object structure (stone) in space and time.

Thus, the "geometrical construction" of an object of the world outside on the footing of the space–time picture alone is not possible within the frame of the projection principle.

This point is important as well when we try to manipulate brain functions within the frame of nanoscience (nanotechnology). In fact, we have to analyze the objects carefully at the nanolevel, which is the ultimate level for the description of object properties.

4.6.3 Nanophenomena from various points of views

For the realistic assessment of nanostructures, the following remarks are instructive: In the transition from basic reality to the picture, an interplay with mind and brain takes place (Section 4.5). The extremity is the picture with its geometrical structures; there is a x, y, z structure at each clock time τ, which is located in the mind of the human observer.

The construction of interaction processes in the picture is "recipe" of what actually takes place in basic reality. For the construction of physically real objects, the "mind" is needed, which is an entity outside space and time. The following scheme is valid:

$$\begin{gathered}\text{entities (light, object, eyes, brain)}\\ \text{in space and time}\\ +\\ \text{mind}\\ \text{(located outside space and time)}\\ \downarrow\\ \text{picture of the object}\\ \text{in space and time}\end{gathered} \tag{4.20}$$

The situation in traditional physics is different: the world before our eyes is considered as physically real, and the mind and the brain are not involved, that is, in traditional physics mind and brain are not coupled to the physical phenomena. Mind and brain can be modeled more or less arbitrarily. In fact, there are a lot of schools of opinion.

Clearly, both points of view, projection principle and the traditional situation, would lead to different valuations in connection with phenomena in nanoscience and nanotechnology.

4.6.4 A general statement

How relevant are the statements of the projection principle? We have to compare its course of action and its information input with what traditional physics says in the same situations.

Within nanoscience and nanotechnology, we go deeper into what we call "reality." This particularly means that we have to formulate the physical problems more general than in micro- and macrotechnology, that is, nanophenomena occupy the physical reality at a level above the usual point of views of traditional physics. This situation requires more general descriptions. With increasing deepness, a standpoint of increasing generality is required.

In this connection, we may state the following: the projection principle does not reflect a model and it is not an approximation of traditional physics. The projection principle reflects a view that is based on a general principle. It is essential to underline again that projection theory does "not" consider space and time as physically real elements. This is an essential condition and is in accordance with Mach's principle, which contains hard but realistic conditions. That is, space and time are here treated more realistically than in traditional physics where Mach's principle is not accomplishable. This situation leads inevitably to the more general projection principle, which is therefore not an approximation of what already exists.

The principles of evolution can and must be incorporated. A separate treatment without the entities "basic reality, mind and brain" makes no sense in projection theory. Everything is interrelated as in nature itself.

4.7 The atomic facet of physically real objects

The "macroscopic facets" of physically real objects are observed "unconsciously." The details in connection with Figure 4.7 refer to this kind of observation and reflect experiences of everyday life, that is, the objects S, D and the stone θ_{STONE} in the picture within the x, y, z, τ frame are directly before the eyes of observer A. Observer A experiences these things in an "unconscious" way and they appear spontaneously without any intellectual help.

The situation in connection with the "atomic facets" is different (see in particular Figure 4.2). The atomic level can only be captured in an intellectual way, that is, within a "conscious" procedure. For the experimental determination of an atomic facet, in most cases, sophisticated methods are needed.

4.7.1 Ideas

What is the procedure? First, the human being must have an "idea" about what he/she wants to know about the object (facet) at the atomic level. This idea defines the experimental method and its associated devices, which the observer needs for the measurement of the properties of the object under investigation, that is, for example the atomic facet of a stone.

In order to get all these facts, the observer needs a theoretical conception, which the observer has to develop, or which must come to the observer's knowledge. All that reflects of course a certain kind of specific intellectual activity.

The theoretical conception is the basis for the experimental determination of the object's properties at the atomic level. Let us choose again the stone as object. What kind of experimental method is adequate for the determination of its atomic-level properties?

The observer decides, for example, to determine the properties of the stone through scattering experiments. As we already outlined, for this purpose he/she needs an electron source *S* and a detector *D*. These devices appear, together with the stone, in front of him/her as macroscopic entities, that is, they appear before his/her eyes unconsciously. This is exactly the situation that is needed and which is represented in Figure 4.7.

The measurement itself requires a clear instruction: the stone and the devices *S* and *D* have to be arranged relative to each other in a way that the theoretical conception dictates, and the arrangement procedure has to be done by the observer in a conscious way. For this purpose, the observer needs the "macroscopic" pictures of the stone and the devices *S* and *D*. Only on the basis of these macroscopic entities before his/her eyes, the observer, it is human being *A* in Figure 4.8, is able to arrange these entities in accordance with the theoretical conception.

The scattering process is described in the picture. The process itself takes place in basic reality but, as we have outlined in the sections before, we may describe the situation in the picture with respect to the frame $\{x, y, z, \tau\}_B$ of human being *B* (see Figure 4.8): the electrons leave the source *S* and are scattered by the stone. The scattered electrons are registered in the detector by *D*. This process is indicated in Figure 4.8 through two arrows between the source *S* and the stone and between the stone and the detector *D*.

4.7.2 The evaluation

The evaluation of the detected spectrum has also to be done in connection with the theoretical conception and leads to the atomic structure and the atomic dynamics of the stone. This is the information about the stone at the atomic level and reflects the picture of its atomic facet.

$$\gamma_B \leftrightarrow \left\{\begin{matrix} \gamma_A \leftrightarrow \gamma_{STONE} \\ \gamma_A \leftrightarrow \gamma_S \\ \gamma_A \leftrightarrow \gamma_D \end{matrix}\right\} \text{(basic reality)}$$

$$\downarrow \text{Tools of evolution } B$$

$$\alpha_A, \alpha_{STONE}, \alpha_S, \alpha_D \text{ (mind of } B)$$

$$\downarrow \eta_B$$

$$\theta_A, \theta_{STONE}, \theta_S, \theta_D$$

(x, y, z, τ structure in the mind of B)

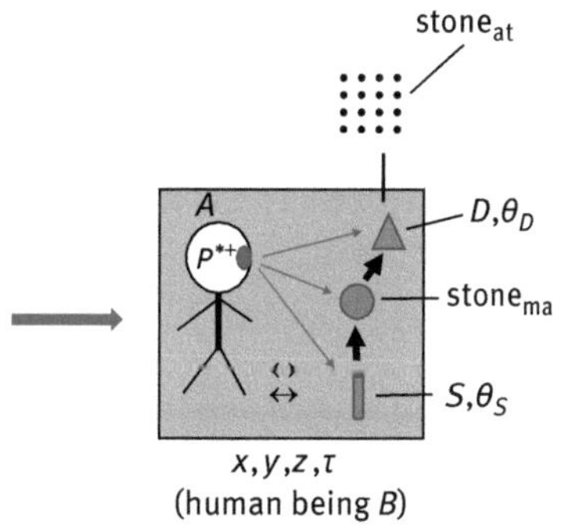

Figure 4.8: For the determination of the "atomic facet" of the stone, we need macroscopic interactions in order to be able to recognize the components of the experimental setup. This is a necessary precondition because only in this case the components can be arranged relative to each other and relative to observer A in a way that is dictated by the theoretical conception. The macroscopic interactions are indicated in the figure by the thin arrows (between A and the stone, between A and S, between A and D). As in Figure 4.7, the entire situation in the figure is a judgement from the point of view of human being B; B observes it within the space–time frame $\{x, y, z, \tau\}_B$. The scattering process, $S -$ stone$_{ma}$ $- D$, based on the correct device arrangement, is indicated in the figure by arrows (between S and stone$_{ma}$, between stone$_{ma}$ and D). The brain P is excited through the two interactions (the macroscopic and the atomic interaction). Thus, the brain is excited twice (simultaneously). The macroscopic interaction leads to P^*. If, in addition, the atomic interaction is switched on, the brain state P^* converts into a further excitation state, denoted here by P^{*+}. In this way, we obtain two pictures of the stone in the form of x, y, z, τ structures: the macroscopic facet stone$_{ma}$ and the atomic facet stone$_{at}$. Both facets exist simultaneously. Furthermore, there are two observation procedures simultaneously effective together with the two interaction types. The connections between the physically real entities γ_B, γ_{STONE}, γ_S, γ_D and γ_A are explained in the text, together with the entities α_A, α_{STONE}, α_S and α_D and the structures θ_A, θ_{STONE}, θ_S and θ_D.

The stone at the atomic level is indicated in Figure 4.8 as point pattern in space and time is denoted by stone$_{at}$. It is the picture of the stone at the atomic level, that is, it is the atomic facet of the stone. On the other hand, the picture of the stone at the macroscopic level (its macroscopic facet) is again indicated in Figure 4.8, as in Figure 4.7, by the full circle and is here denoted by stone$_{ma}$ instead of θ_{STONE}.

The two double arrows in Figure 4.8 indicate that there are two interaction types effective: the interaction at the macroscopic level and the interaction at the atomic level. Both interaction types are simultaneously effective. Once again, we need the information at the macroscopic level in order to be able to deduce the properties at the atomic level. Both facets, stone$_{ma}$ and stone$_{at}$, exist simultaneously. Furthermore, there are two observation procedures, which are simultaneously effective together with the two interaction types, which are simultaneously existent as well.

The pictures stone_{at} and stone_{ma} reflect x, y, z, τ structures in the mind of observer *A*. The entire situation in Figure 4.8 is a judgement from the point of view of human being *B* and he/she has this space–time picture before his/her eyes, that is, *B* observes it. The space–time picture of *B* is based on the frame $\{x, y, z, \tau\}_B$. As in Figure 4.7, human being *B* is characterized in basic reality by γ_B, and γ_B interacts with the stone γ_{STONE}, the source γ_S and the detector γ_D.

4.7.3 The conscious and the unconscious creation of pictures of reality

The pattern stone_{ma} is an "unconscious" picture, which appears spontaneously before the eyes of observer *A*. The pattern stone_{at} reflects a "conscious" picture and comes into existence via the experimental setup consisting of the stone, the source and the detector. For the evaluation of the detected events, a theoretical conception is needed. Furthermore, for the construction of the experimental setup, a theoretical conception is required as well. The evaluation of the experimental results leads to the facet expressed by stone_{at}.

4.8 Mind–brain coupling: two directions

Let us repeat: for the determination of the atomic facet of the stone, we need macroscopic interactions in order to be able to recognize the components of the experimental setup (stone and devices). The determination of the atomic facet itself has to be done in the second step. The entire process is based on unconscious and conscious activities.

4.8.1 Unconscious perceptions

In the case of a sight process (Section 4.6), the light sensors of the body convert the brain into an excited state P^* ($P \to P^*$). Through the coupling between mind and brain the mind is activated, in this case by the brain in the state P^*. The brainstate P^* contains the macroscopic information of the world outside. It is, for example, the macroscopic information about the stone. The sight process in connection with the perception mechanism has been described in Sections 4.3–4.6 in connection with Figures 4.4 and 4.7.

The coupling between mind and brain effectuates that the macroscopic information about the world (stone) is transferred to the mind. In other words, the mind also contains the macroscopic world-outside information and appears "unconsciously" after processing before the eyes of a human being as space–time picture; the processing itself is performed by the mind's processing characteristic η. This process is summarized in scheme (4.21).

$$\begin{array}{ccc} \text{Light sensors} & & \\ \Downarrow & & \\ P \rightarrow P^{*} & \xrightarrow{\text{Mind - brain coupling}} & \text{Mind} \\ & & \downarrow \\ & & \text{Picture} \\ & & \text{(unconsciously created)} \end{array} \tag{4.21}$$

This macroscopic world inside the mind is used for the initiation of "ideas." For example, to learn something about the atomic facet of the stone. For this purpose, the human observer must have an "idea" of how to measure this facet.

4.8.2 Conscious actions

In cases of conscious actions, the notion "intuition" is an essential feature. The intuition together with a theoretical conception are the basis for "ideas" for changing the world outside. For example, in this way it becomes possible to order the components of an experimental arrangement for the determination of the atomic facet of physically real objects, for example, a stone. Clearly, the experimental arrangement has to be in accord with the idea and the theoretical conception. In other words, an atomic facet is determined in a "conscious way." The situation in connection with the stone is described in detail in the sections before.

In the case of "unconscious" perceptions, that is, in the case of observations at the macroscopic level, the information flows from the brain to the mind, which can be recognized in detail through the steps in scheme (4.20). However, within the case of conscious actions, the information flows the other way round: from the mind to the brain. Let us discuss this mechanism in connection with our analysis given in Sections 4.5 and 4.7 together with Figures 4.7 and 4.8. The entire procedure is summarized by the following steps:

In scheme (4.22), the brain state P^{*+} is created through an "idea," which is located in the mind. Mind and brain are coupled (interrelated): the mind creates "unconsciously" a picture of the world outside (environment) and produces a space–time picture of it. The coupling between the mind and the brain effectuates that the brain state P converts into an excited state P^{*}, that is, we have the transition $P \rightarrow P^{*}$.

This picture, created unconsciously, is the basis for a human being to create consciously an "idea" of how to change the world outside.

This is the onset for actions: for the experimental determination of atomic facets, the observer needs macroscopic information about the experimental devices. The observer chooses on the basis of the "idea" and a "theoretical conception," an experimental setup consisting of a source S, a detector D and a stone in Figure 4.7.

This space–time picture with the macroscopic entities is located in the mind. This new picture corresponds to a new mind–state. Through the coupling with the brain, the mind transfers this information to the brain, and the brain converts P^* into the brain state P^{*+} and we get the transition $P^* \rightarrow P^{*+}$, that is, the brain state P^{*+} has been created consciously.

This brain state P^{*+} is the basis for conscious actions, and measurements can be performed, which are regulated by the brain. The handling of the instruments by parts of the observer's body (arms, legs, etc.) are initiated by the brain which is in the state P^{*+}. The evaluation of the measuring results is done by the mind and are stored in the mind. The stored information is then the information about the atomic facet of the object under investigation (it is the stone in this case).

The various steps of the procedure concerning "conscious" actions can be read from the following scheme:

$$\begin{array}{ccc} \text{Conscious actions} & & \\ \Uparrow & & \\ P^{*+} \leftarrow P^{*} & \xleftarrow[\text{coupling}]{\text{Mind–brain}} & \text{Mind} \\ & & \uparrow \\ & & \text{Idea} \end{array} \tag{4.22}$$

The coupling between mind and brain is not one sided. Concerning this feature, brain and mind are equivalent. In both cases, unconscious and conscious actions, the physically real object, the brain, the mind are logically connected with space and time.

Coupling between mind and brain: no direction is preferred

In the case of unconscious actions, the coupling between mind and brain is directed from the brain to the mind [scheme (4.21)]. On the other hand, within the frame of conscious actions the coupling between mind and brain is directed from the mind to the brain [scheme (4.22)]. No direction is preferred. This symmetry can be expressed by

$$\text{mind} \rightleftarrows \text{brain} \tag{4.23}$$

The mind–brain coupling is a necessary condition in connection with observations and actions as well. The space–time pictures before our eyes in everyday life come into existence through mind–brain cooperations. The space–time picture itself is a phenomenon of mind.

4.8.3 The essential transitions: summary

Let us underline that the "idea" is located in the mind of an observer and, due to the coupling of the mind and the brain, the idea is transferred to the brain P^*, and the brain converts into the brain state P^{*+}, that is, we get in analogy to eq. (4.11) the scheme:

$$\begin{array}{c} P^{*+} \\ \uparrow \\ P^* \end{array} \tag{4.24}$$

That is, the mind–brain coupling leads to

$$\begin{array}{c} \text{Idea (mind)} \\ \downarrow \\ P^* \rightarrow P^{*+} \end{array} \tag{4.25}$$

The excited brain state P^{*+} is connected to the parts of the body (legs, arms, etc.) and regulates our actions according to the idea in the mind, that is, the stone and the devices S and D are arranged through the actions of the human being in a way that the idea and the theoretical conception dictate.

An "idea" comes consciously into existence, and also the arrangement of the stone and the devices S and D are done by body actions in a conscious way. On the other hand, the observations in everyday life occur unconsciously; the stone appears spontaneously before a human being.

4.8.4 Interaction types

As we have remarked several times, there are at least two interaction types, which are principally different from each other. Each interaction type reveals a certain facet of a physically real object. As an example for a physically real object, we have chosen a stone. In this section, we would like to underline some formal points, which are necessary to make the details in connection with Figure 4.8 more understandable. Just the difference between the direct observer (human being A) and the external assessor (human being B) is of particular interest. Human being B plays in Figure 4.8 the role of an indirect observer.

First interaction type

In order to be able to determine the atomic facet, certain information at the macroscopic level is required. The direct observer, that is, human being A, interacts via a sight process with the stone, the electron source S and the detector D (Figure 4.8). The sight process belongs to the realm of macroscopic observations and reveals the facet of the object at the macroscopic level. All the statements that are based on five-sense experiences have to be judged as macroscopic in character and take place in everyday life.

These interaction processes take place in basic reality. That is, it is an interaction of the physically real entities γ_{STONE}, γ_{S}, and γ_{D} with observer A, which is characterized as physically real entity by γ_A. This situation is indicated by scheme (4.14). This characterizes the macroscopic interaction type, the first interaction type within our classification. In Figure 4.8 and scheme (4.14) also the indirect interaction is indicated, which is carried out by human being B and which is characterized by γ_B. γ_B interacts with all four entities, that is, with γ_{STONE}, γ_S, γ_D and γ_A.

As we have outlined in Section 4.7, for the determination of atomic facets the participating entities (for example, stone, detector and source) have to be available as macroscopic facets. Human being A must have them as macroscopic objects before his/her eyes. Only on this footing, he/she can plan an experiment for the determination of the atomic facet of the stone. As can be recognized from Figure 4.8, also the macroscopic picture of the stone itself must be known when we want to determine its atomic facet.

The second interaction type

The second interaction type leads to the "atomic facet" of the stone. It is formally defined between all participants: γ_{STONE}, γ_S, γ_D and γ_A. Let us express this situation, in which all entities are interrelated, by

$$\begin{aligned} \gamma_S &\leftrightarrow \gamma_{\text{STONE}} \\ \gamma_{\text{STONE}} &\leftrightarrow \gamma_D \\ \gamma_D &\leftrightarrow \gamma_A \end{aligned} \tag{4.26}$$

This corresponds to what is indicated in Figure 4.8. The interaction part $\gamma_D \leftrightarrow \gamma_A$ refers to the pattern stone_{at}.

The atomic interactions (4.26) and the macroscopic interactions are superimposed and act simultaneously. That is, we have at each clock time τ two interaction schemes, which can be formulated as follows [see also relation (4.14)]:

$$\left\{\begin{array}{l}\gamma_A \leftrightarrow \gamma_{\text{STONE}}\\ \gamma_A \leftrightarrow \gamma_S\\ \gamma_A \leftrightarrow \gamma_D\end{array}\right\} \Rightarrow \left\{\begin{array}{l}\gamma_A \overset{\text{macroscopic}}{\longleftrightarrow} \gamma_{\text{STONE}}\\ \gamma_A \overset{\text{macroscopic}}{\longleftrightarrow} \gamma_S\\ \gamma_A \overset{\text{macroscopic}}{\longleftrightarrow} \gamma_D\end{array}\right\}$$
$$+$$
$$\text{atomic level:}\qquad\qquad (4.27)$$
$$\gamma_S \leftrightarrow \gamma_{\text{STONE}}$$
$$\gamma_{\text{STONE}} \leftrightarrow \gamma_D$$
$$\gamma_D \leftrightarrow \gamma_A$$

Again, the macroscopic part (first interaction type) is needed for the experimental determination of the atomic facet of a physically real object (in our case it is a stone). The entire complex, expressed by the macroscopic part of eq. (4.27), interacts with γ_B, that is, human being B observes the entire complex, which is shown in Figure 4.8.

Interaction particularly means that the entities (stone, source S and detector D), are changed. What is changed? The relative positions of the entities, and this is dictated by the theoretical conception.

Furthermore, the scattering of electrons takes place, which is a physically real process as well and takes exclusively place in basic reality. The scattering spectrum is registered in basic reality and contains the information of the atomic facet of the stone; it is projected in the space time frame $\{x, y, z, \tau\}_A$ of A and observer A obtains the picture of the stone at the atomic level, indicated in Figure 4.8 by stone_{at}.

From basic reality to the picture

All that, that is, the configuration stone source S–detector D but also the scattering protocol, is pulled from basic reality with the tools of evolution of human being A and is transferred to the mind of observer A, which contains the four components:

$$\alpha_S, \alpha_D, \alpha^{\text{ma}}_{\text{STONE}} \text{ and } \alpha^{\text{at}}_{\text{STONE}} \qquad (4.28)$$

α_S and α_D refer to macroscopic level of the source and the detector. The entity $\alpha^{\text{ma}}_{\text{STONE}}$ reflects the information about the macroscopic facet of the stone and $\alpha^{\text{at}}_{\text{STONE}}$ is the information of its atomic facet.

What appears in the space–time picture of observer A, which is based on the frame $\{x, y, z, \tau\}_A$. Information (4.28), which is located in the mind of observer A, is processed by the processing characteristic η_A of human being A and the complex with

$$\theta_S, \theta_D, \text{stone}_{\text{ma}} \text{ and } \text{stone}_{\text{at}} \qquad (4.29)$$

appears as space–time picture in the mind of human being A.

In Figure 4.8, the situation is assessed from the point of view of human being *B*. In this case, everything is projected onto the space–time frame $\{x, y, z, \tau\}_B$. The picture of human being *A* is represented within the frame $\{x, y, z, \tau\}_A$. Concerning the two space–time frame and the position of the objects, condition (4.13) is valid.

Conditions

Let us underline once again that measurements at the atomic level can only be performed with devices, which must be perceivable at the macroscopic level where the five senses are relevant, that is, in the case of everyday life experiences. We work here on the basis of sight processes (Figures 4.5 and 4.7).

The theoretical conception says us what kind of experiments and what kind of instruments are relevant for the determination of certain properties at the atomic level. The realistic choice of devices is necessary and to arrange them correctly relative to each other and relative to the object under investigation (the stone at the macroscopic level). We need in any case instruments that are sensible to the properties at the atomic level, which are, on the other hand, presentable at the macroscopic level.

There arc physically real interactions in basic reality. The transfer to the space–time picture means that there are correlations in space and time instead of physically real processes. Human beings perceive the entities in the space–time picture.

4.9 Changing the environment

We want to assume that human being *A* lives in a certain environment consisting of trees, houses, cars, other human beings, and so on. Human being *A* wants to change the environment through specific movements of the entities within that environment. For this purpose, he/she has to activate his/her arms and eventually his/her legs as well and/or other parts of the body.

All that reflects physically real processes, that is, they exclusively take place in basic reality. The environment is characterized in basic reality by the symbol γ_{env} and the human being *A* again by γ_A, that is, the interaction process is represented symbolically by

$$\gamma_{\text{env}} \leftrightarrow \gamma_A \tag{4.30}$$

The observer *A* has a brain and a physically real body. The difference between the physically real body and the brain is necessary in order to understand the relationship between the mind of *A* and his/her free will for changing his/her environment.

4.9.1 Brain and body

Thus, human being A consists of two components: the brain, characterized by γ_A^{brain}, and the body, expressed by γ_A^{body}. Then, we get

$$\gamma_A = (\gamma_A^{\text{brain}}, \gamma_A^{\text{body}}) \tag{4.31}$$

Instead of eq. (4.30) we obtain

$$\gamma_{\text{env}} \leftrightarrow \gamma_A = (\gamma_A^{\text{brain}}, \gamma_A^{\text{body}}) \tag{4.32}$$

When human being A wants to change his/her environment, he/she has to know what he/she wants to change. The environment is represented as picture before the eyes of observer A. This picture develops a strategy of what is to change and how to do these changes. This strategy reflects an "idea," and this idea is located in the mind of human being A. The notion "idea" has been introduced already in connection with measurements at the atomic level (Section 4.7.1).

How can the idea be transferred to the body of human being A? Here the coupling between the mind and the brain comes into play, which we have already mentioned in Section 4.1.

4.9.2 Daily life

The brain is different from the other parts of the body. Arms, legs, and so on have to be considered as physically real entities without to have contact to the human being's mind. Only the brain is coupled to the mind. The brain is more than a simple physically real entity. The brain steers parts of the body in order to perform certain actions in the environment by human being A.

The tools of evolution pull from basic reality what we have called "physically real object." The pulled information is species dependent. (The notion "species" will still be introduced within the frame of the projection principle.) The tools of evolution pull physically real part of the real objects, which are "useful" for the members of the species under investigation. Only this part is relevant in everyday life. This is the message of the principles of evolution or more precisely, it is an attribute of the principle of usefulness.

But what kind of interaction is useful for human beings? Only the world of the "five senses" is perceivable for human beings, and only these facts are relevant in daily life. Not the knowledge about the "absolute truth" is important but that is what is useful. This situation is in fact fulfilled through the projection principle, which connects physically real objects (systems) and the observer's mind under the assistance of the brain and the space–time (representation frame in the case of a human being).

4.9.3 Features of the processing characteristic η

The tools of evolution connect basic reality with the mind of human being A and this holds of course for all the other human beings. But there is a further connection tool, which couples the mind with space and time. This second coupling is accomplished by the processing characteristic η (Section 4.1).

The processing characteristic η is basically a feature of the mind but its features can be transferred to the brain; η is primary not defined in connection with the brain or with brain functions but enables the possibility for conscious actions through the observer.

The brain steers parts of the body and initiate body actions. These actions must be in accordance with the space–time pictures, which human being A has before his/her eyes. The pictures are directly created by the mind without the brain. In Section 4.8 we already treated the coupling between mind and brain and the action through the brain. In the following, we will deepen that and we will give more details and conditions concerning the mind–brain coupling.

The characteristic function η contains the information about what human being A wants to change and creates a space–time picture on the basis of this information. The function η collects the pulled information, which we denoted earlier by α, and we have

$$\eta = \eta(\alpha) \tag{4.33}$$

Since the actions must be in accordance with this space–time picture, created by the mind through η, also the brain, which is responsible for body actions, must contain the information expressed by (4.33). This feature is accomplished through the fact that mind and brain are coupled. The processing characteristic η is therefore also a coupling constant.

The pictures are directly created by the mind without brain; on the other hand, the body actions are directly initiated by the brain without mind. In other words, there is a certain independence of the two lines.

4.9.4 Features of the reality picture process

In principle, the idea initiates conscious body actions and it contains the entire process of change with respect to the human being's environment due to the body actions, which take place through human being A. The process of change takes place within a definite time interval $\Delta\tau$ and starts at clock time τ_1 and ends at time τ_2. We have $\Delta\tau = \tau_2 - \tau_1$. The idea contains information about possible actions, which take place in the time interval $\Delta\tau$ and are based on instantaneous perception processes.

The reality picture process takes place instantaneously. The experience shows that this feature is reflected by our daily experiences. When we touch an object before our eyes with a finger, we "feel" it at once without delay, that is, the physically real process, which takes place in basic reality, and the touch of the object in the space–time picture takes place instantaneously without delay. Clearly, within the frame of the projection principle, the touch in space at time τ is a touch of two geometrical objects, whereas the physically real process takes place in basic reality.

Both phenomena take place simultaneously and are connected through various characteristic points: tools of evolution, mind and brain. That is, matter and space–time are not directly connected as in the case of the container principle but are interrelated through various intermediate stations. However, as we already said, the signals do not start in basic reality and move over the tools of evolution toward the mind and the brain, and finally reach the space–time picture. But all that takes place instantaneously. Nevertheless, the changes in the environment take place within a certain time interval $\Delta\tau$, as we underlined earlier.

The reality-observer system is not composed of various components but form "one block" and the components are not separable. The one component cannot be effective without the other. The components exist, but are not separable from the block. The reality-observer system is a perception system and is not determined by physically real processes. It is a system for the representation of physically real processes, which occur in basic reality.

The projection principle reflects primarily a reality-observer principle and is for the perception and the use of what is before the eyes of human beings in everyday life.

Observation and its primary stages

In this connection it must be underlined once again that within projection theory the notion "observation" exclusively refers to perception procedures in space and time. Human beings are not able to take another standpoint. Because they cannot leave space and time, as we remarked several times, human beings are caught in space and time. This kind of observation provides a physically real interaction process in basic reality. We come to an observation of the facts in basic reality through the tools of evolution, the mind, and mind-processing transactions. The observation itself is reflected through an unconscious picture before the eyes of the human. But this direct impression has a detailed primary stage, expressed in Figure 4.9 through block *E*.

4.9.5 Some general statements about the coupling

The processing characteristic η is primary a feature of the human being's mind who observes the environment and tries to change it with respect to his individual

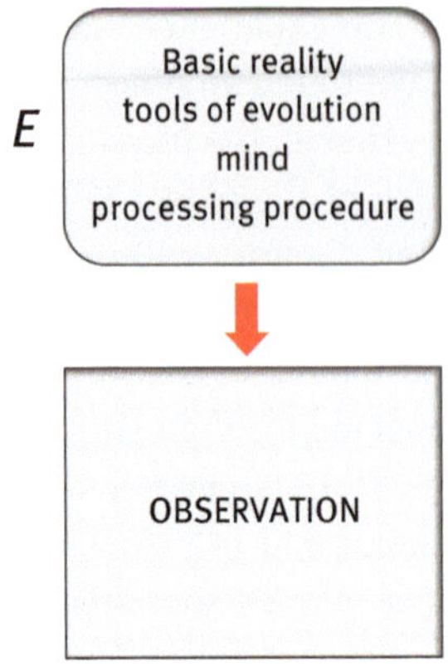

Figure 4.9: Observations take place in space and time, but there are primary stages outside space and time. The primary stages, summarized in block *E*, are the physically real interactions in basic reality and the action of the tools of evolution. In the center is the mind with its processing procedure. Block *E*, which does not exist in traditional physics, is essential in connection with observations in projection theory.

needs. The individual needs are determined through the principles of evolution and here the principle of usefulness is in the foreground. Since this principle is basic, it has been introduced at the outset in the theoretical formulations.

From ideas to actions

The characteristic η is basically not defined in connection with the brain, but the brain is coupled to the mind, and, as we already remarked, η has to be considered as the coupling parameter. In Section 4.9.3 we explained why? In the case of body actions, the brain comes into play, and the coupling to the mind becomes necessary. The characteristic function η couples the mind and the brain, and this coupling is the basis for the body actions. What is here the mechanism, that is, the general procedure?

The behavior of human beings is well characterized by their "body actions." A human being can only act within his/her environment (the world around him/her), if he/she knows the motivation for an action. For this purpose, the human being observes his/her environment and forms a picture of this environment, which appears directly before his/her eyes in an "unconscious" way. This picture gives rise to "conscious" reflections about the environment around him/her, that is., he/she thinks about the world outside and, in particular, he/she thinks about possibilities to improve his/her individual situation with respect to the environment. In other words, he/she develops an "idea." This idea is the basis for "body actions."

The idea is an entity of the mind and must be transferred to the human being's body, which does the necessary actions for the improvement of the environment in favor of his/her life. How this activity comes into existence? How is the idea transferred to the human being's body? At this point, the "brain" comes into play. In fact, between the idea and the observer's action the brain is needed.

A human being develops an "idea" on the footing of the picture of the world outside, which appears unconsciously in front of him/her. Within the frame of the idea, the human being develops consciously an action plan for changing the individual situation in his/her environment in order to improve his/her life conditions.

The coupling between mind and brain

The picture itself is created through the mind without to use the brain [Section 4.1 and scheme (4.5)]. However, for the human being's body actions, the brain functions are needed. The brain functions steer the movements of specific parts (arms, legs, etc.) of the human being's body. This is necessary when the observer himself/herself wants to change his/her environment. Therefore, a "coupling" between the mind and the brain becomes necessary in order to enable the transfer of the idea to the body functions of the human being.

The steerage of the body movements through the brain functions require a mechanism, which has to be in accordance with the processing characteristic η of the mind in the creation of the picture. The reason for this requirement is simple and has already been indicated in Section 4.9.3: the picture, created through the mind, is involved in the development of the "idea" for changing the environment. On the other hand, this idea is involved in the laws for the body steerage through the brain. The reason here is also obvious: the brain and the mind are coupled. Without such a coupling, the idea, developed by the human being, could not lead to the body movements for changing the human being's environment.

This coupling, which we observe extensively in daily life (Section 4.1), has to be imagined as follows: because of this coupling, the idea, located in the mind, converts the brain P into an excited state P^*. This excited brain P^* is able to steer the body movements in accordance to the idea located in the mind. These actions are then consistent with the picture, if their processing features are exactly the same, that is, if the processing characteristic η_a for the body actions, steered by the brain, and the processing characteristic η_b for the creation of the picture by the mind are identical. In conclusion, it is in both cases essential to use the same parameter η. Then, we get

$$\eta_a = \eta_b = \eta \tag{4.34}$$

The various steps that are needed for the transfer of an idea (mind state) to the human being's body actions are summarized in scheme (4.35):

$$\begin{array}{c}\text{Mind }(\eta)\\ \downarrow\\ \text{Picture of environment}\\ \downarrow\\ \text{The picture is used for}\\ \text{the creation of an idea}\\ \downarrow\\ \text{The idea couples to the}\\ \text{brain}\\ \downarrow\\ \text{The brain steers the body movements}\\ \text{for changing the environment}\end{array} \tag{4.35}$$

Let us underline once again that the following condition for scheme (4.35) has to be considered: the laws for the steerage process of body movements through the brain have to be in accordance with the creation process of the picture by the mind. The picture created by the mind is applied in the body-steerage process by the brain. Thus, in the laws for the steerage process, the same parameter η for the processing characteristic has to be used. Otherwise, inconsistencies come into play. That is, in the coupling of the mind with the brain, also the brain has to be dependent on the parameter η:

$$\text{brain} = \text{brain}(\eta) \tag{4.36}$$

Then, we get in the case of actions the following relation:

$$\text{Mind}(\eta) \rightarrow \text{brain}(\eta) \tag{4.37}$$

Within the frame of the projection principle, the notions "observation" and "body action" belong together and are not separable. This particularly means that the brain is permanently coupled to the mind, also in reversed cases. This is expressed through the coupling constant η and we have

$$\text{Mind}(\eta) \leftarrow \text{brain} = \text{brain}(\eta) \tag{4.38}$$

and

$$\text{Mind}(\eta) \rightleftarrows \text{brain}(\eta) \tag{4.39}$$

Relation (4.39) is valid, independently of whether there is an observation or an body action or nothing of both.

Equation (4.39) is similar to relation (4.23), which refers to observations and body activities without argument on the footing of the processing characteristic η. Equation (4.23) refers to specific cases (macroscopic and atomic facets of physically real objects).

Outside space and time

The brain couples the mind and physically real objects. If there are no such objects, this kind of coupling is not definable and the brain becomes superfluous. In this case, also the physically real part of the brain does not exist. The mind, on the other hand, can exist without brain. However, the observation line from mind to space and time does not exist without physically real object states.

There are reasons to assume that the brain only exists in the physically real realm. The mind can occupy states that are not physically real in character, for example, in the case of metaphysical states. Within projection theory, metaphysical states may in principle exist, but they are not representable in space and time. Only physically real object states appear in space and time, that is, only pictures of physically real objects appear spontaneously before us in everyday life.

The general physical laws of nature can also not be represented in space and time (also not within frame of traditional physics). They describe physically real objects, but the laws themselves occupy nonphysically real object states and belong to the realm of mind.

4.9.6 The two poles: basic reality and picture

We would like to summarize and supplement the main fact that we have discussed in connection with body actions through mind states. Let us consider again human being A, who wants to change his/her environment, that is, the world around him/her. Such a "change" is a physically real process and takes place in basic reality. We have symbolized the human being A in basic reality by the letter γ_A and the environment by γ_{enc}; we assume again that γ_{enc} describes the entire environment. The planned changes of human being A are conscious actions and, therefore, in projection theory the mind of A becomes relevant. In fact, we exactly experience the world in this way. We think and change the environment on the footing of what we have intended to change.

The brain of A must couple to the mind of A, and this is realized within the frame of the projection principle through the coupling constant, which refers, as we have several times underlined, to the processing characteristic. Let us denote the coupling constant with respect to human being A by the letter η_{A}. In principle, another human being, say B, could have another coupling constant, which we want to denote by η_B. Concerning η_A and η_B with $\eta_B \neq \eta_A$, the following general relationships are valid [see in particular eq. (4.39)]:

$$\begin{aligned} \text{Mind}(\eta_A) &\rightleftarrows \text{brain}(\eta_A) \\ \text{Mind}(\eta_B) &\rightleftarrows \text{brain}(\eta_B) \end{aligned} \tag{4.40}$$

The relations given by (4.40) are permanently effective (Section 4.9.5); the brain and the mind have the same processing characteristic, in this case η_A for observer A and η_B for observer B.

The coupling constant (processing characteristic) η is more than a simple number but it contains all the items, which are necessary to produce a space–time picture from the physically real raw information α, which reflects an objective quantity.

Before we start with the description of "activities" through human being A, we would like to repeat some general essentials. We already discussed body actions in Sections 4.1–4.8, but η independent and only in rudimentary form.

We try to understand how the information, which is located in basic reality, is transferred to space and time. Therefore, the two poles are "basic realty" and "space and time." Let us briefly characterize them.

Is basic reality describable?

It must be mentioned once again that for the facts in basic reality, that is, for physical real entities and processes, analytical descriptions in the sense of theoretical physics are not possible. As we have outlined earlier, a human being has principally no access to basic reality and the entry to the absolute truth remains hidden. Human beings are caught in space and time and these elements do not exist in basic reality. Also the processing in mind with the usual tools is ruled out; the processing in mind, that is, the conversion of the raw data pulled from basic reality to space–time information, is not describable in a traditional way. The reason is obvious: the mind creates space and time and such a creation is hardly settable here. Therefore, all the procedures taking place in connection with mind, brain and matter are merely captured "qualitatively." This suggests to assume that we deal here with fundamentals aspects; each theoretical description, that is, a mathematical description, is based on a qualitative information [74].

Within projection theory, mind states definitely exist. However, these states do not appear in space and time. The principles of evolution give space and time a restricted functionality, namely, to represent only "physical real object states" and not "states of the mind."

This is in fact our permanent experience and this is the reason why some people actually believe that states of the mind do not belong to reality or do not exist. The projection principle gives an answer on this kind of questions. In particular, the theory reveals that the mind creates pictures of physically real objects, that is, the objects appear as space–time experiences. Furthermore, it explains why such objects are not perceivable in its basic form, namely in basic reality where the absolute truth is located.

No empty space and no empty time

The conversion of the raw physically real data, pulled from basic reality, to a space–time information, is done by the mind. Since the space–time elements x, y, z, τ are not physically real, they do not belong to basic reality. Then, space and time are elements of the mind. Space and time are created by the mind in connection with the picture processing.

The fact that single space–time points, which we denoted by x, y, z, τ, are not observable as physically real quantities, finds its explanation in the fact that the mind does not create space and time without basic reality information, that is, without physically real objects. The mind could possibly do it, but an empty space–time would reflect a pure mind state, which is not representable. The mind creates space and time and the mind itself cannot be projected on this space–time structure. This point underpinned the projection principle.

In particular, we have the reality-observer principle to keep in mind (Section 4.9.4). This principle is characterized through the feature not to be composed of various components but form "one block," and the components are not separable. We underlined in Section 4.9.4 that the one component cannot be effective without the other.

No picture of the picture

At clock time τ, we have a space impression before us, which reflects, as we know, a certain x, y, z structure. The impression suggests that the world outside is embedded in space. However, we cannot observe its elements x, y, z as physically real elements, that is, the points x, y, z are not accessible to measurements. In particular, we cannot describe the elements x, y, z in the picture, as we can in the case of physically real objects. This would in fact mean that we construct a "picture of the picture." Such a construction makes however no sense and is therefore forbidden within the frame of the projection principle, and it is relevant to mention that the construction of a "picture of the picture" is actually not possible. We can experience space and time only in connection with physically real objects, which reflect geometrical structures in projection theory.

4.9.7 Objective but observer dependent

There are only those items in space and time that are located in basic reality, and these items are independent from the observer's properties. This information defines an "objective" fact. However, the envisaged information is "observer dependent."

This is because space and time are observer dependent and are created through the mind of the observer.

The transition from basic reality to space and time is a projection, and the projected information is objective in character. The information could be projected onto another representation frame, different from space and time, but the objective character of the information would remain.

4.9.8 Feelings

We observe the world outside as space structure at clock time τ, and we "feel" the objects of the environment in certain situations. Traditional physics works with the container model; here matter is embedded in space and time. Within projection theory, the physically real world is projected onto space and time. Both views are, from the conceptional point of view, very different from each other. However, they lead superficially to the same impression in the case of observations. Let us discuss this point in somewhat more detail.

Impressions: container model and projection principle

What do we experience? We touch an object of the world outside with our fingers and "feel" it at once. Therefore, it is often concluded (traditional physics) that the object before our eyes is physically real in character, but the projection principle delivers exactly the same impression, although it is different from the usual traditional view (container model).

Within the frame of the projection principle, there is a direct contact between geometrical structures, the geometrical structures of the fingers and the object. The same process takes place "simultaneously" in basic reality as physically real process. There is a strict correlation between the geometrical-structure situation, which is directly before our eyes, and what is given through the physically real situation in the world outside (basic reality), which is not directly before us but is reflected through a certain "feeling."

In other words, a physically real process in basic reality is proved through this feeling and the space–time picture. The feeling itself does not produce a space–time picture. In fact, there is no geometrical x, y, z, τ structure defined for what we call pain or similar phenomena. The pain is located in the mind and is possibly processed further but it does not appear in the space–time picture.

In conclusion, the world outside is experienced through the container model and the projection principle in the same way, but the mechanisms have to be assessed differently [7].

Pain and other feelings: no physically real aspects

A human being touches a geometrical structure (as for example a stone) with his/her fingers, which are given as well in the form of geometrical structure in space and time. He/she acts on the stone so that the stone moves with a certain velocity from one position to another. The human being experiences this process in space and time, that is, with respect to the picture, but the actual physically real process takes place simultaneously in basic reality to which the human being has no access but he/she "feels the stone," which can even be a pain, depending on the impact. The pain does not belong to the class of physical effects, and this is because pain is a subjective phenomenon. In physics, we deal with objective facts that are free of subjective elements, and pain is a subjective element.

Various connections between mind and basic reality

The tools of evolution connect the world outside (basic reality) with the mind of a human being. We have further mind–reality couplings, which exist independently from the tools of evolution; they are not influenced by them. In Section 4.1 we stated that an idea, that is, a thought, can directly be coupled to the physically real world outside. Example: we look at a tree and get an idea due to this impression. The tree is an entity of the world outside and the idea is an element of the mind. The idea due to the tree is independent of other mind states. This is an effect independently from the tools of evolution.

Within projection theory, similar connections between mind and basic reality can occur. The mind can interact "unconsciously" with basic reality (direction: mind → basic reality), and there is a response and a certain "feeling" in mind that comes into existence (direction of the response: basic reality → mind). This is in fact the mechanism when a finger touches a stone.

No subjective aspects in space and time

However, feeling (pain) is a "subjective" aspect and does not appear in the space–time picture, which is exclusively based on what the tools of evolution deliver. Only physically real objects are selected by the tools of evolution and "objective" facts appear in space and time. Subjective facts, like feelings, pains and so on, do not appear in the space–time picture. There are no measuring instruments for the registration of pain or similar phenomena.

Nature has designed the reality-observer conception in this way, and the projection principle is adapted to this conception. But we experience the feelings like pain and there can be only one solution: The feelings are phenomena in the mind, they are exclusively located in the mind; they may be processed further or not but never in the form of a space–time picture. Also here we work without the tools of evolution.

The pain itself comes into existence in basic reality and is due to the physically real interaction process between the finger and the stone assessed from the point of view of the space–time picture.

The mind is in the center

The mind is responsible for “conscious and unconscious” events. The mind is also involved in “objective and subjective” facts. That is, the mind is in the center of everything. In fact, there is no other unit, which would have the ability to treat such a diversity of essential features. The brain is obviously not qualified for that; it connects the mind with body functions.

4.10 Summary and final remarks

Matter is that what a human being “unconsciously” experiences with his/her sense apparatus (five senses) and which is directly before the observer’s eyes. This situation reflects the macroscopic level. In other words, matter is primarily an “observed” feature and not an objectively given fact of the world outside. “Observed” here means that the sensory apparatus of an observer defines essentially what we call matter.

There are other levels, for example, the atomic level. The atomic level is the “ultimate” level of nanoscience (nanotechnology) at which the properties of usual matter and those of biological systems (living systems are included) emerge.

Matter

Matter is an observer-dependent attribute. We know that there are various types of observers. The environment consists of species, and their members are essentially different from each other; their sensory apparatuses can strongly differ from each other. From this fact, we have to conclude that the attribute “matter” is different for different types of observers.

Objects of the world outside have the feature “matter” in the physical sense if they have constant properties. When do the objects have constant properties? This question must be answered in connection with evolution: the systems (objects) developed in the course of time and we assume that these evolutionary developments reach a “stationary” state after a certain time period. Then, matter has to be classified with respect to objects in a stationary state.

That is, physically real objects have the property “matter.” Matter is defined through objects with constant properties and which behave constantly in the environment. Evolution moves toward a stationary state, and this state reflects the "constant" properties and the constant behavior of objects in the environment. Projection theory created the presuppositions for this assessment.

The constant properties and the constant behavior effectuate that the objects always behave unchangingly even when their environment is changed. The objects behave exactly in the same way in repeated experimental situations. This reflects a definite law. In this particular situation the introduction of "physical laws" becomes meaningful.

In connection with projection theory, matter is that what a human being (observer) pulls from an objectively existing world outside (basic reality) which is physically real. It is the observer's view of the world and this defines his/her physical conceptions of the things. The true, observer-independent objects are located in basic reality and are not accessible to human beings.

The observer produces what we call "physically real objects." In particular, it is a species-dependent point of view. We assume here that also the members of other species judge their own world in accordance with evolutionary states where constant properties and constant behavior are the seminal features.

A certain species is organized through the "principle of usefulness." Not cognition is in the foreground but the differentiation between "favorable toward survival" and "hostile toward survival" is important in nature. This is in fact the case at least at the early phase of evolution and everything, which has been developed evolutionary, is based on the features developed earlier.

The reality-observer principle

For the observation of physically real entities in basic reality, the entire block consisting of "observer, tools of evolution, processing characteristic and the picture of reality" is involved and must be applicable at "each" time τ. Therefore, the procedure from basic reality to the picture occurs instantaneously, that is, without any delay.

The entire block consisting of the four entities "observer, tools of evolution, processing characteristic and the picture of reality" belong together and act simultaneously. This block with these specific features defines the notion "observation." It is "one" unit and is not separable. None of the entities can exist without the others.

This is the reality-observation principle. This reality-observation principle reflects a "compressed state."

Observation in particular means that a biological system interacts at the macroscopic level of everyday life using sensory apparatus, which is typical for it. In the case of human beings, the sensory apparatus is identical with the five senses.

An observation at the macroscopic level without the five senses is not possible, that is, observation requires an interaction process between the observer and the system that the observer wants to observe. This interaction is a physically real process taking place in basic reality.

Chapter 5
Matter

5.1 The black box

The raw information α, selected from basic reality, contains information that is distributed on various levels. The levels reflect certain "facets" of α. In Section 4.3.2, we have discussed two levels: the macroscopic level and the atomic level (Figure 4.2). Let us pick up the macroscopic level again. At this level, the notion "matter" has its source.

Let us consider a human being, denoted again by the letter A, and let us assume that A interacts with the environment (world outside). What kind of interactions do we expect? There is a mutual interplay between human being A and his/her environment, which is "evolutionary" in character. The evolutionary developing physically real states are pulled by the tools of evolution and appear in the mind of the human being (Section 4.1) in the form of space–time pictures. After the evolution processes, "stationary pictures" of reality are observed.

Within the projection theory, the effects due to evolutionary processes can be included at the outset. This is an important point since the principles of evolution are reflected in all activities performed by human beings because they themselves are developed in accordance with the principles of evolution. In the formulation of the projection principle, the laws of evolution can be included at the outset.

The evolutionary processes effectuate that the interaction between human being A and the world outside is strictly concerted and coordinated, respectively. Human being A interacts with the world through species-dependent "sensory tools." Let us discuss the details in this section and let us draw conclusions from the facts.

5.1.1 Interactions and the five senses

The mind of a human being and the objects do not interact in basic reality and are independent of each other and they can be introduced as separate entities. The interactions between the human being and an object of basic reality are physically real as well and take place in basic reality.

Physically real

What does "physically real object" mean in this connection? Answer: All that, which has the ability to interact with the human being's sensory apparatus and measuring instruments, we want to consider as "physically real object (system)." This statement is in accordance with the formulation "The term physically real object means that the objects have the ability to interact." This reflects a general definition.

https://doi.org/10.1515/9783110525595-005

Effects of evolution

The effect of evolution can be formulated within the frame of the projection principle as follows: The sensory tools of human beings and the information from the world outside (basic reality) are strictly adapted to each other, and they are developed together in this sense during certain time periods. In principle, the environment can remain constant during this evolutionary time period. In connection with human beings, the sensory apparatuses are given by the five senses and can possibly be considered as a final state in the evolution process of human beings.

The sensory apparatuses of living biological systems are strictly adapted to their environment. The sensory apparatuses of human beings developed during a certain time period. The human being, that is, the observer of an object, perceives only such information from the physically real entities, which the sensory apparatus allows.

At the end of the evolutionary process, the sensory apparatus remains constant if the environment remains unchanged. Then, the sensory apparatus is given by the five senses and contains the information about the environment at the macroscopic level. This is the complete information about the world outside, but this statement merely refers to what the five senses see, that is, it is a five-sense world.

5.1.2 The observer's mind and the objects in basic reality

Through the interaction processes in basic reality, the independence between the human being and the physically real object are abrogated. The principles of evolution effectuate that there is a strict correlation between the five senses of a human being and the world outside (environment).

Due to the evolutionary processes, the observer only perceives a part of the world outside, and this part is tailor-made to his/her needs. This peculiarity enables the observer to perform certain actions in the environment with "certainty," and refers here to the observation of physically real objects at the macroscopic level, where the five senses are effective.

The sensory apparatuses define the system

The evolutionary selection processes with respect to the human being's sensory apparatus develop in the course of time. After a certain time period, at the end of the evolutionary process, the sensory apparatus reaches the human being's five-sense system if the evolutionary process is actually finished, but we do not know.

The evolutionary selection processes between the human being's five-sense system and the world outside cause that the "observed world" contains just the information, which the five senses define, no less, no more. This observed world (system) is species specific, and this is because the sensory apparatuses are species specific.

This defines the macroscopic level, which reflects everyday life experiences. It is a species-specific kind of evolutionary adaption. This allows the members of a species to perform an effective and certain handling with the world outside. Such processes reflect the principle of usefulness (Section 4.1.4).

5.1.3 Products of mind

The biological states of human beings are developed with respect to the facts within the world outside after the principle of usefulness. The sensory apparatuses of human beings (the five senses) are strictly adapted to the physically real structure of the environment of the human being. Such sensory apparatuses developed during certain time periods can be very long. During this time period, the world outside developed in general as well.

Clearly, the actual basic reality does not change and appears to be "absolute" in character, but only that is changed what the human being is able to perceive. We will analyze this situation in more detail later. The "observed world," different from basic reality, is defined through the sensory apparatuses and is therefore species specific as well.

The selected information by the five senses is processed by the processing characteristic η of the mind and its contents are represented as space–time picture. This picture occupies the mind and the feelings (pain, etc.) as well. However, we know that "feelings" do not appear in the space–time picture. But they are, on the other hand, strictly correlated with the geometrical positions in space and time: When we touch in the picture a stone with a finger, that is, we do it within the geometrical x, y, z, τ structure before our eyes, there is simultaneously an unconscious process in basic reality. The stone is touched by the observer's finger in the example above and the observer feels it immediately (Chapter 4).

The feelings (pain, etc.) remain in the mind and cannot be depicted as geometrical formation. This is a strict consequence of the reality-observer principle. Only objective facts are viewable as space–time structure, but not subjective elements, and the effect of feeling belongs to the realm of subjective facts. For example, a geometrical shape of headache is not known, it is simply not defined. All that has already been outlined in Section 4.9.

Also the products of mind cannot be represented as space–time picture and do not appear unconsciously before our eyes, as we do in connection with the physically real environment. The products of mind reflect subjective features and remain therefore in the mind of the human being. In certain cases, the products of the mind can be pictured (constructed) consciously. But these constructions do not belong to the realm of the five senses and only these appear as geometrical structures before us in an unconscious way.

5.1.4 Laws

The principle of usefulness is valid for all human beings and probably also for other living (biological) systems. It should therefore be considered as a general principle. General facts are usually expressed in the scientific realm through certain "laws," qualitatively and/or quantitatively.

What "useful" evolutionary fact can be clothed by a law? In this connection, the following is relevant: It is obviously easier to handle the environment if it represents instead of an open system a "closed" unit. It is therefore useful to work with closed units and not with open systems. Thus, we may assert that nature has the tendency to form "closed systems."

This law refers to human beings and possibly also to other creatures. Let us call the tendency to form in the world outside closed units "closed system law." However, such a law can only refer to the "observed world." Only the observed world can be recognized and assessed by human beings; this statement should also be valid for the members of another kind of species.

It is "useful" for human beings to work with or within isolated systems, just when we try to formulate mathematical expressions for the physical appearances in nature. We will discuss this point later, and we will connect the reality-observer principle to theoretical physics.

5.1.5 Definition of a system

How can we define a system? A physically real system or object is defined through the sensory apparatus of the observer. The system (object) introduced here has the features in analogy to what we called "observed world." The system (object) is therefore an "observed system (object)." Since the sensory apparatus develops evolutionary, the system also develops evolutionary. It changes its properties continuously. Such developing systems are not wrong but difficult to handle, that is, the system is not in a useful state.

At the end of the evolutionary process, a closed system or closed object is reached with constant properties. This behavior is of course due to the fact that the sensory apparatus becomes a closed unit at the end of the evolutionary process. That is, we have reached a stationary state.

"Stationary" means with respect to the observed environment that no new objects and/or new phenomena appear in the space–time picture. These kinds of effects come into existence during the evolutionary processes and are exclusively projections onto space and time. The objects may come and go, that is, they may enter or leave space and time. After the evolutionary processes, no new aspects appear in space and time. The pictures before the eyes form a closed system (object).

One point needs to be mentioned: The changes of the sensory apparatuses feign the effect as if the world itself would change.

5.1.6 Hidden reality

The environment remains unchanged but the observing, living system, that is, in our case a human being. During the evolutionary process, the human being develops in accordance with the closed system law. No doubt, the environment may develop as well, but we would like to assume here that the environment remains constant, at least approximately.

What is changed? The sensory apparatus of the human being is changed. Since the sensory apparatus is connected to the entire body, the entire body will be changed during the evolutionary process. It changes in the course of time until a closed system is reached, and this is obviously the case when the sensory apparatus is given by the five senses.

Then, the "observed environment" is just that what the five senses generate. In the case of members of another species, we expect other "observed worlds," that is, the members of this species deal in such cases with other observed environments, and this is because there are other sensory apparatuses, different from the five-sense system of human beings.

All takes place in basic reality but appears in this form in space–time picture. Only the facts in the picture allow to assess what we call "evolution." The picture contains just that what is useful for human beings. Basic reality plays in this case the role of a "black box." We actually do not know what in basic reality happens. There is a strict barrier (Section 4.2). The situation is summarized in Figure 5.1.

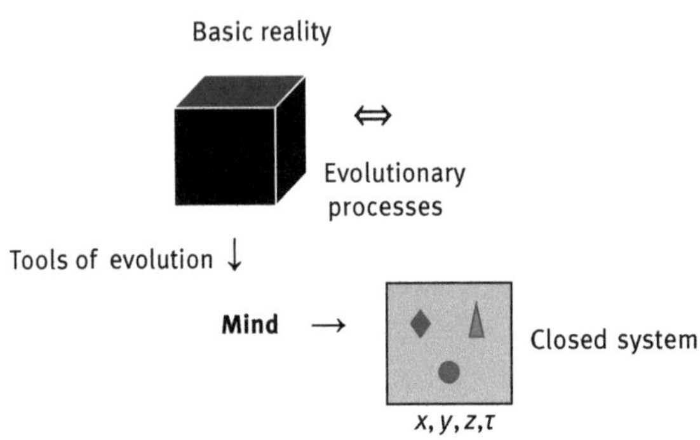

Figure 5.1: For the development of living systems (human beings, etc.), the principle of usefulness is relevant. These are evolutionary processes and take place in basic reality, which has to be seen, however, only as a black box. Therefore, everything has to be assessed from the point of view of the space–time picture. The principle of usefulness and the closed system law refer therefore to space–time pictures. The space–time pictures behave like "closed systems" if the evolutionary processes are finished. The physically real objects interact in basic reality and are pulled by the tools of evolution leading, after processing, to the space–time picture, which represents a closed system, that is, the x, y, z, τ structure in front of the observer represents a closed system. The interaction in basic reality is just so that in the space–time picture the closed system law is fulfilled. Since human beings are caught in space and time, the observer can only assess situations from the point of view of space–time pictures.

Nanolevel

Biological evolution takes/took place at the "nanolevel" where the atomic processes happen. It is the ultimate level, as we have outlined in the chapters earlier. The closed system law is basically valid at this nanolevel, but it has an impact on the macroscopic phenomena where the five senses are effective.

Evolution at the level of everyday life refers to physically real objects and processes. The effects of evolution come into existence through interaction processes. Only physically real objects have the ability to interact with the entities of their environment. The closed system law can only be reached via interactions between the human being and the environment.

5.1.7 Conclusion

Only that is useful for an observer what the five senses perceive. The observed information about the environment is a five-sense information, no less, no more. The sensory apparatuses are adapted to the environment and change in the course of the evolutionary process; in principle, the environment can remain constant during this evolutionary time period. The evolutionary process itself is steered by the closed system law. It stops when the system is closed, that is, if the sensory apparatuses behave stationary. Then, we have the final version of the sensory apparatus. Again, in the case of human beings, the sensory apparatus is identical with the five senses.

The tools of evolution pull only what is needed and just that is recognizable for an observer. Thus, the tools of evolution pull only that part of the environment, which the five senses can generate. Only this part actually appears in the space–time picture.

5.1.8 Theoretical developments

As we underlined, the term "closed system" refers to space and time. The contents of the space–time system cannot be changed: No entities can leave space and time and no entities can enter space and time. In particular, also those with basically new properties cannot enter space and time. That is, we have definite systems located in space and time, which have in particular constant features without surprises. This conception is useful for daily actions.

The behavior of closed systems

These constant features are characterized by the following attributes: The appearances of the physically real objects in space and time, which come into existence together with the feeling of the space–time impressions, are the basic information at the macroscopic level, that is, at the level of everyday life. Space–time impressions and the feeling of them occur simultaneously.

Closed systems have obviously constant features: Effects always take place in the same manner, that is, arbitrary repetitions of the experiment lead to exactly the same result and the same effect. Even when the structural configurations and compositions are changed, the effects remain explainable on the same footing, namely, on the basis of general laws. In other words, the findings with respect to closed systems are reproducible. This holds for five-sense observations and for macroscopic observations with measuring devices as well; all configurations lead to the same observed laws.

A world, consisting of a constant number of objects, is experienced on the basis of visual impressions (x, y, z, τ structures) and specific feelings. If the sensory apparatuses behave stationary, this world (observed world) appears as "closed system." For such a world, physical descriptions are possible. Isaac Newton introduced with large success on the footing of space–time impressions the notions "force F, momentum $\mathbf{p}$ and energy E." These features define that what we call "material world" and the entities are denoted as matter. The physically real objects themselves are assigned by a mass parameter m. Even quantum systems can be treated on the basis of this conception.

How are the physical laws connected to what is located in the mind? In the mind, the entity α is in the foreground. The physically real entity α, which is again the raw information pulled from an object γ of basic reality, corresponds to mathematical (physical) laws. In projection theory, these laws are based on the two variables $\mathbf{p}$ and E for the momentum and the energy. This $\mathbf{p}, E$ information, which defines reality, is projected onto space and time leading to the picture of reality. This "theoretical" picture $(\text{space} - \text{time structure})_{\text{theory}}$ can be compared with the corresponding "observed" picture $(\text{space} - \text{time structure})_{\text{observed}}$. The situation is summarized in Figure 5.2.

Open systems

During the evolution process, the space–time system is open, that is, space and time are open for the input and output of physically real entities. After the process, the "space–time doors" are closed and the environment becomes a closed system. However, we have always to keep in mind that all these space–time phenomena are exclusively projections onto space and time.

On this footing, we may construct evolutionary processes, in particular those at the nanolevel, where self-organizing processes are of basic relevance. The container model of traditional physics does not offer such kind of mechanism.

$$\begin{array}{ccc} \gamma & & \\ \downarrow & & \\ \alpha & \leftarrow & \text{Theoretical description} \\ \downarrow \text{Projection} & & \downarrow \text{Projection} \\ (\text{space–time structure})_{\text{observed}} & \leftrightarrow & (\text{space–time structure})_{\text{theory}} \end{array}$$

Figure 5.2: Closed systems are the basis for theoretical descriptions. Theoretical descriptions in the form of physical laws are expressible. How are the physical laws connected to what is located in the mind? In the mind, the physically real entity α is located. α is the raw information pulled from an object γ of basic reality. The entity α corresponds to mathematical (physical) laws. This information, which defines reality, is projected onto space and time leading to the picture of reality. This "theoretical" picture $(\text{space–time structure})_{\text{theory}}$ can be compared with the corresponding "observed" picture $(\text{space–time structure})_{\text{observed}}$.

What develops evolutionary and what not?
The biological evolution refers to the physical real processes in basic reality and not to the states of the mind, which do not belong to the class of physically real entities. That is, within this view the states of the mind do not develop in the course of time to which also belong that what we call space–time feeling. The elements x, y, z, τ of space and time, which characterize this space–time feeling, belong to the realm of mind and do not develop evolutionary of the mind. However, the structures within the x, y, z, τ frames, which reflect physically real entities, develop evolutionary.

5.1.9 The complete world

It is alluring to assume that such a world, which is an observed world and which behaves as closed systems, defines in this form the "complete world." It is in fact assumed in traditional physics that there exists nothing else than just those things that are defined through the "observed world," and it is particularly assumed that there is nothing outside space and time.

Our physically real actions and observations always refer to space and time, and also here we are seduced to maintain that everything takes place within space and time. This leads to a world view, which has been used with large success in traditional physics. The projection principle corrects this view. The selection processes, an item within the principles of evolution, with its law "as little outside world as possible" feign this impression of a complete world which is embedded in space and time. Selection means to isolate but this isolation does not mean that there is nothing else. Just this seemingly complete world can be handled optimal. But a "selected system" is not identical with a "complete system."

Human beings are merely caught in space and time and this fact feigns a complete world. But this is obviously a fallacy when we measure with the yardstick of the projection principle. The physically real world was developed toward a "closed system." In this case, we cannot look beyond space and time and, therefore, the world outside must appear as complete system.

5.1.10 The main items

The projection principle conveys that the world before our eyes appears as closed system and all processes seem to be taken place within a strict impermeable space–time block. In particular, the following points have to be underlined:

1. Changes in connection with the world outside (environment), done through individual actions, belong to the realm of macroscopic phenomena, where the five senses are effective.
2. We can call the world, which are based on five-sense phenomena, as "material world." Mind states are not registered by the five-sense world, as we also know from the traditional view. But mind states exist within projection theory. All x, y, z, τ structures in the pictures of reality belong to the material realm but are, on the other hand, states of the mind. The mind states are not representable through x, y, z, τ structures. The tools of evolution transfer exclusively physically real (material) information to the mind.
3. The effect of evolution can be formulated within the frame of the projection principle as follows: The sensory apparatuses of human beings and the information from the world outside (basic reality) are strictly adapted to each other, and the sensory apparatuses are developed during certain time periods. In connection with human beings, the sensory apparatus are given by the five senses. Measuring devises refine the statements at the five-sense level but do not open the door for additional senses.

Once again, at the macroscopic level, the five senses are effective, that is, it is the level of the five senses. Clearly, specific measuring devices can be constructed and produced on the footing of five-sense experiences.

5.2 Classification

In this section, we would like to classify the components, which are used within the frame of the projection principle. This principle reflects a reality-observer principle. The following components are relevant:

Basic reality
Mind
Brain
Space and time

All these units belong strictly together, and the function of one component cannot resign on the existence of the others. The tight binding between these components is demonstrated in the earlier sections.

Let us consider human beings. For the perception of the physically real world (basic reality), an observation process is necessary. Within the frame of the projection principle, the above four components are needed.

The mind is here in the center. In basic reality, matter (physically real entities) is located. The brain regulates body functions and transfers information from the world outside to the mind and vice versa. The picture of reality is the only unit to which human observers are directly connected. In fact, human beings are caught in space and time. The physically real world is projected onto space and time and a picture is created.

5.2.1 Traditional view

In traditional physics, we have no qualified model for the connection of these components. The world within traditional physics is defined through what the world before our eyes dictates. That is, this world is simply given (represented) by the structures, which we have directly before the eyes in everyday life. This defines a complete system from the point of view of traditional physics. The material objects are embedded in space and time. As we remarked several times, this conception is the basis of the container model.

For the understanding of the material processes, the mind, the brain and the basic aspect of reality are not needed. This is the naive perspective of the situation. Here, the true nature of space and time is not considered within this approximation.

5.2.2 Projection principle

The relation of the above elements (mind, brain, basic reality) to space and time is different from that what traditional physics teaches. What are the relationships to space and time?

Basic reality

Basic reality does not contain space and time. Space and time are not physically real and such types of elements are not suited to embed physically real (material) objects into them.

Mind

The mind creates the entities space and time. It is therefore not describable in space and time. The mind cannot create itself. The processing characteristic η of the mind creates a space–time picture of the physically real (material) raw information α, which is pulled from basic reality.

Brain

The brain of human beings has the ability to interact with the mind. It records information of the mind and transfers it to the parts of the body. In this way, the human being is able to perform body actions. The brain is also able to transfer information from the world outside (basic reality) to the mind. Thus, the brain is not completely composed of matter. The brain does not create space and time, but it has the ability to include products of the mind, depending on the situation. It is involved in conscious and unconscious actions initiated by the mind. Only the material part of the brain is representable in space and time.

Tools of evolution

The tools of evolution have nothing to do with space and time; they are not representable in space and time. They connect basic reality with mind. We do not know the mechanism how the tools evolution pull information from basic reality, but we know the necessity for their existence.

5.3 Local interactions and block interactions

Physically real objects are located in basic reality; here "located" does not mean that the objects can be ordered with respect to space and time. As we often underlined, space and time with the elements x, y, z, τ do not belong to basic reality; the elements x, y, z, τ are not physically real.

We assume that there are individual objects

$$\gamma_k, \quad k = 1, 2, \ldots, N \tag{5.1}$$

These objects are assumed not to be isolated, but there are certain relations between them. In principle, all N objects are involved. Also these relations define physically real processes, which we usually call "interaction."

What does the interaction effectuate, locally and globally? Locally with respect to "single objects" and globally in connection with "groups (blocks) of objects." Both features are a matter of fact: We observe them, that is, we observe single objects but also biological blocks, which are usually called species. Just in connection with living biological systems, these blocks (species) are important and belong to the field of nanoscience (nanotechnology). We will specify this point in more detail later.

5.3.1 Group formations

The interaction between the individual objects is defined through all observable effects and, in particular, the tendency for the generation of "group formations" (creation of species) has to be included here. Facets come into existence (local effects, Chapter 4) and group states are formed as global effects. Through these group states, specific biological blocks (species) are created with characteristic properties and, as we know, there are various of them in nature.

The specific facets are selected through observation methods. In Section 4.3, we discussed the facets at the macroscopic level, which are selected by the five senses. On the other hand, there is the atomic level that needs sophisticated measuring instruments. All that is analyzed in Section 4.3.

However, the classification of the cosmos into an ensemble of interacting objects must not be the basic state. The cosmos is possibly in its fundamental form a "unified whole"; in this case, the assumption of an ensemble of more or less individual objects would merely be an approximation [37].

5.3.2 What is pulled?

The tools of evolution are species specific. However, they pull information from species-independent facts located in basic reality. This seems to be a contradiction, but it is not. The following point deepens that:

To a human being belongs "human being-specific tools of evolution," which we want to denote by $(\text{tools of evolution})_{\text{hum}}$. On the other hand, we have other species as, for example, a turkey is a member of another species. A turkey is clothed with "turkey-specific tools of evolution," which we want to denote by $(\text{tools of evolution})_{\text{turkey}}$. We already discussed the case of a turkey in Chapter 1. Turkeys show quite another behavior than human beings; they obviously developed in the course of evolution a world view, which is quite different from that of human beings.

All these species-dependent tools of evolution pull information from "all" blocks (species), but merely a restricted information from each of these blocks. This is due to the principles of evolution that dictate the maxim "as little outside world as possible" (Section 4.1). This amount of restricted information constitutes a mixed information where all blocks participate. Here, in general, "all" objects are concerned but a restricted amount of information with respect to each object as well.

Not the complete information in connection with "one" block is registered through the tools of evolution, but a reduced part from "each" block, that is, from each species. This is actually what a human being observes. Such situations reflect just that what we understand under the maxim "as little outside world as possible." In the picture of reality of a specific species the information of "all" species appear, but only a part of them. This part is selected in accordance with what is "useful" for the observer.

5.3.3 Useful portions of all aspects

The members of a species notice “useful” portions of “all” aspects of the world outside (environment). This species-specific information is assessed from the point of view of species-typical pictures of reality. As we already know, these pictures come into existence through the mind of the observer (human being, turkey, etc.). The mind processes the useful portions, pulled from basic reality, and it creates the picture within a certain representation frame, which is expressed by space and time in the case of human beings.

Clearly, the mind with its processing characteristic, which we denoted by η in Section 4.1, are individual entities and are therefore also observer specific and species specific, respectively. In particular, it has to be assumed that the representation frames is species specific as well.

In conclusion, the members of a species deal with “useful” portions of “all” aspects of the world outside (environment) and create a species-typical picture from this information. Thus, the point of view varies from species to species and must be considered in the assessment of living biological systems and all the other entities in the world outside.

5.3.4 Interaction types

The interaction processes are physically real and therefore take place in basic reality. We may assume that there is a superposition of interaction types, corresponding to the types of observed effects.

We remarked earlier that we have to distinguish between local and global interaction types: local with respect to “single objects” and global in connection with “groups (blocks) of objects,” which define species. We assumed that there are N objects in basic reality. Then, the local interaction is expressed by

$$\begin{gathered} y_i \leftrightarrow y_j \\ i, j = 1, 2, \ldots, N, \quad i \neq j \end{gathered} \tag{5.2}$$

We suppose that this local interaction type is superimposed by a global interaction part. What does “global” mean in this connection?

Let us assume that there are ℓ species in the cosmos. We denoted a species also through the terms “biological state” or “block.” Each of these ℓ blocks consists of block_k objects and this number varies from species to species, where k labels the species, that is, we have

$$\text{block}_k, \quad k = 1, 2, \ldots, \ell \tag{5.3}$$

The interactions themselves with respect to the specific blocks $k = 1, 2, \ldots, \ell$ are expressed by

$$k = 1:$$

$$\gamma_i \xleftrightarrow{\text{block interaction 1}} \gamma_j \Rightarrow i, j = 1, 2, \ldots, \text{block}_1, \; i \neq j \tag{5.4}$$

$$k = 2:$$

$$\begin{aligned} &\gamma_i \xleftrightarrow{\text{block interaction 2}} \gamma_j \Rightarrow i, j = h_{m2} + 1, h_{m2} + 2, \ldots, h_{m2} + \text{block}_2, i \neq j \\ &\text{with} \\ &h_{m2} = \text{block}_1 \end{aligned} \tag{5.5}$$

$$\vdots$$

$$k = s:$$

$$\begin{aligned} &\gamma_i \xleftrightarrow{\text{block interaction } s} \gamma_j \Rightarrow i, j = h_{ms} + 1, h_{ms} + 2, \ldots, h_{ms} + \text{block}_s, i \neq j \\ &\text{with} \\ &h_{ms} = \text{block}_{s-1} \end{aligned} \tag{5.6}$$

$$\vdots$$

For the starting condition of the next block $k = s + 1$ we have

$$h_{ms} + \text{block}_s = h_{m(s+1)} \tag{5.7}$$

Since the number of objects in the world remains constant, the following condition holds:

$$\sum_{k=1}^{\ell} \text{block}_k = N \tag{5.8}$$

Not all objects of the world, denoted by N_{world}, belong to the realm of biological systems. Therefore, we have $N < N_{\text{world}}$, but in principle N may comprise all objects in the world: $N = N_{\text{world}}$.

The block interactions vary from species to species, and we have

$$\begin{aligned} &\text{block interaction 1} \\ &\neq \text{block interaction 2} \cdots \\ &\neq \text{block interaction } s \neq \cdots \end{aligned} \tag{5.9}$$

In conclusion, a specific biological state (species) comes into existence through physically real interactions between physically real objects, which are located in basic reality. During the evolutionary process for the formation of a certain species, say k, the objects i, j within a block, expressed by

$$i, j = h_{mk} + 1, h_{mk} + 2, \ldots, h_{mk} + \text{block}_k, \quad i \neq j \tag{5.10}$$

interact with each other, where $h_{mk} + 1, \ldots$ labels the objects inside block k. The blocks contain temporary a certain number of objects, which varies however in the course of time. Finally, we have ℓ species (blocks or biological states) consisting of physically real objects, whose numbers inside a block are in general different of each other and we have

$$\text{block}_1 \neq \text{block}_2 \neq \cdots \neq \text{block}_\ell \tag{5.11}$$

It is important to underline that to each block (species), observers of the same kind are associated. We have therefore species-dependent point of views with specific frame of references, which is given in the case of human being by space and time having the elements x, y, z, τ.

5.3.5 Stationary states

To a specific biological block belongs an observer type (human being, etc.) with a specific sensory apparatus. All species develop evolutionary and of course their observing members as well. It is an "observed block (species)" with "observing members," quite in analogy to that, what we have remarked earlier in connection with "observed world."

In other words, in the course of clock time τ, the block (species) changes and also the observers with their sensory apparatuses, which belong to the biological block (species). They develop in mutual dependence:

$$\text{observed species } (\tau) \leftrightarrow \text{sensory apparatus } (\tau) \tag{5.12}$$

The ℓ biological blocks and their members develop evolutionary, and we suppose that all blocks behave similarly:

$$\begin{aligned} &\text{species}(\tau_a)_k \leftrightarrow \text{sensory apparatus } (\tau_a)_k \\ &\text{species}(\tau_b)_k \leftrightarrow \text{sensory apparatus } (\tau_b)_k \\ &\qquad\qquad\vdots \\ &\text{species}(\tau_{\text{final}})_k \leftrightarrow \text{sensory apparatus } (\tau_{\text{final}})_k \\ &\text{with} \\ &\tau_a < \tau_b < \cdots < \tau_{\text{final}} \end{aligned} \tag{5.13}$$

The time dependence is effective until the closed system law is fulfilled. Then, the species and the sensory apparatus do not develop further and become stationary.

Let us assume that this case is given at $\tau_{\text{final}\,k} = (\tau_{\text{final}})_k$; at clock time $\tau_{\text{final}\,k}$ the evolutionary development is finished. The various species reach their stationary state at different clock times:

$$\tau_{\text{final}\,1} \neq \tau_{\text{final}\,2} \neq \cdots \neq \tau_{\text{final}\,\ell} \tag{5.14}$$

For times $\tau > \tau_{\text{final}\,\ell}$ the biological block as well as the sensory apparatus have the properties

$$\begin{aligned}&\text{sensory apparatus}\,(\tau_{\text{final}\,k}) = \text{sensory apparatus}\,(\tau)\\ &\tau > \tau_{\text{final}\,k}\end{aligned} \tag{5.15}$$

and

$$\begin{aligned}&\text{species}\,(\tau_{\text{final}\,k}) = \text{species}\,(\tau)\\ &\tau > \tau_{\text{final}\,k}\end{aligned} \tag{5.16}$$

The species, that is, the biological blocks, and the sensory apparatuses behave stationary for $\tau > \tau_{\text{final}\,k}$. Then, the closed system law is fulfilled. The evolutionary processes are finished. This is definitely the case for clock times $\tau > \tau_{\text{final}\,k}$, unless we put a member of a species into a new environment.

Remark

In connection with the definition of blocks (species), the interaction between the environment and the sensory apparatus is fundamental. During the evolutionary processes, the state of the complex "sensory apparatus and environment" is changing. That is, the species-creating interaction refers to the complex "sensory apparatus and environment," but may change as well in the course of time τ.

5.3.6 Species-dependent world views

An observer, for example, a human being, which belongs to a specific species, is clothed with a sensory apparatus adapted to this specific biological block. He/she experiences the world exclusively on the footing of this sensory apparatus. He/she assesses all the other species with respect to this sensory apparatus. Another observer, for example, a turkey, which is based on another biological block, that is, which belongs to another species, experiences the world in another way. The turkey's sensory apparatus is different from the sensory apparatus of human beings, and the entire world before its eyes is seen differently from that what a human being sees in the same situation. The experiment by Wolfgang Schleidt demonstrated that impressively (Chapter 1).

The evolutionary selection processes between the human being's sensory apparatus (five-sense system) and the world outside cause that the "observed" world outside contains just the information, which the five senses define, no less, no more. This is of course also true for turkeys: The evolutionary selection processes between the turkey's sensory apparatus and the world outside cause that the "observed" world of the turkey contains just the information, which its sensory apparatus defines, also here no less, no more.

5.4 Species-dependent features

We have a lot of species on the earth and the members of each species construct a picture of the physically real world (basic reality). The structures of the pictures are identical for the members of a certain species. What picture type of what species is realized in basic reality? None of them! None of them is preferred in basic reality. This is in fact solved through the conception itself: The structures in basic reality are not identical with those in the picture. This is only what we can state about basic reality. The reason is obvious: As we remarked many times, there can be no space and no time in basic reality because space and time do not belong to the class of physically real entities. This fact is valid for all kinds of species if the reality-observer principle is valid.

5.4.1 Interaction processes determine the evolutionary facts

We assume that the unit γ_{env} (Section 4.9) does not vary with time τ but the evolutionary processes are reflected through the picture that is selected from the raw information α_{env} which is, on the other hand, pulled from γ_{env}. The entity α_{env} is dependent on the species under investigation, and this is because the tools of evolution are species dependent.

The members of a certain species can merely know a specific form of the world outside, which we called in Section 5.1.2 "observed world" or "observed environment." This observed world develops evolutionary in the course of clock time τ. Thus, biological evolution is an observer-specific effect and refers to developments with respect to clock time τ. It is through the interaction in basic reality; the "order" in connection with evolutionary effects is determined by the physically real interactions in basic reality, but it is "deciphered" through the time τ. This statement needs more attention.

The processes are observed through members of various species, and such a member is located as physically real entity in basic reality. Let us study the situation in connection with human beings and let us denote a human being in basic reality by γ_{human}. On the other hand, the physically real environment is denoted by γ_{env}. In order to be able to learn something about the environment, the human being has to interact with the environment, that is, there is an interaction process between γ_{env} and γ_{human}, which is symbolically expressed by $\gamma_{human} \leftrightarrow \gamma_{env}$. Since evolution belongs to the category of physically real processes, the interaction must produce these evolutionary effects. How is that expressed in basic reality? We have principally no access to basic realty and cannot say something about the interaction mechanism, but general statements are possible.

5.4.2 Compressed information

The system (world) can be assumed to be stationary with respect to the clock time τ. We always argue in terms of τ; our intuition is tailor-made to τ, and we have no direct access to other standards. The world behaves stationary in basic reality, that is, outside space and time. This point will be deepend in Section 5.13.5.

Stationary means that the world does not develop. The entire (complete) world or system is given at each time τ, that is, all at once. The characteristics of the world and other systems are given all at once and are given in "compressed" form. In this way, the term "compressed form" is defined.

The world exists all at once

However, the world and the systems are not experienced all at once, but piecewise, part by part. None of the parts of the world is preferred concerning their existence at each time τ, and none of them is excluded within the observation procedure. In other words, all parts have to be seen as equivalent.

In conclusion, in the course of time τ the parts of the world are not existent one after the other, but all at once. If, for example, the world consists of k parts, each of them is in principle tangible at clock time τ; again, all the parts are equivalent and none of them is preferred concerning their existence at τ.

Probability distributions

Then, we are confronted with a statistical problem. There must be a probability distribution for the k parts at τ: the probability aspect is an observer-induced property. The statistical behavior is coercive when the entire information about the world or any other system cannot be experienced all at once.

From this point of view, the statistical behavior is not an objective category of nature, but is observer dependent. An observer type who would be able to register the entire system (world) all at once would not experience the world as an entity with statistical behavior.

5.4.3 Three facts

In Section 5.4.1 we discussed the interaction between a human being and his environment and denoted it by $(\gamma_{\text{human}} \leftrightarrow \gamma_{\text{env}})$. The complete interaction takes place "instantaneously" and is stationary in character and is not a function of time τ, but is, because of the instantaneousness, an information in "compressed" form (Section 5.4.2).

The facts in basic reality have no relatedness to space and time. The elements x, y, z, τ do not appear in basic reality, that is, the facts (the interaction processes) in basic reality are independent on x, y, z and also on τ. In conclusion, the facts are unrelated to x, y, z, τ.

On the other hand, human beings are caught in space and time and they can assess the situation only from their point of view. The facts concerning $\gamma_{\text{human}} \leftrightarrow \gamma_{\text{env}}$ are therefore constant for a human being independently from the clock time τ. This is a fact when we work within the reality-observer principle. There are three facts that are relevant in the assessment of processes within the frame of the projection principle.

First fact

The information about the interaction process $\gamma_{\text{human}} \leftrightarrow \gamma_{\text{env}}$ is given in "compressed" form in basic reality. This is the first fact.

Second fact

The second fact is that this complete information concerning α_{env} is projected onto space and time. The complete information is then "distributed" on a time interval $\Delta\tau$ with respect to the clock times τ, which is defined by the complete information $\alpha_{\text{env}}$with $\tau_a \leq \tau \leq \tau_b$, where τ_a is the initial time and τ_b is the final time, that is, at clock time τ only a part of α_{env} is observable.

Third fact

The third fact is that the human being observes only x, y, z structures at clock time τ. These structures (houses, trees, cars, etc.) appear before the eyes of the human being. Also this fact belongs to the realm of evolution, because we deal here with a selection process.

5.4.4 Species-selected information

The facts, α_{env}, are constant at each time τ between the initial time τ_a and the final time τ_b $(\tau_a \leq \tau \leq \tau_b)$. This constant information at each time τ in $\Delta\tau = \tau_b - \tau_a$ defines, as we already quoted, the information in compressed form. It is an assessment from the point of view of a human being who is caught in space and time. The complete information about the interaction process $\gamma_{\text{human}} \leftrightarrow \gamma_{\text{env}}$ is expressible in terms of space and time. The time-dissolved form only contains a part of the complete information.

The environment, expressed by γ_{env}, contains the evolutionary order in compressed form and the members of a certain species have access to it through the selecting tools of evolution. The entity α_{env} is the species-dependent part of the basic information $\gamma_{\text{env}}\,(\gamma_{\text{env}} \rightarrow \alpha_{\text{env}})$, that is, also α_{env} contains the evolutionary order in compressed form.

Compressed facts and deciphered information

This behavior is due to the feature that the interaction processes take place instantaneously in basic reality, that is, it contains at "each time τ" the complete interaction terms (possibilities). The observer perceives the complete information α_{env}, but not as compressed fact but in deciphered form.

As we have already outlined earlier, the complete information is "distributed" on a time interval $\Delta\tau$ with respect to the clock times τ, which is defined by the complete information α_{env} with $\tau_a \leq \tau \leq \tau_b$, where τ_a is again the initial time and τ_b is the final time. That is, at clock time τ only a part of α_{env} is observable.

The reason for that is obviously also here explainable by the "principle of usefulness." It is less burdened for a human being to treat the environment piecewise and not as a compressed block. Such a procedure is more useful for survival.

Therefore, the members of "each species" have their own species-dependent law of evolution, depending in particular on the frame of representation and the tools of evolution.

The species-dependent law of evolution is based on the compressed interaction law between the observer of a species and the environment in basic reality, and this exists τ independent but in the form of instantaneous information. In Chapter 6, we will formulate mathematically the laws for this behavior quantum-theoretically in connection with a quantum aspect of time, which is not known in traditional quantum theory. In this way, compressed states as well as states in deciphered form can be described convincingly.

5.4.5 Varying the species

Let us consider a human being and an animal. What do we expect for two species that are different from each other? We want to assume that the reality-picture principle is valid for both, for the human being and for the animal. Here we have to assume that the complete perception procedures for both species are different from each other.

Then, the physically real part of the animal γ_{animal} and that of the human being γ_{human} are different from each other. This is trivial. However, also the tools of evolution, the processing characteristic η as well as the pictures of reality together with representation frame of both biological systems are differing. In particular, the frame of representation, which is given in the case of the human being by the elements x, y, z, τ, takes for the animal other features and are characterized, for example, by the elements r, s, p, μ.

Then, in the transition from a human being to an animal, the following items are changed:

$$\begin{gathered}\gamma_{\text{human}}, \text{tools of evolution}_{\text{human}}, \eta_{\text{human}}, \text{picture}_{\text{human env}}(x, y, z, \tau) \\ \downarrow \\ \gamma_{\text{animal}}, \text{tools of evolution}_{\text{animal}}, \eta_{\text{animal}}, \text{picture}_{\text{animal env}}(r, s, p, \mu)\end{gathered} \tag{5.17}$$

with

$$\gamma_{\text{animal}} \neq \gamma_{\text{human}} \tag{5.18}$$

$$\text{tools of evolution}_{\text{animal}} \neq \text{tools of evolution}_{\text{human}} \tag{5.19}$$

$$\eta_{\text{animal}} \neq \eta_{\text{human}} \tag{5.20}$$

$$\text{picture}_{\text{animal env}}(r, s, p, \mu) \neq \text{picture}_{\text{animal env}}(x, y, z, \tau) \tag{5.21}$$

The various elements are connected through the reality-picture principle and we obtain scheme (5.22):

$$\begin{array}{ll} \text{HUMAN BEING} & \text{ANIMAL} \\ \gamma_{\text{human}} \leftrightarrow \gamma_{\text{env}} & \gamma_{\text{animal}} \leftrightarrow \gamma_{\text{env}} \\ \quad \downarrow \text{tools of evolution}_{\text{human}} & \quad \downarrow \text{tools of evolution}_{\text{animal}} \\ \alpha_{\text{human env}} & \alpha_{\text{animal env}} \\ \quad \downarrow \eta_{\text{human}} & \quad \downarrow \eta_{\text{animal}} \\ \text{picture}_{\text{human env}} & \text{picture}_{\text{animal env}} \\ x, y, z, \tau & r, s, p, \mu \end{array} \tag{5.22}$$

Because of these transformation characteristics, the picture of the environment, which the animal experiences, is different from the picture that the human being has before his eyes, although the same environment is concerned.

This is an assessment from the point of view of projection theory and should be valid in this form when we go from species to species.

5.4.6 The situation for human beings

We do not know the situation in connection with animals, but we are able to make statements with respect to human beings. Human beings are caught in space and time. Thus, the physically real interaction $\gamma_{\text{human}} \leftrightarrow \gamma_{\text{env}}$ can only be assessed from the point of view of the picture, which is a x, y, z structure at each clock time τ. We already analyzed the situation in Section 4.5 for a stone, but we would like to extend that through an environment consisting of three entities: a car, a house and a tree (see Figure 5.3).

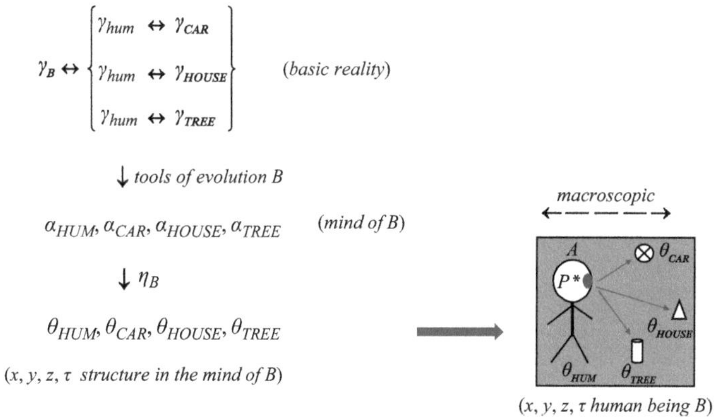

Figure 5.3: Human being A (symbolized by γ_{hum}) interacts with his environment γ_{env}: $\gamma_{\text{hum}} \leftrightarrow \gamma_{\text{env}}$. It is a sight process, taking place at the macroscopic level. We assume that the environment is composed of a car γ_{CAR}, a house γ_{HOUSE} and a tree γ_{TREE}. The picture contains the x, y, z, τ structures θ_{CAR}, θ_{HOUSE}, θ_{TREE} of the car, the house and the tree that are represented in the space–time frame $\{x, y, z, \tau\}_B$ of human being B. Human being A with eyes is represented in the picture as well and is denoted by θ_{HUM}. As in Figure 4.5, the entire situation is again assessed from the point of view of human being B. The "observation" of the car, the house and the tree through human being A must lead to pictures X_{CAR}, X_{HOUSE}, X_{TREE} in the mind of A, which are represented within the space–time frame $\{x, y, z, \tau\}_A$ of human being A. Due to the interaction between the brain P of human being A and the three entities car, house, tree, the brain itself is changed, and it converts into an excited state indicated by P^*, as in connection with Figure 4.5. Other entities of the environment (photons, etc.) are not indicated in the figure. The position of the pictures X_{CAR}, X_{HOUSE}, X_{TREE} inside human being A has to be identical with the positions of θ_{CAR}, θ_{HOUSE}, θ_{TREE}, which is indicated in the figure by the arrows going from A to θ_{CAR}, to θ_{HOUSE} and to θ_{TREE}.

5.5 Variants

We denote again the physically real environment in basic reality by γ_{env}. The ℓ species experience the environment γ_{env} differently. We have ℓ variants of the same thing. These variants are expressed through pictures. However, all variants must be simultaneously stick in γ_{env}. It is a certain kind of superposition, but we have to be careful since a picture is different from that what is located in basic reality; due to the evolutionary effects the picture is a reduced form of basic reality and reflects a projection as we know.

The variants must be simultaneously stick in γ_{env}. How that is accomplished cannot be analyzed from the point of view of a human being because there is no access to that what happens in basic reality. This is probably also the case for members of other species.

It is even difficult to construct a metaphor for that. The rainbow is certainly a bad example, but let us nevertheless quote it here: The light of the sun is composed of many colors as we can perceive when a rainbow appears. The white sun light corresponds to γ_{env} and the various colors correspond to the pictures created by the various species.

5.5.1 Various pictures of the same thing

The physically real entity γ_{env} is composed of many species-dependent information and each of them represent the same thing, that is, they represent the environment γ_{env}. These species-dependent information are pictures:

$$\text{picture}_{\text{env}\,k}, \quad k = 1.2, \ldots, \ell \tag{5.23}$$

$$\gamma_{env} \begin{cases} \text{picture}_{\text{env}\,1} \\ \text{picture}_{\text{env}\,2} \\ \vdots \\ \text{picture}_{\text{env}\,\ell} \end{cases} \tag{5.24}$$

with

$$\text{picture}_{\text{env}\,1} \neq \text{picture}_{\text{env}\,2} \neq \cdots \neq \text{picture}_{\text{env}\,\ell} \tag{5.25}$$

Even the representation frames can be different from species to species. The representation frame of human beings is given by space and time, but other species may in principle not work on the footing of space and time.

5.5.2 Local and global effects

Concerning evolutionary developments, we have to distinguish between two cases, which come into existence through various interaction types. Let us briefly discuss both cases:

1. In Section 5.1 we studied an observer with a certain sensory apparatus. This observer changed the environment, from environment 1 to environment 2. The

sensory apparatus fulfilled the closed system law for environment 1, but not for environment 2. This situation gives rise to an evolutionary process with respect to the sensory apparatus. The observer interacts with the new environment 2 and his sensory apparatus changes until it is adapted to environment 2. Within the frame of this interaction type, objects are observed at various levels. Facets come into existence that are based on "local" effects (Section 4.3).

2. The second case refers to the evolutionary development of biological blocks (species). There is a tendency for the generation of "group formations" (creation of species, see Section 5.3). These group states reflect "global" effects. Through these group states specific biological blocks (species) are created with characteristic properties and, as we know, there are many of them in (biological) nature.

In summary, in connection with the projection principle we have to distinguish between two interaction types: local and global interactions, leading to levels (facets) and group formations (species). The facts are represented in Figure 5.4.

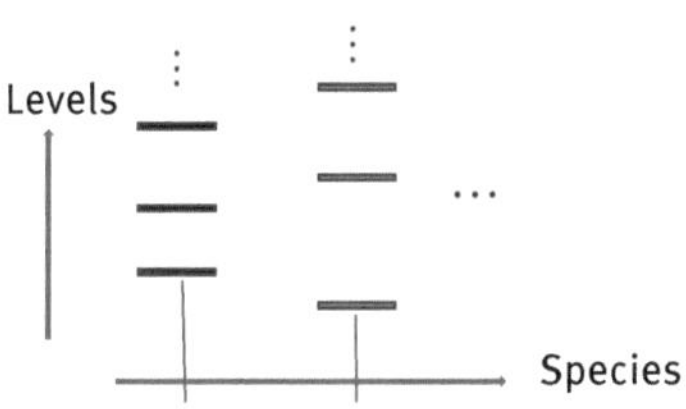

Figure 5.4: The physically real objects of the world are organized in terms of two interaction types: local and global interactions. Within the frame of local interactions, facets come into play, which are distributed on "levels." On the other hand, the global interactions lead to group formation, which are known under the notion "species".

Remark

The formation of specific biological blocks (species) must not come into existence through species-specific interactions between physically real objects. In principle, the creation (formation) of species could also be established in a collective way without block-specific interaction types. There is possibly a certain analogy to the cluster formation with respect to the relatively simple many-particle systems in solid-state physics. In the formation of species, the physics of becoming is probably essential.

The main problem in connection with the evolutionary development of biological species and blocks, respectively, is the fact that within the development of quantum structures the "time" cannot be described quantum theoretically in traditional physics since the time is here still a classical parameter. Thus, in future descriptions, the quantum aspect of time has to be introduced adequately. This point has been underlined in Chapter 1. The projection principle is able to treat the phenomenon "time" quantum theoretically. Everything is interlocked: mind states, matter and quantum aspect of time. More details are given in Chapter 6.

5.6 Nanoengineered brain functions

The members of a certain species (let us denote it by the letter F) can only assess the situation of other species from their point of view, but cannot take the point of view of a member of another species. The members of species F can be, for example, defined by human beings and the other species could be an animal (e.g., a turkey, Chapter 1). The members of species F live in "world F." The members of another species, say D, live in "world D." According to Section 5.5, world F is different from world D:

$$\text{world}\, D \neq \text{world}\, A \tag{5.26}$$

The reason is simple. The members of species D interact with the environment in another way than the members of species F, and they developed their own species-specific sensory apparatus. But there is a further point: The members of species D create in general pictures of reality on the basis of another frame of representation, which can in principle quite different from space and time, that is, the elements x, y, z, τ do not appear in this case. That is, for the members of species D space and time with x, y, z, τ are unknown and, on the other hand, the frame of representation, valid for species D, is unknown for the members of species F, which we assume to be human beings.

5.6.1 No higher ordered point of view

The members of each kind of species are caught in their own world; in this case, with respect to world F and to world D. World D cannot be assessed from the point of view of world F and vice versa. There is no higher ranking point of view for the assessment of all these possible worlds. Such a higher ordered state is not defined within projection theory and is even not thinkable.

This situation has consequences: If a human being (assumed to be a member of world F) changes body functions of an individual of world D, he/she will have only a restricted possibility to assess the behavior of the individual of world D after a brain-functional (nanotechnological) manipulation. The human being does the changes from the point of view of world F, but the effect takes place in another world, namely, with respect to world D. This is a remarkable and far-reaching statement and demonstrates that a nanotechnological impact, which takes place at ultimate level, can lead to uncontrolled situations.

Within the frame of the projection principle we come to an important conclusion: The members of species F assess the world outside differently when we compare it with the world view of species D.

Are such constructed situations in nature thinkable? Are they observable? Yes, it is. The experiments by Wolfgang Schleidt within the realm of behavior research (we called it "chick experiment" in Chapter 1) demonstrate that turkeys behave in fact differently from human beings and both belong to different species (biological states). These experiments are dramatic and had in fact not been expected by the scientific community.

5.6.2 The ethical standpoint

In this respect, the usual body changes through surgeries will not be so dramatic. But with the manipulation of brain functions, we go a step further. Here we enter the brain–mind realm. The chick experiment belongs to this category and this experiment has led to an uncontrolled situation (Chapter 1). Here the brain functions are influenced through the manipulation of the environment of the turkey, indicating that the world view of a turkey is different from the world view of a human being.

When a human being (world *F*) changes nanotechnologically the brain functions of a member, which belongs to world *D*, he/she changes the brain functions at the "ultimate level" (Chapter 1). Thus, the effects due to such changes can be expected as fundamental. The human being cannot know the effect of the brain-functional changes on the life conditions in world *D*, and this is because the human being, member of world *F*, is caught in space and time, which is unknown with respect to world *D*. A member of world *F* has assess merely on what happens in world *F*.

Such kind of brain-functional experiment in the nanorealm has to be performed very careful, and its possible effects have to be assessed "before" the experiment starts. However, such an assessment will hardly be possible and, therefore, such far-reaching experiments should be barred; we should relinquish on them. The ethical standpoint dictates that the chick experiment has in fact be considered as questionable when we measure with the yardstick of ethics.

The chick experiment demonstrates that the existence of species (biological states, characteristic blocks) is fundamental. Far-reaching consequences for the members of the species are inescapable, just when the members are confronted with the individuals of other species.

The projection principle allows the development of a conception, which allows to understand the main features of such a scenario. Here, physically real objects and the mind and the brain of an observer can be ordered systematically. This is important since all these entities are obviously interrelated as the experiments reveal. In particular, the frames of presentation (space and time in the case of human beings) have to be considered as species specific as well. Concerning the interrelations, more details are given in the next section.

5.6.3 What is changed within nanotechnological alterations?

What kind of effects do we expect when we change brain functions through a nanotechnological impact? Here not only the members of other species are concerned, but the members of the human species as well. That is, what can we principally expect when we change brain functions of a human being at the nanolevel?

In Section 4.9.5 we have underlined that there must be a coupling between the brain and the mind. The processing characteristic η has to be considered as the coupling parameter. The coupling constant η is not a simple number but reflects a peculiarity; the mind of the human being processes through the η information from basic reality (world outside) and creates the picture of reality that we have directly before us in everyday life.

The coupling between brain and mind is expressed within the frame of the projection principle by the peculiarity that the mind and the brain have the same processing characteristic η: If the characteristic η of the brain is changed also the characteristic η of the mind is changed and vice versa. In other words, we have not only the relations $\text{mind}(\eta) \rightarrow \text{brain}(\eta)$ but also the law $\text{mind}(\eta) \leftarrow \text{brain}(\eta)$ [eq. (4.39)].

This feature is well founded in Section 4.9. In the case of actions (Section 4.9.5), the mind influences the brain. On the other hand, the brain influences the mind if we manipulate the brain as, for example, through a nanotechnological impact; in this case, the brain is changed at the ultimate level and this is fundamental.

In this section, we discuss the possibility for nanotechnological changes of the brain (brain functions). There is in fact the principal possibility that through a nanotechnological impact on the brain functions, the processing characteristic η is changed from η to η_{NANO}. Then, from our statements made above directly follows that the brain and the mind are changed correspondingly, that is, we get in the case of the transition:

$$\eta \rightarrow \eta_{\text{NANO}} \tag{5.27}$$

Also the transition

$$\text{brain}(\eta) \rightarrow \text{brain}(\eta_{\text{NANO}}) \tag{5.28}$$

Equation (5.28) effectuates that the mind is changed simultaneously and we get

$$\text{brain}(\eta_{\text{NANO}}) \rightarrow \text{mind}(\eta_{\text{NANO}}) \tag{5.29}$$

That is, also the mind occupies a new state with η_{NANO}:

$$\text{mind}(\eta) \rightarrow \text{mind}(\eta_{\text{NANO}}) \tag{5.30}$$

The mind is responsible for the picture of reality, that is, the x, y, z structure at clock time τ. The details are given in Figure 5.5.

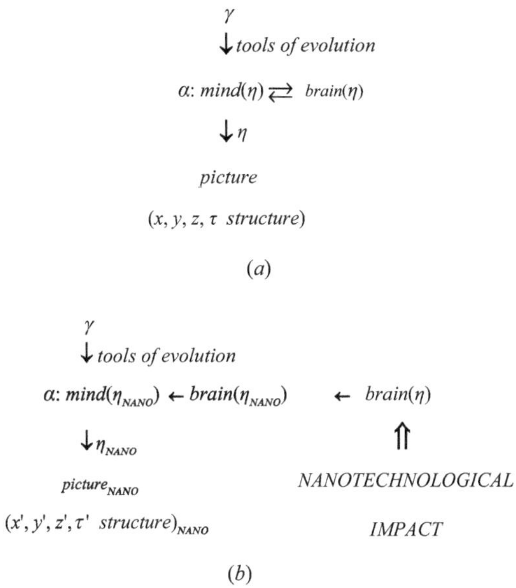

Figure 5.5: The brain functions of a human being are nanotechnologically changed. What are the consequences? How does the impact propagate? The evolutionary selected information α ($\gamma \to \alpha$) is not changed through the nanotechnological impact, and the tools of evolution are also not influenced. (a) The situation without nanotechnological changes and (b) with such kind of changes. (a) If the processing characteristic η of the brain is changed, also the processing characteristic η of the mind is changed and vice versa. In the case of "actions," the mind influences the brain. On the other hand, the brain influences the mind and can be manipulated through the brain as, for example, by a nanotechnological impact; in this case, the brain is changed at the ultimate level and this is fundamental. (b) The brain is changed nanotechnologically, that is, the processing characteristic η with respect to the brain is changed: $\eta \to \eta_{\mathrm{NANO}}$. Then, also the mind and the representation frame are changed from x, y, z, τ to x', y', z', τ'.

Changing the frame of representation

In general, with varying processing characteristic η ($\eta \to \eta_{\mathrm{NANO}}$), the representation frames are also changed. As we already discussed, the processing characteristic η is responsible for the picture structure created by the mind. For η, we assumed that the elements x, y, z, τ are effective, and for η_{NANO} we get the elements x', y', z', τ' (Figure 5.5) with

$$x, y, z, \tau \neq x', y', z', \tau' \tag{5.31}$$

Then, we have

$$(x, y, z \text{ structure at time } \tau) \\ \rightarrow (x', y', z' \text{ structure at time } \tau')_{NANO} \tag{5.32}$$

The time τ (clock time) is an "external" parameter. The basic experience, on which everything in the world is based, says the following: at time τ a human being has a certain x, y, z structure before his/her eyes. Within the frame of the projection principle, the notation "external" here means that τ is located outside the space structure, that is, outside the x, y, z structure. The clock time τ is in any case an element in connection with the picture (x, y, z structure) and, therefore, it is created by the mind on the basis of η. Therefore, the clock time τ is dependent on the processing characteristic η as well and not only the coordinates x, y, z. In principle, there can be drastic changes depending on the degree of the change.

Internal compatibility but not with other human beings

In the case of brain manipulations, the concerned human being is in general influenced basically. When we measure with the yardstick of projection principle, we may state that even the individual's world view can be changed. That is, the world before his/her eyes in daily life has possibly no longer similarity with what he/she saw before the brain manipulation.

In other words, due to specific manipulations of brain functions, the human being produces in general a new world view. This can, however, only be happen if the brain functions are changed at the fundamental level, and the changes due to a nanotechnological impact are in general done at the fundamental ultimate level where the properties emerge.

The world's view of the concerned human being is changed, but it must be underlined explicitly in this connection that the world's view of other human beings is not changed. This can of course lead to serious problems for the brain-manipulated human being.

It is easy to recognize that for a certain human being, the "observation" of the world outside and the simultaneous "actions" within this new world view would be compatible. This is because exactly the same processing characteristics hold for "observation" and "action." However, this new world view, which is based on η_{NANO}, would not be in accordance with the world of the other members of the same species, which is based on $\eta \neq \eta_{NANO}$. Such a creature with η_{NANO} would probably not be able to survive.

Furthermore, the transition $\eta \rightarrow \eta_{NANO}$ would take place at once, that is, not within a certain time interval. However, evolutionary processes need a sufficiently large time interval, that is, no evolutionary processes can be involved. No doubt, this could lead to additional problems, because the sense apparatus of the concerned human being is not adapted to the environment.

5.7 The reality-picture principle

In Section 4.3.1, human being B "observes" an interacting unit consisting of human being A and a stone. All three entities are physically real and belong to basic reality. They are denoted by γ_B, γ_A and γ_{STONE}. Human being A interacts with the stone, which is symbolically expressed by $\gamma_A \leftrightarrow \gamma_{\mathrm{STONE}}$. Only human being B is involved in an "observation process."

What does human being B observe? He observes γ_A and γ_{STONE} in their "interacting" states, that is, observer B interacts with the stone and with A. The observation of the interaction process itself in $\gamma_A \leftrightarrow \gamma_{\mathrm{STONE}}$ will not be captured because an "interaction of interaction" (suggested by $\gamma_A \leftrightarrow \gamma_{\mathrm{STONE}}$) makes no sense and has to be excluded.

5.7.1 Objective representations

What does observation mean within the frame of the projection principle? We need a physically real observer; it is in this case human being B denoted by γ_B. The entity γ_B interacts in basic reality with the object or complex, which human being B wants to observe (here $\gamma_A \leftrightarrow \gamma_{\mathrm{STONE}}$). This is performed by the sensory apparatus of observer B.

The notion "observation" only makes sense if, beside the interaction trough the sensory apparatus, an observation result is actually expressible. This condition is fulfilled if a "picture of reality" is existent and defined, respectively.

That what is located in basic reality as an "objective" physically real unit must appear in the picture as an "objective" unit as well, which is given at time τ as x, y, z structure. That is, the "subjective" features belonging to the observer (in our case human being B) are not of interest here. The picture itself is in our example in the mind of human being B.

The observations are done within the macroscopic realm, where the sensory apparatus is given by the five senses. They always appear in an unconscious way before our eyes (before the eyes of human being B). The corresponding interaction process in basic reality is unconscious as well. In general, human being B observes the entire environment and not only human being A and the stone. These everyday life experiences are merely limited through the possibilities of the five senses.

5.7.2 Ideas and conscious actions

All that belongs to the realm of "unconscious" observation that we have spontaneously before us, as x, y, z structure at clock time τ. In this way, we perceive unconsciously trees, houses, cars and all the other things we meet in everyday life.

But there can also be a certain intention to observe the environment. Then, a conscious "idea" is behind the observation and the observation itself does not reflect no longer simple unconscious observation. The idea to change the environment often comes into existence through the conscious observation of the environment. In other words, in the case of actions (Section 4.9), when we want to change the environment, "conscious" states come into play.

The idea comes into existence on the footing of the picture before the human being's eyes. The idea itself initiate the mind to perform changes in basic reality, which, on the other hand, are instantaneously transferred to the picture which belongs to the observing human being. This picture is located, as we know, in the mind of the observing human being. The observation of the picture is necessary in order to be able to create the idea.

5.7.3 Observer's interaction and the space–time picture

The notion "observation" is characterized in projection theory through an interaction process in basic reality and its representation in space and time as a picture. It is an interaction between an observer and that what the observer wants to observe. This process is tightly linked to the corresponding picture, which reflects the features of the interaction process. Without the observer's interaction in basic reality, no picture in space and time, and without picture in space and time no interaction of the observer in basic reality. This is the "reality-picture principle." A picture in space and time always means "observation."

5.7.4 From basic reality to the picture of reality

From basic reality to the picture, various interstages are involved: The tools of evolution of human being B as well as the mind's processing characteristic η are essential. The following scheme is relevant:

$$\begin{array}{c} \text{basic reality (observer } B) \\ \qquad\qquad \longrightarrow \text{picture of reality (mind of } B) \\ \qquad\qquad \Uparrow \\ \qquad\qquad \text{tools of evolution of } B, \\ \qquad\qquad \text{processing characteristic } \eta_B \text{ of } B \end{array} \tag{5.33}$$

The observer (human being B) makes statements about physically real objects and processes in space and time. Scheme (5.34) summarizes the main facts: At each clock time τ, observer B has x, y, z structure before his eyes. The x, y, z, τ structures come into existence through a procedure starting in basic reality over the tools of evolution and the processing characteristic η_B:

$$\tau: \left\{ \begin{array}{c} \text{observer} \longrightarrow \text{objects, processes} \\ \Uparrow \\ \text{tools of evolution} \\ \text{processing characteristic} \\ \text{picture of reality}: x, y, z \text{ structure} \end{array} \right\} \qquad (5.34)$$

All the interstages, indicated in eq. (5.34), do not take place one after another, but they occur simultaneously at each time τ. In the next section, we will explain the situation.

5.7.5 Instantaneous transitions

Again, the x, y, z, τ structures before the observer's eyes come into existence through a procedure starting in basic reality over the tools of evolution and the processing characteristic η_B. This procedure takes place instantaneously. Let us justify that.

We may not consider the interaction process $\gamma_A \leftrightarrow \gamma_{\text{STONE}}$ in basic reality and its space–time picture as separate events that take place successively. Both events belong to the same procedure, which we call "observation." Each observation takes place at a definite clock time τ. The interaction process $\gamma_A \leftrightarrow \gamma_{\text{STONE}}$ is not the "cause" for the picture appearance, which would then play the role of the "effect." Such a "cause–effect" procedure is not reflected by the observations taking place within the frame of the projection principle.

We have to distinguish between two cases. Case 1: In the cases of usual observations, the interaction processes in basic reality, which reflect the observation, and which appears as picture in space and time, seem to be a cause–effect procedure: First, the interaction and after that the appearance of the space–time picture. This argumentation is, however, not correct, because there is the second case. Case 2: A finger touches the geometrical structure in the picture and due to this touch a physically real process takes place in basic reality and the picture is produced via the mind of the observer. This example has already been discussed in Section 4.2.3.

The touch of the finger in connection with the picture cannot be the cause for the physically real process in basic reality. If there is no interaction process in basic reality, there can be no picture of reality. The problem is only solvable if the two processes (interaction in basic reality and the appearance of the picture) take place simultaneously.

The process in basic reality has its counterpart in the space–time picture. There is a strict connection between the physically real action between human being A and the stone in basic reality and the movement of the stone in the picture. These connections between the process in basic reality and the equivalent process in the picture occur instantaneously. If the process in basic reality takes place at clock time τ'_a or at τ'_b and the equivalent process in the picture at clock time τ_a or at τ_b we must have

$$\tau'_a = \tau_a \tag{5.35}$$

$$\tau'_b = \tau_b \tag{5.36}$$

This means the term simultaneousness is in connection with basic reality and the picture.

In conclusion, cause and effect take place simultaneously and, therefore, the observation procedure cannot be assessed in terms of the traditional cause–effect view with a certain time difference with respect to cause and effect. The traditional cause–effect views refer to physically real processes where matter is involved. This is, however, not the case when we consider the standard observation procedure $y \rightarrow$ picture. Clearly, an object in basic reality, here denoted by y, reflects a physically real entity, but the transition from y to the picture is not material in character; here, the states of the observer's mind are essentially involved and not matter.

5.7.6 Movements

In Figure 4.3, the space–time picture is shown that human being A has before his/her eyes. In the picture, there is a stone, which can be moved by observer A with his/her arms. Human being A can observe parts of himself/herself through a self-observation as, for example, the arms of A, which is however not indicated in Figure 4.3. Let us suppose that that human being A moves the stone. We have to distinguish between two situations: The process in basic reality and the equivalent activity in the space–time picture.

The physically real action between the stone and human being A takes place in basic reality, but we do not know how it works. We have no access to basic reality, but merely to space and time: We are caught in space and time, as we have remarked earlier several times.

The situation is assessed from the point of view of the picture as follows: Human being A moves the stone relative to its environment (also not indicated in Figure 4.3) from position $a = (x_a, y_a, z_a)$, at time τ_a, to position $b = (x_b, y_b, z_b)$ that it reached at time τ_b. We assume for simplicity that the movement is in accordance with classical mechanics.

At time τ_a, the state of the system under investigation is denoted by γ_a. γ_a interacts and the state is changed. At time τ_b, the state of the system is denoted by γ_b. We have

$$\gamma_a \neq \gamma_b \tag{5.37}$$

and

$$\tau_a \neq \tau_b \tag{5.38}$$

The equivalent process in the picture leads a movement of the geometrical structure of the object (stone) from position $a = (x_a, y_a, z_a)$, starting at time τ_a, to position $b = (x_b, y_b, z_b)$, which is reached at time τ_b. We have

$$\{x_a, y_a, z_a\} \neq \{x_b, y_b, z_b\} \tag{5.39}$$

On the other hand, the transition time interval from γ_a to x_a, y_a, z_a, τ_a is zero, and the transition time interval from γ_b to x_b, y_b, z_b, τ_b is zero as well (see Figure 5.6). Again, we do not deal with material processes in this case, and the cause–effect principle in the traditional sense has no basis here.

$$\tau_a : \begin{cases} \gamma_a \\ \downarrow \\ \alpha_a \\ \downarrow \\ x_a, y_a, z_a \end{cases} \Rightarrow \tau_b : \begin{cases} \gamma_b \\ \downarrow \\ \alpha_b \\ \downarrow \\ x_b, y_b, z_b \end{cases}$$

$$\Delta\tau = \tau_b - \tau_a \neq 0$$

Figure 5.6: Physically real processes take place in basic reality. At time τ_a, the state of the system under investigation is denoted by γ_a. γ_a interacts and the state is changed. At time τ_b, the state of the system is denoted by γ_b. The process in basic reality has its counterpart in the space–time picture. The system (stone) moves within the time interval $\Delta\tau = \tau_b - \tau_a \neq 0$ through the action of an observer (human being *A*) from position $a = (x_a, y_a, z_a)$ to position $b = (x_b, y_b, z_b)$. Whereas the transitions from γ_a to x_a, y_a, z_a and from γ_b to x_b, y_b, z_b occur instantaneously, the transition from γ_a to γ_b takes place within the time interval $\Delta\tau = \tau_b - \tau_a \neq 0$.

5.7.7 The indivisible observation procedure

What does "observation" mean within the frame of the projection principle? Observation means that certain things of the world outside become perceivable. We already know that reality (world outside, environment) is defined as species-specific unity and its properties are determined through the sense apparatus of the members of a certain species (Section 5.1). But these processes do not yet define that what we call "observation."

When is an observation complete?

The perception of the species-specific unity is only possible under inclusion of space and time, and this is because human beings are caught in space and time and cannot leave space and time. The characteristic processes take place in basic reality and are projected onto space and time, that is, the observation is only complete through the inclusion of a space–time picture.

In other words, the notion "observation" can only then be a fact if the observer interacts in basic reality with a species-specific object and if, in addition, there is a representation in space and time. That is, the space–time picture and the interaction of the observer with a species-specific object are not separable. Both characteristics exist as "one" unity. It is an indivisible observation procedure. In Chapter 4, this specific situation has been analyzed in more detail.

Species dependent and observer dependent

The world outside is not only a species-specific unity, but it is observer dependent as well. Each observer lives in his own world, whereby the worlds of the members of the same species are identical. The world outside is represented within the observer-dependent frame of representation. Again, the frame of representation of human beings is given by space and time. Other species may work on the basis of other frames of reference.

Absolutely given objects

The objects of basic reality are existent as absolute entities and are, in the case of biological systems, also absolute in character within species-dependent formations. They are all defined as absolute entities in basic reality, provided they are not observed. The members of the various species experience the same physically real objects, which are given in basic reality in absolute form, differently. The various situations reflect a superposition (Section 5.5.1).

These entities of basic reality contain information in absolute form and are more than the species-dependent information produced through an observer. That is, the absolute world outside exists but only if there is no impact by an observer. As soon as an observer is coupled to the absolute world, the absolute character gets lost.

When an object, say y_i, interacts with an observer's sensory apparatus, the observer perceives it simultaneously as space–time picture, that is, he/she represents the interaction result in a picture.

Clearly, the N objects y_i, $i = 1, 2, \ldots, N$, in basic reality interact with each other, that is, there are processes $y_i \leftrightarrow y_k$. However, such processes have primarily nothing to do with what we called "observation." An observation is initiated through the application of the sensory apparatus. Then, the unit "interaction in basic reality and the space–time representation" becomes active.

Remark

The forming of species through global interaction processes (Section 5.1) can be understood as a general evolutionary selection activity. On the other hand, the tools of evolution select only those things from the environment of an envisaged processing procedure that are useful for the acting observer. As an example for such a processing procedure, we have treated in Chapter 4 the sight process. The sight process itself is not (essentially) changed in the transition from basic reality to the picture.

5.7.8 Space and time are not subjects of observation

We do not know how basic reality is structured. Due to the evolution barrier (Section 4.2.1), a human being is not able to assess the actual situation in basic reality. But the basic reality (world outside) can be observed in restricted form as a picture in space and time. The space–time is the frame of representation. Can this frame be considered as physically real unit? We have remarked several times that space and time cannot be physically real. But is there still a backdoor to make space and time to physically real entities like matter? Let us briefly discuss this point in the following.

This transition from basic reality to the picture cannot be done directly because a transformation is necessary. Here the "mind" of the observer is in the center. What is the procedure?

The tools of evolution pull the species-dependent information from basic reality and transfer it to the mind. No space and no time are involved. The mind processes this information further with its processing characteristic η and creates the picture in space and time. These processes due to the tools of evolution, and the creation of the picture cannot be represented in space and time and are therefore not imaginable for a human being.

In fact, we want to observe physically real objects and not space and time, that is, the elements x, y, z, τ. The objects are "represented" in space and time using the elements x, y, z, τ; the entities x, y, z, τ themselves are not subject of observation. The representation of x, y, z, τ in space and time would not be possible, and would in fact make no sense because such a procedure reflects a "representation of representation." In Section 4.9.6, we underlined that we cannot observe the elements x, y, z, τ as physically real elements and, therefore, it makes no sense to picture them like physically real (material) objects.

The information of the world outside is projected onto space and time. However, space and time themselves are created in the course of the projection procedure, that is, space and time are not existent in the mind of the human being as permanent entities. Space and time are created through the mind if there is something to represent. Space and time are superfluous if there is nothing to represent. Physically real objects and the space–time are tightly linked, and one cannot exist without the other. The picture itself is located in the mind.

5.7.9 The relevance of everyday life experiences

The world outside appears spontaneously before us, that is, it appears unconsciously before our eyes. This is in fact the most basic impression we have from the world outside. Our theoretical conceptions are based on these impressions. It is the orientation for all our developments.

In fact, there is nothing, which would be more important for human beings than everyday life experiences. The relationship to the environment has to be perceived and judged realistically through the observer, and this is essentially based on that what is spontaneously as picture before his/her eyes. A misjudgment in connection with the structure of the picture can in fact be life-threatening. This situation makes everyday life experiences to the most relevant symptom in the assessment of the world outside [74].

So, for human beings these direct observations are the basis for everything; they are even the yardstick for developments in theoretical physics. If a theory is not in accordance with what we have before our eyes in everyday life, the theory has to be discarded. This statement is confirmed through John Ziman. He wrote: "Failure to accord with reliable 'commonsense' evidence is quite as discreditable as falsification of a theory by a contrived, abstruse experiment" [74].

The spur to the measuring of natural observations arises at the level of everyday life; without it, no measurable fact would be known [75]. In other words, the five-sense world reflects the most basic facts and are the starting point for all more sophisticated developments.

That particularly also means that in the scientific field one cannot do without the visual experiences of everyday life. The verbal description without the visual picture is incomplete [74]. In any case, verbal descriptions themselves nearly always refer to other similar visual patterns. John Ziman wrote:

> How would one define the adjective "serrate", except to say it was "like a saw"? Try to imagine a purely verbal account of such an object, without actually drawing a picture or seeing an actual specimen. We need not to led astray by the numbers that occur in such descriptions. These merely summarize, in consensible language, the outcome of other sensory experiences, such as counting and measuring and will seldom be amenable to the logical transformations of any mathematical theory [74].

After that, a mathematical description of everyday life observations is only possible if they are really imaginable, and things are imaginable exclusively in space and time. We know that human beings are caught in space and time and this means that only the things in space and time are imaginable. The statement by John Ziman distinctly underlined [74] that only things are accessible to mathematical descriptions if they are presentable in space and time. That what is not presentable in space and time is not imaginable and cannot be expressed through mathematical laws. The verbal description without the visual picture is, as John Ziman stated, incomplete.

In other words, we base our descriptions essentially on that what we experience in everyday life, that is, the things that are faced in daily life play an essential role also in the scientific realm.

5.8 Are space and time given twice in nature?

The pictures before us in everyday life are pictures of material objects in space and time. The objects in the picture reflect physically real objects of basic reality in the form of geometrical structures.

These geometrical structures are strictly interwoven at each time τ with space. In other words, the objects and the space–time are not separable. This is a requirement of projection theory. Space and time are in fact needed to express the geometrical structures, which reflect the material objects as picture. This is the view of projection theory.

Separability

The peculiarity of the projection principle that the material objects and the space–time are not separable is really observed in nature. No effect is known in nature where the material objects appear separate from space and time. This is a relevant statement when we judge effects within the frame of the projection principle. What about traditional view?

In traditional physics the situation is different. The material objects are located at each time τ in space; the notion "basic reality" is unknown here. Also here the objects have the relevant property not to appear without space and time, but this statement is based on experimental experiences. That is, in traditional physics we observe the material objects without exception in space and time, but we believe intuitively that the space–time and the objects are separable and that they can exist independently from each other. But there is no case observed where such a separability and independency can really be observed.

The projection principle explains this behavior and requires in fact that the space–time and the objects cannot be considered as separable entities. In traditional physics, such a possibility is principally allowed and would in fact correspond to our intuitive demands when we assume that the physically real objects are embedded in space and time. The projection principle does not allow separability, and this in fact observed.

5.8.1 The features of space and time are the key

Human beings developed in the course of evolution not to perceive the "absolute truth," but their biological structure developed after other criteria, which can be summarized under the term "principle of usefulness." We already applied this term several times in this book. In other words, human beings (and probably also other living biological systems) perceive only those things of the world outside that are "useful" for them. What is the deeper reason for that?

The projection principle reveals the cause why we do not perceive the absolute truth, which we called above basic reality. Let us discuss more details in this section.

Space and time are the elements of the picture

The specific features of space and time are the reason for the occurrence of basic characteristics in the material (physically real) world outside. As we have underlined in Chapter 1, space and time, that is, the elements x, y, z, τ, are not physically real. They are not accessible to empirical tests and can principally not be measured by devices.

That what cannot be measured does not belong to the class of physically real objects. Therefore, space and time cannot belong to basic reality, where exclusively physically real entities are embedded and, as we underlined several times, space and time do not belong to this class of manifestation.

This must have consequences: Space and time (x, y, z, τ) are the elements of the frame of representation for the picture of reality, which is located in the mind of human beings. Space and time (x, y, z, τ) cannot belong to basic reality where exclusively physically real entities are embedded.

Basic information and perceivable facts

Human beings are caught in space and time and perceive only what is given in terms of space and time. Basic reality is not accessible to human beings, but it contains the basic information about the world. This means that information must be transferred from basic reality to the picture.

Thus, a transformation takes place from basic reality to the picture of reality. Such a transformation only makes senses if it is connected with a specific purpose. A "one-to-one-transformation" of information would be a dead and superfluous action; nature would not be organized in this way. Why not an observation directly in basic reality, in this case, without space and time? This is, however, not the way and is not imaginable. On the other hand, the way over a transformation must have a reason and purpose, respectively.

No one-to-one transformation

A one-to-one transformation would be the maximum transfer of information. Due to the specific τ-selections (Section 5.4.4), it must be less than the maximum. Only a part of the maximum information is used through a human being, a part that is relevant for human beings. This is the requirement of the projection principle, which is based on realistic space–time features.

What part of the maximum information (absolute truth) is it, which is transferred? What is the cause for it? What is the criterion and the law behind that? Also here the projection principle contains in principle the answer. We may argue as follows: Only the notion "usefulness" can be the reason for the deviation from the ideal way of maximum information. The human being developed with respect to this characteristic. This is in fact that what is usually reached under the phenomenon "evolution," that is, human beings developed in the course of time in the direction of perfect usefulness.

In our case here, the "principle of usefulness" is a requirement of the projection principle. The time-developing process is in this connection a relatively unimportant detail.

Conclusion

In conclusion, the specific features of space and time emerge not only the projection principle but the essential characteristics of evolutionary phenomena as well, particularly the principle of usefulness. In particular, we recognized in this connection that the absolute truth (basic reality) is not comprehensible but it definitely exists.

A lot of scientists are firmly convinced to have the final laws of nature (absolute truth) in their hands soon, and there are prominent physicists among them. But this is obviously only a dream and not reachable when we measure with the yardstick of the projection principle.

5.8.2 Beyond the material level

The principle of usefulness could be considered as absolute feature if there is nothing else in the world outside but merely the material reality. This is in fact the case if the container model of traditional physics is considered. Projection theory offers a more sophisticated view. The principle of usefulness refers here to the material level but there is more than the material world.

Usefulness means within the projection principle that with the tools of evolution only the helpfulness is selected, that is, the useful from the general states of basic reality, which are not accessible. If also the general state would reflect only useful facts, the tools of evolution would not be necessary.

The tools of evolution exist because there are qualitative differences with respect to the information and we may assume that not only useful characteristics are transferred from basic reality to the mind. The useful part for human beings is defined through the five senses at the macroscopic level. But there are other levels (the atomic level, see Figure 4.2). The projection principle opens the door for more sophisticated views.

No doubt, the material part, experienced by human beings in everyday life, is dominant, and this is because the material level developed very early in the course frame of evolution. But it is the most simple level and served at the beginning of the development of human beings in nature, mainly for survival. In the course of time, states of the mind developed more and more, but the material level could not be passed and it is still dominant. We do not know how the states of the mind are influenced by evolution. Nevertheless, through the conscious impact on mind states, thinking with all its facets can be moved more into the foreground.

5.8.3 Where are the tools of evolution positioned?

The members of a species pull only a part of the maximum information from basic reality. It is the raw information α, which is a part of the physically real (material) entity γ in basic reality (Section 4.2). α is the part that is "useful" for the individual in his/her actions in everyday life. The remaining information $\gamma - \alpha$ included in γ is permanently ignored.

The entity α is located in the mind of the observing human being and is that what we called "useful" information. This phenomenon has already been discussed within traditional views. In Section 4.1.4 we cited in this connection Hoimar von Ditfurth; let us repeat the essential point here. He wrote:

> No doubt, the rule "As little outside world as possible", only as much as is absolutely necessary is apparent in evolution. It is valid for all descendants of the primeval cell and therefore for ourselves . . . In principle only those qualities of the outside world are accessible to our perception apparatus, which, in the meantime, we need as living organisms in our stage of development. Also our brain has evolved not as an organ to understand the world but an organ to survive. [73]

It is remarkable that the principle of usefulness is obviously a consequence of specific space–time properties. The τ-selection (Section 5.4.4) reflects this principle.

The stationary limit

The quantity α is a stationary information, which developed in the course of evolution within a certain time interval $\Delta\tau = \tau_b - \tau_a$, where τ_a is the starting point of the evolutionary process and at τ_b the evolutionary development is finished. The selection mechanism remains unchanged for $\tau \geq \tau_b$ and behaves stationary; the information package α refers to the stationary case.

The physically real (material) object γ in basic reality is not a subject of evolution and is a compressed information (Section 5.5), which is a stationary unit as well and is completely given at each clock time τ in complete form, also within the time interval $\Delta\tau$, that is, we have: $\tau_a \leq \tau \leq \tau_b$ (Figure 5.7).

Basic reality
γ
↓ ⟵ Tools of evolution
Mind
α

Figure 5.7: The physically real (material) object γ in basic reality is not a subject of evolution and is a compressed information, which is not dependent on time τ. The quantity α is a stationary information, but it developed evolutionary in the course of time τ and reaches finally the constant information package α. The entity α is located in the mind of the observing human being and is called "useful" information. The mind is observer dependent and is therefore species specific, but the complete information γ not. The transferred information from basic reality to the mind has to be a selected information. This in particular means that the tools of evolution are positioned between basic reality and the human being's mind.

There are evolutionary processes in basic reality but not with respect to evolutionary selection processes. To this kind of evolutionary process belong species formations that we have treated in Section 5.3. Furthermore, there are also no such selection processes in the observer's mind. The mind can only contain physically real information in selected form. Why?

Whereas the complete information about the world is embedded in basic reality, the mind is observer-dependent and is therefore species-specific. Thus, the transferred information from basic reality to the mind has to be a selected information. It must reflect the situation from the point of view of a member of a certain species. In other words, the tools of evolution are positioned between basic reality and the human being's mind. They pull selected information from basic reality and transfer it to the mind. The situation is represented in Figure 5.7.

5.9 Preconditions for observations and descriptions

Once again, human beings observe and act in space and time, that is, only those things are imaginable and directly observable that are located in space and time. In accordance with John Ziman, we may state that physically real entities can only be described mathematically if they are imaginable. After that, a mathematical description is only possible if they are really imaginable and things are imaginable exclusively in space and time.

With John Ziman we have to accentuate that a mathematical description of physically real facts outside space is not possible. If entities are located outside space and time, they cannot be described mathematically but merely qualitatively. To these entities belong basic reality, the mind, the tools of evolution and the processing characteristic η as well; these units are necessary for the observation of material objects and physically real processes as well, and to these entities belong space and time.

5.9.1 Geometrical structures and the physically real effects

Space and time, in particular the elements x, y, z, τ, are not physically real. We cannot put the elements x, y, z, τ on the table and there do not exist measuring devices, which would be able to detect the elements x, y, z, τ, that is, x, y, z, τ are definitely not observable. No doubt, it is important to observe new effects, but it is equally important to recognize what is principally not observable.

It is remarkable that the space–time features, discussed in connection with the projection principle, are also reflected by the ideas of philosopher Immanuel Kant in the nineteenth century. This point has already been underlined in Chapter 1. However, the projection principle reflects basic physics and has primarily nothing to do with philosophy. The basic properties of space and time are the pre-condition for the formulation of realistic physical theories.

We have geometrical structures before us (tress, houses, etc.), and not physically real (material) objects. But human beings feel them when they touch them, for example, with the fingers that are geometrical structures in space and time as well. The geometrical structures are correlated with a characteristic feeling, which reflects a physically real process that however takes place in basic reality, that is, outside space and time [77].

We have the illusion that all these effects take place in space and time in connection with our bodies. In other words, it is throughout believed that there are material objects embedded in space and time. Fact is, however, that our bodies and the other objects are geometrical structures in space and time; their physically real (material) manifestation is located in basic reality. All these facts have been discussed in Ref. [77] and in Section 4.9.

5.9.2 Interactions with space and time?

We have to argue carefully. It is simply no possible to embed physically real (material) objects into the space–time, which is definitely not physically real. The points x, y, z, τ are not observable in isolated form and they can only exist in connection with objects, which must however appear in this case as geometrical structures. The existence of material objects in space and time is only possible if also the elements x, y, z, τ represent physically real entities.

The reason is obvious: A material object can at clock time τ_m only be connected to a certain space–time point, say x_m, y_m, z_m, if the object interacts at τ_m with this space–time point x_m, y_m, z_m. However, such an interaction process presumes that the elements x_m, y_m, z_m, τ_m are physically real, which is however not the case, as we underlined several times.

The conclusion is straightforward: In space and time there are no material objects but merely geometrical structures of them, as picture so to speak. This is actually fulfilled within the frame of the projection principle. The material manifestations of objects are here exclusively outside space and time. As we know, within frame projection theory "outside space and time" means that the material objects are embedded in basic reality, that is, outside space and time.

5.9.3 Relationships

We can only put things together if relationships between these things are definable. In the case of material objects and space–time, there must exist a relationship between objects and space–time. Otherwise, the space–time and the objects "know" nothing of each other. Both entities are side by side, and there is no possibility to liaise them. Such a construction, however, makes not much sense in physics.

As we have accentuated earlier, there can be no relationship between the material objects and the space–time, if the objects are material in character. This is because the space–time is not physically real, and a relationship with the material objects without interaction is not thinkable. In this case, the material object takes at clock time τ a certain position in space and the object itself does not reflect a geometrical structure within the same space having the elements x, y, z.

We have therefore only one possibility to express a relationship between objects and the space–time and this is given by the projection principle: For human beings, any object is exclusively perceivable as picture in space and time. It is the projection from basic reality onto space and time.

In the case of object pictures, the relationships between the projected objects and the space–time are explicitly given, because the objects themselves are expressed (defined) at clock time τ by the elements x, y, z. The picture of an object, and only a picture is perceivable, is defined through space and time, and this is because its individual structure is given at each time τ in terms of the elements x, y, z, τ. In Ref. [77], more details concerning material objects and their relationship to space and time are discussed.

5.9.4 Facts outside space and time

Since human beings exclusively work in space and time, we cannot imagine how an object is formed outside space and time, that is, in basic reality where the physically real (material) objects have their most basic features.

Moreover, a theoretical (mathematical) description outside space and time would not be possible and would in particular not directly be verifiable. There is obviously no basis for that. We can merely develop theoretical conceptions, which are compatible with the space–time structures before our eyes and those that are obtained in connection with measurements in space and time. In basic reality, the objects appear as "raw" information without having the possibility to express their properties in terms of our theoretical laws of physics.

5.9.5 Traditional view

In traditional physics, the physically real (material) world is embedded at each clock time τ in space. The effect of "inertia" demonstrates that there must be at each time τ an interaction between the material objects and the space. This is of course forbidden when we take the space–time properties seriously. Newton's classical mechanics works in many cases excellently, but it is burdened with physically real space–time effects (inertia). This shortage could not be eliminated through the theory of relativity, and this elimination was in fact a requirement by Ernst Mach. Albert Einstein was an advocate of Mach, at least in his early days, but he was not able to fulfill Mach's principle. This famous principle is a yardstick up to the present day. There are many theoretical investigations in this field and are intensively discussed in the scientific literature.

The solution of this serious problem has obviously been based on the following fact: We have not the physically real world before us, but it is a "picture" of the physically real world.

5.9.6 What is outside?

We assess the world on the footing of space and time. The properties of space and time dictate what units outside space and time are necessary for the observation of the physically real (material) world. Their relative functional order is determined through the properties of space and time as well. However, the units outside space and time can only be captured qualitatively because a mathematical formulation is not possible outside space and time. This is in strict accordance with the statement of John Ziman [74].

The entire line given through the item's basic reality, mind and so on define the projection principle. The projection principle is, on the other hand, the basis for the mathematical formulation of physical effects. The deduced theoretical laws can then not be used to describe theoretically (mathematically) basic reality, mind, tools of evolution and the processing characteristics η. For these units, only "qualitative" statements are possible. This complex of units defines the projection principle and, as we remarked, this complex can only be captured in qualitative form and cannot be described by the theoretical-physical equations. The existence of this complex (basic reality, mind, etc.) belongs to the preconditions for the formulation of the basic theoretical laws.

Again, the complex (mind, tools of evolution and the processing characteristics η) is needed for the observation of physically real (material) objects and processes. Only those entities are investigated in physics, which are in particular expressed through our spontaneous observation in daily life, where we have the material world directly before our eyes. These are the well-known unconscious observations at the macroscopic level on the basis of the five senses. As we know, the physically real (material) world itself is embedded in basic reality.

Let us repeat the transition from basic reality to the picture. Three units are intercalated: the tools of evolution, the mind and the processing characteristic η. They are necessary and enable the transition from basic reality to the picture. The situation is summarized in Figure 5.8.

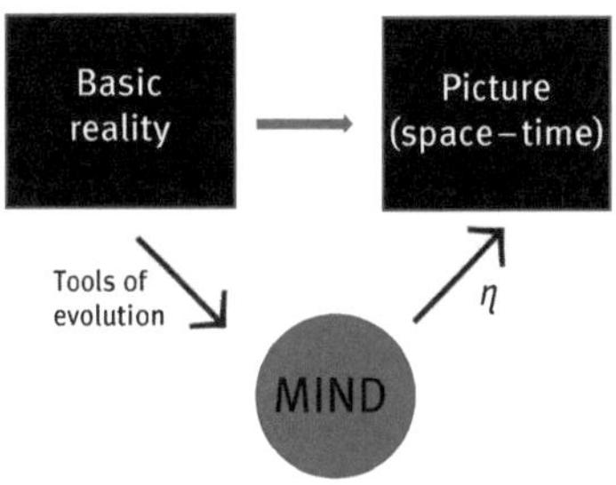

Figure 5.8: Within the frame of the projection principle, the observation takes place in space and time. The physically real (material) objects of basic reality are projected onto space and time and appear as picture before the eyes of human beings. The transition is from basic reality to the picture and three units are intercalated: the tools of evolution, the mind, and the processing characteristic η. This complex of three units are necessary for the observation process, and this complex is located outside space and time and is definitely not describable by the theoretical physical laws. The three units are merely captured through "qualitative" arguments. Human beings perceive the world exclusively in space and time, and also the theoretical physical descriptions of the objects are restricted to space–time formulations.

All that defines the projection principle. This is the precondition for the perception of the material world and its formulation in terms of theoretical physical laws. However, these three units cannot be physically real in character, that is, they cannot be made of material substances. Thus, they are not representable in space and time, but they are used for the representation of material entities in space and time. The three units are needed to describe material entities.

In fact, in the construction of the recognition system, these three units are located outside space and time, and this situation also applies for basic reality and the mind. They are all not describable through the theoretical laws of physics and are merely understandable in terms of "qualitative" arguments. Exclusively physically real objects and processes are describable through the theoretical physical laws. The situation can be summarized as follows: We describe and understand material objects in space and time through units outside space and time.

Because the existence of the complex, consisting of the tools of evolution, the mind and the processing characteristic η, is the "precondition" for the observation and the description of the material world, the complex itself cannot consist of matter. In fact, the existence of the complex is defined outside space and time, where only qualitative statements are possible. The statements by John Ziman (Section 5.7) are seminal here.

The choice of other variables, different from x, y, z, τ, is not possible without reference to the usual space–time variables. In Chapter 6, we will demonstrate that the choice of another set of variables is possible when we base its construction on x, y, z, τ. In any case, human beings are caught in space and time.

5.9.7 Mind-matter interrelations

Within the projection principle, we have the following strict order: Mind and matter are independent of each other. Matter does not create the mind, and the mind does not create matter. However, the mind processes the matter information pulled from basic reality and produces a picture of the physically real (material) objects in space and time.

5.9.8 Raw information and its corresponding theoretical formulation

An object γ in basic reality and its pulled part α are outside space and time and are therefore not describable in terms of theoretical physics. Only the processed form of α through the observer's mind and its representation in space and time are observable, and the observer has this space–time picture (space–time representation) spontaneously and directly before his eyes; only these space–time representations are describable with the theoretical (mathematical) means of physics.

In Chapter 6, we will introduce the wave functions $\Psi(\mathbf{p}, E)$, $\mathbf{p} = (p_x, p_y, p_z)$ and $\Psi(\mathbf{r}, t)$, $\mathbf{r} = (x, y, z)$, where $\mathbf{p}$ is the momentum of an object (system) and E is its energy. The letter t refers to the quantum time, which is not known in traditional quantum theory. The important point is here that the function $\Psi(\mathbf{p}, E)$ corresponds to the raw information α. But $\Psi(\mathbf{p}, E)$ does not describe α; it merely corresponds to it. This statement is based on the following fact: The raw information α is processed (transformed) in the mind leading to a picture in space and time; it is the x, y, z structure at clock time τ that human beings experience in front of them. The picture and the entity α have exactly the same information (e.g., see Section 4.9). The same is the case for $\Psi(\mathbf{p}, E)$ and $\Psi(\mathbf{r}, t)$: $\Psi(\mathbf{r}, t)$ and $\Psi(\mathbf{p}, E)$ contain exactly the same information, and $\Psi(\mathbf{p}, E)$ is projected (transformed) onto space and time leading to $\Psi(\mathbf{r}, t)$, which also describes the picture before the eyes of the observing human being.

5.10 Structures in space and time

5.10.1 Static structures

The spontaneously experienced impression reflects a static structure. This is the reason why these impressions can be seen as orientation and standard, respectively. A certain structure, say σ, remains constant if there is no influence in the sense of "cause and effect." Under this condition, the space structure is frozen and there does not appear spontaneously and unconsciously another structure, say σ', before us with

$$\sigma' \neq \sigma \tag{5.40}$$

For example, a tree remains a tree if there are no external influences on it. This is the basic experience on which everything in our life is based. On the other hand, if a theoretical conception σ' has been developed leading to eq. (5.40), the developed structure σ' has to be assessed as wrong. This is completely in accordance with Ziman's statement given in Section 5.7.

Again, the unconsciously experienced impressions before us in daily life are fundamental. There is in fact no other phenomenon in nature, which could be considered as equivalent and with such a relevance.

It must be underlined once again that the phenomena before us reflect just that from the world outside which comes into existence through the sensory apparatus, and these are defined by the five senses in the case of human beings.

5.10.2 Representations

A material object, for example a tree, is within the projection theory always a geometrical structure in our observations. The tree does not exist without the space–time elements x, y, z, τ. A tree is therefore characterized through

$$\text{tree:}\, x, y, z, \tau \tag{5.41}$$

We cannot separate the tree from space and time, that is, the tree and also all the other objects are caught in space and time like human beings; the observable objects do not exist without the elements x, y, z, τ. The tree and all the other physically real objects are not detachable from space and time. But they exist in basic reality without x, y, z, τ.

In other words, the mind of a human being never creates objects without representation frame:

$$\begin{aligned} &\text{tree:} - \\ &\quad \rightarrow \text{not existent} \end{aligned} \tag{5.42}$$

Moreover, the mind of a human being does not create elements u, v, w, r that are different from x, y, z, τ and we have

$$\begin{aligned} &\text{tree: } u, v, w, r \\ &\qquad \rightarrow \text{not existent} \end{aligned} \tag{5.43}$$

This view is species dependent. Biological systems, which are different from human beings, might have other criteria for the assessment of the world outside. The frames of representation can in particular quite differently organized than in the case of human beings.

5.10.3 An important point: no space–time in the world outside

In daily life, we have a space–time picture of the outside world directly and spontaneously before us that is a picture of basic reality, which is not accessible to human observers. There is a space–time in the mind of human beings, that is, we have an inner picture, but there is no space–time in basic reality (in the world outside). The basic properties of space and time do not allow such a view.

Nevertheless, it is assumed throughout that there is besides an "inner" space–time, also a space–time in the world outside. In this case, we would have also a space–time "outside." Then, we would have two spaces and two times. However, we have to be careful: Nature would not choose such a construction. Why? Because this question is important to answer, let us analyze this point on the basis of fundamental features.

"Two" space–times for "one" observation procedure?

Assume that there are "two" space–time structures for "one" observation procedure, which is naive and probably too simple. In Chapter 1, we cited the well-known psychologist C. G. Jung. Because of the relevance of his view, let us repeat his statement: "When one thinks about what consciousness really is, one is deeply impressed of the wonderful fact that an event that takes place in the cosmos outside, produces an inner picture, that the event also takes place inside" [8]. This comment represents the typical view and reflects the usual conception that human beings have about the whole scenario.

The statement by C. G. Jung suggests just that what we have already pointed out: The image as well as the real world outside are embedded in space and time with exactly the same structure: Outside for the "material" objects, inside for "geometrical" object-positions. These are in both cases space structures at clock time τ.

Is such a solution an elegant way with respect to a scientific yardstick? Here the principles of evolution are essential and come into play. Evolution does not produce twice; the principle of usefulness does not allow to produce "two" identical structures for "one" observation process.

Nevertheless, C. G. Jung's statement with respect to "outside world" and "inside world" is correct, but the assumption that an event takes place twice is not provable.

Only the inner world with its geometrical positions for the objects is directly before our eyes. We merely "believe" the outside world is structured in exactly the same way. The geometrical object positions are only replaced outside through the material objects, but we do not detect this structure outside. The reason is obvious: Human beings are caught in space and time. We do not detect the material world outside directly because there is no space and no time in the world outside. This is a general statement and is in particular fulfilled in the case of the projection principle.

Then, we know why basic reality is not observable: There is no space and no time in basic reality. This fact is compatible with the feature that space and time do not belong to the class of physically real entities. In basic reality, there are exclusively physically real entities.

What C. G. Jung called "inner picture" is identical with our spontaneously created impression before the eyes. Again, all human beings are caught in space and time. We have the inner space–time picture directly before our eyes, but we have no direct access to the world outside, that is, we do not observe the world outside, which is identical with basic reality.

We know that there is a basic reality (world outside) but it remains hidden. From our remarks quoted earlier, we have to conclude that only the inner picture contains space and time, but the world outside is not. In other words, we have no direct access to the world outside because there is no space and time in the world outside.

In summary, we have to state that the reason why we do not observe the world outside is the circumstance that there is no space and no time in the world outside, and human beings can only make statements about physically real entities in connection with space and time. This is the reason why we have only the ability to observe the world outside as inner picture. A human being does not perceive the world outside because there is no space and time in the world outside (basic reality). Thus, we can only recognize the world outside as an inner picture.

All that is compatible with the fact that space and time do not belong to the category of physically real entities (see also Chapter 1), that is, a physically real world outside cannot be embedded within a frame (space and time) that is not physically real.

5.10.4 Interactions and correlations

Basic reality (the world outside) consists of individual objects, which are mutually dependent, that is, there are relationships between the various objects. Within the frame of the projection principle, the relationship is expressed in basic reality through physically real "interaction processes" between the physically real objects. The relationships in the picture are reflected by "correlations" between the geometrical structures (pictures of the objects) in space and time.

There are relationships between the entities in the picture as well as in basic reality. The relationships in the "picture" are given by correlations, but an equivalent form of relationship is thinkable by interaction processes in "basic reality." There are two versions of relationships.

5.10.5 Only the information about the material objects is relevant

After C. G. Jung, we have a picture in space and time, but there is an equivalent space–time structure in the world outside. However, we do not need a x, y, z, τ structure outside when it is already inside. In particular, who watches at clock time τ the x, y, z structure outside?

All that is too naive. We can be sure that nature does work in this way. In particular, we have to be taken into account that space and time are not accessible to empirical tests, that is, they are not observable. There are even no measuring instruments thinkable, which would have the ability to measure (detect) the elements x, y, z, τ.

How can elements be transferred from outside (basic reality) into the inner of the human being when these elements are not physically real like space and time? Because of their features, space and time cannot interact with the physically real world. They are in particular not accessible to the sensory apparatus, which reflects a certain kind of device ensemble. In other words, the inner space–time states do not come from outside but are elements inside the individual. It is intended to learn something about the physically real objects and they have to be observed but not the elements of space and time that are not physically real.

Only the information about the objects are transferred from the world outside to the inside world of the observer. The physically real objects are pictured through space and time. In the world outside there are no observers, that is, representation frames are not needed and no space and no time are necessary in the world outside (basic reality). In the next section, we will deepen the situation in connection with space and time.

5.10.6 Separability

Human beings intuitively believe that the space–time and the objects are separable and that they can exist independently from each other. But this is obviously a fallacy from the point of view of projection theory. The projection principle teaches that such kind is a separability and independency is not possible. This is in fact in accordance with our daily experiences. In nature, no effect is known where the objects appear separately from space and time.

Newton's theory, theory of relativity

Nevertheless, Newton's mechanics is essentially based on separability. The situation is illustrated in Figure 5.9. In Newton's physics, an object is characterized by the mass m. However, the coordinates x, y, z are not dependent on m, and the mass m is not dependent on x, y, z, and this property is valid at each time τ. It is relatively easy to recognize that this kind of separability is not essentially changed within theory of relativity.

Despite this shortage, both theories work well. Allowedly, the daily impressions actually convey the conception of separability. However, we have to be careful because such a principle can have an essential influence on the entire theoretical background.

Projection principle

Within projection theory, the situation is different. Here the objects are geometrical structures and are strictly interwoven with space and time. That is, the objects are given in the form of geometrical structures and these are expressed at each time τ through the space elements x, y, z. In other words, separability is explicitly excluded within the frame of the projection principle. Only this item is in accordance with observation.

Space–time experiences

Human beings have the peculiarity to perceive the material world in space and time.

This peculiarity should not be restricted to the special case of everyday life experiences but has to be considered as a basic property, that is, it is a property, which is not dependent on the specific situation, for example, on everyday life experiences. This property "the perception of the world in space and time" should be seen as independent on every type of experience and it reflects therefore a general feature.

Space and time normally appear in everyday life experiences in an unconscious way. But space and time can also come into existence in a conscious way through thinking. For example, an architect develops a building, represented in space and time through thinking. The architect would be able to develop the details with closed eyes.

5.10.7 The main features

Let us summarize the essential facts in connection with physically real (material) objects and space and time. In particular, from all that follows that basic reality does not contain space and time, and this statement is in accordance to what we have outlined in Section 5.10.3. We may formulate the situation in projection theory as follows:

1. Everything, which is physically real, is represented in space and time, no matter what kind of source is responsible for the appearance of the space–time itself. Observations are exclusively done in space and time, that is, human beings are caught in space and time.
2. Physically real entities, which are not observable in space and time, do not exist. The sensory apparatus has been evolutionarily developed with respect to this property (Section 5.1). The sensory apparatus is given by the five senses and extract the complete information about the environment at the macroscopic level, that is, this is the complete information about the world outside, from the point of view of human beings. This also means that the existence of any physically real space–time phenomena is accessible to human observers.
3. Physically real entities, which are not embedded in space and time, are not observable. Since basic reality cannot be embedded in space and time, it is not observable, that is, basic reality is not accessible to human beings.
4. All that belongs to the statement “human beings are caught in space and time.” Since human beings do not observe basic reality, we have to conclude that basic reality does not contain space and time.
5. Why are human beings caught in space and time? The answer is simple: They are caught in space and time because the necessary information for coping daily life is completely arranged in space and time. Nothing of the physically real (material) world is located outside space and time.

Remark

In accordance with our statement earlier, basic reality with its physically real (material) entities would directly be observable if it would be furnished with a space–time. In this case, we would observe a world without selection processes, and the resulting impression would be different from all the other observation types. However, these basic reality observations do not occur, and we have to conclude that there is in fact no space and no time in basic reality. (We stated earlier that the existence of any physically real space–time phenomena is accessible to human observers.)

Furthermore, such basic reality observations would imply that the absolute truth would be accessible to human beings. This would be against the projection principle and leads in fact to a contradiction when projection theory is the yardstick.

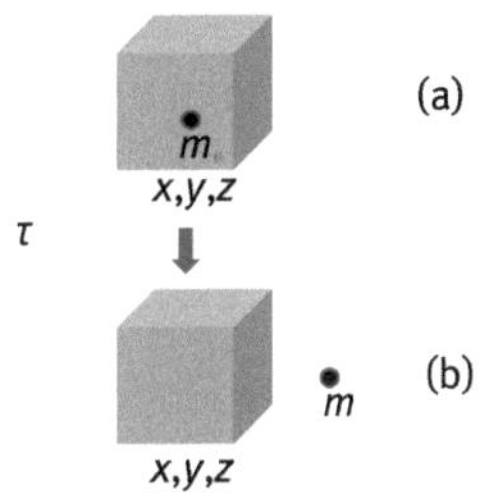

Figure 5.9: Within the frame of the projection principle, space and time and the objects are firmly interwoven and are not separable. This is in accordance with the observations. In fact, we never observe in nature that the objects appear separately from space–time. The peculiarity of the projection principle that the material objects and the space–time are not separable is really observed in nature. No effect is known in nature where the material objects appear separate from space and time. This is a relevant statement when we judge effects within the frame of the projection principle. What about traditional view? In traditional physics, the situation is different. The material objects are located at each time τ in space; the notion "basic reality" is unknown in traditional physics. Also here the objects have the relevant property not to appear without space and time. Nevertheless, it is intuitively believed that the space–time and the objects are separable and that they can exist independently from each other. That is, the step from figure (a) to figure (b) is conceivable in traditional physics at each time τ. However, separability and independency in this sense have never been observed in the world.

5.11 Mechanisms for interaction processes

In traditional physics, the interaction processes take place in x, y, z space and is effective at each clock time τ. But there are serious problems in connection with this conception, and it reveals that the assumption of physically real processes in space and time is a questionable view. We already discussed this point, but let us accentuate this statement by further arguments, just with respect to interaction processes.

5.11.1 Newton's law

Within traditional physics, the material objects are at each time τ embedded in space. Assume that physically real interactions (processes) that take place in space and time lead to the strange situation. Such kind of processes cannot be explained and described.

In the case of Newton's gravitational law $m_1 m_2/r^2$ (r is the distance between two interacting masses m_1 and m_2), we have merely the mathematical formula $m_1 m_2/r^2$, but we do not know where it comes from. As we know, this formula works excellently, but since we do not know the mechanism behind it, we are not able to develop this line further. (The theory of relativity offers a principal new view, but here we have similar problems.)

Why can the situation not be seen as satisfactory? Richard Feynman's comments to this problem are quoted in Refs. [12, 76]. Let us resume the comments together with the statements given in Ref. [12].

Impinging masses

The material world within Newton's physics consists of individual bodies and there is an attractive force between them. As we remarked, Newton introduced the formula $m_1 m_2/r^2$ for the description of this attraction. Within the mechanical corpuscular world view, it was expected that this formula can be explained by an ensemble of other bodied (masses) and we would like to denote these other masses by the letter m_{gi}. That is, within this view one tried to describe the attraction between m_1 and m_2 by means of the masses m_{gi}. This situation can be symbolically expressed by the scheme

$$\begin{array}{c} m_1 \leftrightarrow m_2 \\ \Uparrow \\ \{m_{gi}\} \end{array} \tag{5.44}$$

For example, within this model, the attraction between the Sun and the Earth is explained through impinging particles of mass m_{gi}. But it turned out that such a procedure does not work. The formula $m_1 m_2/r^2$ could not be explained in this way. It was a fallacy!

In this connection, a valuable comment by Richard Feynman is instructive. He wrote [76]:

> Maybe I could invent a better one. Maybe you can, because nobody knows the ultimate. But up to today, from the time of Newton, no one has invented another theoretical description of the mathematical machinery behind this law Which does not either say the same thing over again, or make the mathematics harder, or predict some wrong phenomena. So, there is no model of the theory of gravitation today, other than the mathematical formula.

In other words, we have a formula for Newton's gravity, but we do not know where it comes from.

Is the mechanical corpuscular world view incomplete or has the basic conception of matter embedded in space and time (container model) to be replaced? We already know that the container model of traditional physics represents a limited view, but does it also influence the interaction processes? We will discuss this point later, and the projection principle will also help further in this case.

Let us remark that there are certain plausible interpretations with respect to gravity that are known under the denomination "Action-at-a-Distance" and the "Proximity Effect" (e.g., see Ref. [12]), but these interpretations cannot be used to derive the famous mathematical expression $m_1 m_2/r^2$. We will discuss these interpretations in another context.

Innate attraction

Newton based his physics on the "mechanical corpuscular world view," but this conception could not explain the interaction processes. The conception of an "innate" attraction between every pair of matter bodies was added to the corpuscular view. However, this conception has to be considered as a stranger within the frame of corpuscularism. It is a "dissimilar" conception, dissimilar to the principles of corpuscularism.

Thomas Kuhn gave good comments on this topic. In his opinion, the conception of an innate attraction between material bodies reflects an "occult" quality. Kuhn remarked the following [10]:

> Yet, though much of newton's work was directed to problems and embodied standards derived from the mechanical-corpuscular world view, the effect of the paradigm that resulted from his work was a further and partially destructive change of the problems and standards legitimate for science. Gravity, interpreted as an innate attraction between every pair of particles of matter, was an occult quality in the same sense as the scholastics tendency to fall.

In other words, the interaction process becomes an occult quality when we measure with the yardstick of the mechanical corpuscular world view. In fact, only those elements have to be used for the description of the interaction that are given through the basic formulation of the theory itself. Innate properties of matter do not belong to this class of elements and can be considered as occult in character, as Thomas Kuhn did. Such elements appear to be artificial.

Instead of the impinging particle model [see scheme (5.44)], which definitely does not work, the introduction of "innate" attractions between every pair of matter bodies leads to scheme (5.45):

$$\begin{gathered} m_1 \leftrightarrow m_2 \\ \Uparrow \\ \left\{ \begin{matrix} \text{innate} \\ \text{property} \\ \text{of } m_1 \text{ and } m_2 \end{matrix} \right\} \end{gathered} \tag{5.45}$$

The transition from eq. (5.44) to eq. (5.55) is a drastic step and the term "innate property of matter" reflects nothing else than an "occult" item, as correctly remarked by Thomas Kuhn [10].

Newton himself was not convinced of this solution [expressed here in eq. (5.45)]. Concerning this point, Thomas Kuhn remarked:

> Therefore, while the standards of corpuscularism remained in effect, the research for a mechanical explanation of gravity was one of the most challenging problems for those who accepted the Principia as paradigm. Newton devoted much attention to it and so did many of his eighteenth-century successors. The only apparent option was to reject Newton's theory for its

failure to explain gravity., and that alternative, too, was widely accepted. Yet neither of these views ultimately triumphed. Unable to practise science without the Principia or to make that work conform to the corpuscular standards of the seventeenth century, scientists gradually accepted the view of gravity was indeed innate [10].

Critical remarks

To explain satisfactorily, the origin of Newton's gravity is a principal and important task. The final view seems here to be given by the conception of an innate property of matter. No doubt, we can go this way, but any other view, arbitrarily chosen, would be possible, without changing the law $m_1 m_2/r^2$.

Even Newton declined to declare the phenomenon of gravitation as an innate property of masses; he declined to deviate from the corpuscular view [represented in scheme (5.45)], as is indicated in the comments by Thomas Kuhn outlined in Ref. [10].

The explanation of gravity through innate properties of matter is not only a dissimilar conception with respect to corpuscularism, but it does not allow to deduce Newton's mathematical law $m_1 m_2/r^2$. On the other hand, the mechanical corpuscular world view through impinging masses does not work, and there are hardly other particle constructions conceivable. Therefore, we have to think about other possibilities.

Alternative way

As an alternative for the explanation of gravity through innate properties of matter, we have to ask whether the underlying "conception" of traditional physics is realistic enough.

This conception can be summarized as follows: "Physically real (material) objects are at each clock time τ embedded in x, y, z space." Is this conception really applicable in the case of physically real interaction processes? This is hardly possible especially the projection principle offers a more realistic conception, just in connection with interaction processes. We will give the basic construction in the next section.

5.11.2 Momentum and energy chunks

The projection principle is obviously more suitable to explain interaction processes. We have here not to recourse to occult conceptions. In particular, the formulation of interaction processes can be done more generally than in the case of the traditional view.

In the analytical treatment within the frame of the projection principle (Chapter 6), the physically real processes take place in (p_x, p_y, p_z, E) space and the information is projected onto (x, y, z, t) space, that is, there are no masses, no momentums and no energies located in (x, y, z, t) space but only geometrical structures. Here the (p_x, p_y, p_z, E) space is a fictitious realty and substitutes the selected part τ of basic reality γ.

The interactions themselves are given through quantum fluctuations $\Delta p_x, \Delta p_y, \Delta p_z, \Delta E$ of the entities p_x, p_y, p_z, E. All thinkable fluctuation models are possible, which fulfil the conservation laws for momentum and energy. Moreover, within the frame of this description a variable for a system-specific time, which we denoted by the letter t, has to be introduced as well (see also Section 1.7), as in the qualitative discussion in connection with the units α and γ . The projection principle requires that explicitly.

The projection effect and also the existence of the system-specific time change the situation with respect to the understanding of interaction processes basically. The impinging particle mechanism (5.44) appears to be a naive conception and is not useful for the description of Newton's gravitational law. But to explain the interaction as innate property of matter is also not adequate and the notion "interaction" becomes an occult character, as Thomas Kuhn remarked [see scheme (5.45)].

What are the principles? How are interaction processes formulated within the projection principle? The details are given in Chapter 6, but let us already indicate, in this section, the basic peculiarities.

Two objects, having the variables $\mathbf{p}_1, E_1$ and $\mathbf{p}_2, E_2$ in $(\mathbf{p}, E)$ space, interact with each other. Interaction here means that they exchange chunks of momentum and energy. $\Delta \mathrm{p}_i$ and ΔE_i, that is, instead of the impinging masses m_{gi} of traditional physics, we have within projection theory quantum fluctuations $\Delta \mathbf{p}_i$ and ΔE_i, which are incessantly exchanged between the two objects:

$$\begin{gathered} p_1, E_1 \leftrightarrow p_2, E_2 \\ \Uparrow \\ \{\Delta \mathbf{p}_i, \Delta E_i\} \end{gathered} \tag{5.46}$$

In $(\mathbf{r}, t)$ space we have exclusively geometrical positions, which is the result of the projection from $(\mathbf{p}, E)$ space. The various geometrical positions in $(\mathbf{r}, t)$ space are correlated through the physical real processes in $(\mathbf{p}, E)$ space that are projected onto $(\mathbf{r}, t)$ space.

The momentum and energy chunks $\Delta \mathbf{p}_i$ and ΔE_i (the fluctuations) are not dissimilar to the basic elements of the theory, which are given in $(\mathbf{p}, E)$ space by the variables $\mathbf{p}$ and E.

5.11.3 Comparisons

Let us briefly compare the features of the impinging mass model, the innate property description and the interaction by quantum fluctuations of projection theory.

Impinging mass model

As indicated in scheme (5.44), the theory only uses entities on which the mechanical corpuscular system is based, that is, here only masses appear: m_1, m_2 and m_{gi}. That is, we do not leave the mechanical corpuscular world view, and no dissimilar elements appear. But, unfortunately, the model does not work. The model is more or less fixed on “one” kind of mechanism. There are obviously no other mass configurations possible that are essentially different from the model of impinging masses.

Innate property description

The interaction processes through the introduction of innate properties between matter represents an ad hoc invention. It is a dissimilar conception with respect to the mechanical corpuscular view.

Quantum fluctuations

The quantum fluctuations $\Delta\mathbf{p}_i$ and ΔE_i are not dissimilar to the basic elements of the theory, which are given in $(\mathbf{p}, E)$ space by the values $\mathbf{p}_1, E_1$ and $\mathbf{p}_2, E_2$. We do not leave the basic conception when we introduce the quantities $\Delta\mathbf{p}_i$ and ΔE_i. We are here not fixed on “one” kind of mechanism, as in the case of the impinging mass system, but each mechanism is possible for which the conservation of momentum and energy.

Further features

In traditional physics, the physically real interaction processes take place at each time τ in x, y, z space, whereas in projection theory the physically real interaction processes are at each time τ outside x, y, z, t space and take place within $(\mathbf{p}, E)$ space.

5.11.4 Observability and nonobservability

Concerning interaction processes in traditional physics, there is Newton’s gravitational law $m_1 m_2/r^2$ and there is also Coulomb’s law qQ/r^2 . Details are given, for example, in Ref. [12]. Both formulas have been applied with success and are still actual formulations. We have two formulas, but we do not know where they come from. They work excellently, but we cannot assume that these laws reflect final solutions. However, we are not able to develop them further, because we do not know the mechanism behind these physical laws.

Let us give some further comments in connection with the gravitational law $m_1 m_2/r^2$. This law has been interpreted by means of some physical conceptions. As we already remarked in Section 5.11.1, here two interpretations are of particular relevance: “action at a distance” and the so-called proximity effect. The meaning of these interpretations is given in Ref. [12]. Here we only want to remark one essential point.

It is of principal interest to note that the "action at the distance" and the "proximity effect" are merely interpretations of the gravitational law to m_1m_2/r^2. However, Newton's force law m_1m_2/r^2 cannot be derived on the basis of these interpretations.

The "action at the distance" and the "proximity effect" are qualitatively different from each other, but they are nevertheless exactly equivalent conceptions. It is not possible to make a choice between these two conceptions, since there is no experimental way to distinguish between them, as they both have the same consequences. The implements used in the formulation of the interpretation conceptions cannot be observed.

Then, the following question arise: Are these conceptions scientific at all? This is a very principal question with respect the world view. What is the mind-set dictated by traditional physics and what is the mind-set within the projection principle? With respect to the "action at the distance" and the "proximity effect," the world view of traditional physics is concerned, and let us discuss this (usual) view first.

Mind-set in traditional physics

In traditional physics, the standard view is expressed through the container model, that is, matter is at each clock time τ embedded in space. It is throughout supposed that the "complete" material world is given as space structure. This must have consequences for the scientific assessment of the phenomena around us. Just the believe that the complete material world is located in space and time, is incisive: Only those entities are considered to be existing for which actually observation methods exist or which are thinkable. If that is principally not possible, the concerning entities are judged to be metaphysical in character, that is, they are considered to be nonscientific.

Measured with this yardstick, the conceptions "action at a distance" and the "proximity effect" have to be considered as metaphysical in character. Both notions have therefore to be classified as nonscientific. This is no problem with respect to the law m_1m_2/r^2 because this formula remains conserved in its form and has not to be discarded.

Mind-set in projection theory

The situation is different in projection theory. The notion "observation" gets through the projection phenomenon new aspects. Here two points are essential and should be the initial facts in the analysis:

1. That what is located in space and time is not the "complete" world.
2. Human beings are caught in space and time.

We have to analyze the situation with respect to these peculiarities. Here human beings have not the possibility to observe the complete number of units and entities, respectively. These units and entities are embedded as physically real quantities in basic reality and appear in selected form in space and time.

Because in projection theory only a part of the material world is accessible to human beings, this restriction of observation possibilities has nothing to do with metaphysical effects in the above sense, but this kind of nonobservability is due to the restricted possibility for human beings to observe the material world. Once again, note that only a part of basic reality can be captured. There are things outside space and time to which human beings have no access because they are caught in space and time.

Conclusion

The term "existence and nonobservability" in traditional physics means "metaphysical," that is, nonscientific. In projection theory, the situation is different, and "existence and nonobservability" means here "scientific" even when the existence is outside space and time.

In particular, the mind and basic reality are located outside space and time. They are definitely existent but not observable. However, they do not belong to the class of metaphysical elements.

The brain is partly material in character and, in fact, we know a space–time picture of this part. But we also know that the brain is more than that. In projection theory, we have a clear assignment between all these entities. All that has been analyzed in the earlier sections.

Remark

There is no way to apply the notions "action at a distance" and the "proximity effect" within the frame of the projection principle. These notions are constructions within traditional physics.

5.12 Nanoscience and projection phenomena

Newton's law of force m_1m_2/r^2 is at each clock time τ expressible through the potential energy $V(r) = V(x, y, z)$. The function $V(x, y, z)$ describes in traditional physics the physically real interaction processes in x, y, z space. Also, the projection principle works with a similar function but is extended through the system-specific time t and we get instead of $V(x, y, z)$ the term $V(x, y, z, t)$, which is also valid at each clock time τ. In other words, in the transition from traditional physics (tp) to the projection principle (pp), we have

$$V_{\text{tp}}(x, y, z) \to V_{\text{pp}}(x, y, z, t) \tag{5.47}$$

There is a fundamental difference between $V_{\text{tp}}(x, y, z)$ and $V_{\text{pp}}(x, y, z, t)$: Whereas $V_{\text{tp}}(x, y, z)$ describes "physically real processes" in x, y, z space, the function $V_{\text{pp}}(x, y, z, t)$ describes "correlations" between the (geometrical) objects in $V_{\text{pp}}(x, y, z, t)$ space [$(\mathbf{r}, t)$space].

5.12.1 Outside of space and time

Once again, the physically real processes in projection theory take place in $(\mathbf{p}, E)$ space (outside $(\mathbf{r}, t)$ space), and instead of $V_{\text{pp}}(x, y, z, t)$ we get in $(\mathbf{p}, E)$ space the momentum–energy operator

$$V_{\text{pp}}\left(i\hbar\frac{\partial}{\partial p_x}, i\hbar\frac{\partial}{\partial p_y}, i\hbar\frac{\partial}{\partial p_z}, -i\hbar\frac{\partial}{\partial E}\right) \tag{5.48}$$

where

$$i\hbar\frac{\partial}{\partial p_x},\quad i\hbar\frac{\partial}{\partial p_y},\quad i\hbar\frac{\partial}{\partial p_z} \tag{5.49}$$

are the operators for the space coordinates in $(\mathbf{p}, E)$ space. The term

$$-i\hbar\frac{\partial}{\partial E} \tag{5.50}$$

is the operator for the time coordinate also in $(\mathbf{p}, E)$ space (see also Chapter 1). The operator rules can be derived easily in projection theory [12]. More details are given in Chapter 6.

5.12.2 Considerations with respect to nanoscience

In nanoscience, we deal with atoms and molecules, which consist of charged elementary particles (electrons, etc.). Here gravity does not play a role, but the Coulomb interaction is relevant. The description of many-particle systems, in particular in the nanorealm, interactions between electrons and their screening are relevant and influence the material properties critically. In particular, the atomic (molecular) structure and dynamics are very much influenced by the interaction for which the Coulomb interaction is the starting point.

We have the following situation: The force between the two charges q and Q is proportional to qQ/r^2 where r is the distance between the two charges. The term qQ/r^2 looks like the corresponding law m_1m_2/r^2 for Newton's gravity and, in fact, the same problems appear in connection with qQ/r^2: Also here we do not know where it comes from. It works excellently, but we cannot also in this case assume that the law qQ/r^2 reflects the final solution. We are not able to develop this law further, because we do not know the mechanism behind the formula qQ/r^2.

All that is important for nanoscience because the entities of nanosystems are described at the ultimate level by atoms and molecules for which in traditional physics Coulomb's force law qQ/r^2 is basic. On the other hand, reliable force laws are necessary since nanoproperties are very sensitive to small variation in the force laws (interaction potentials).

As in the case of gravity, Coulomb's force law qQ/r^2 is at each clock time τ expressible through the potential energy $V_{\mathrm{Coul}}(r) = V_{\mathrm{Coul}}(x,y,z)$. The function $V_{\mathrm{Coul}}(x,y,z)$ describes in traditional physics the physically real interaction processes between two charges in x,y,z space. In projection theory, the term $V_{\mathrm{Coul}}(x,y,z)$ has to be extended through the system-specific time t and we get instead of $V_{\mathrm{Coul}}(x,y,z)$ the term $V_{\mathrm{nano}}(x,y,z,t)$, when we concentrate ourselves on the nanorealm. In analogy to eq. (5.47), we get

$$V_{\mathrm{Coul}}(x,y,z) \rightarrow V_{\mathrm{nano}}(x,y,z,t) \tag{5.51}$$

Whereas $V_{\mathrm{Coul}}(x,y,z)$ describes "physically real processes" in x,y,z space, the function $V_{\mathrm{nano}}(x,y,z,t)$ describes "correlations" between the (geometrical) charges in x,y,z,t space [$(\mathbf{r},t)$space].

Also in the case of nanosystems, the physically real processes in projection theory take place in $(\mathbf{p},E)$ space and, as in the case of eq. (5.48), we get in $(\mathbf{p},E)$ space the momentum–energy operator

$$V_{\mathrm{nano}}\left(i\hbar\frac{\partial}{\partial p_x}, i\hbar\frac{\partial}{\partial p_y}, i\hbar\frac{\partial}{\partial p_z}, -i\hbar\frac{\partial}{\partial E}\right) \tag{5.52}$$

Drastic changes in the theoretical description can be expected when we work within the frame of the projection principle.

A realistic description of nanophenomena is necessary because changes at the nanolevel, which is the ultimate level in the description of material properties, will influence our daily life essentially. Just in connection with food and medicine developments, where it is even planned to change brain functions, realistic descriptions are necessary. For all that we need reliable world views and theoretical conceptions. We must know what we do!

5.12.3 Further statements

The container model of traditional physics and the projection principle are views that are essentially different from each other. The container model reflects a hard approximation, which is obviously compensated through other approximations. But this cannot be the final way for solving scientific problems. The projection principle is more realistic, and this is because this physical view treats space and time more realistically.

No doubt, in the nanorealm we need reliable theoretical conceptions for the treatment of the various phenomena at this level, and this is because nanotechnological changes can be very far-reaching and can in particular modify our world drastically and sustainably. For example, through nanotechnology we have in principle the possibility to intervene in the biological evolution and this is dangerous if such intervention is not done with care.

5.12.4 Intervention through evolution

Evolutionary processes take place at a level where nanoscience has its ultimate level; we know that at this level the properties of usual matter and biological systems emerge. This is fundamental. Evolutionary processes are nevertheless large-scale developments in nature, where large space–time regions are concerned. When we change a local, but characteristic structure in the nanorealm with nanotechnological means, we not only influence the local situation, but the global large-scale environment as well.

The entire scenario is coupled at the ultimate level, at least the proximate environment. For example, if we change a biological system targeted at the nanolevel, which is however still under evolutionary development, we possibly change its evolutionary course simultaneously through this nanotechnological intervention. In other words, the nanotechnological intervention is fundamental and we cannot exclude that it gets out of control. It is therefore necessary to simulate the intervention "before" the changes are done. For this purpose, reliable theoretical conceptions have to be on hand. We should not forget that such kind of changes are interventions in the evolution itself.

5.12.5 Raw space–time impressions

The material world outside is projected onto space and time. There are geometrical positions described at each clock time τ by the variables x, y, z, t. Clearly, no physically real entities are projected onto space and time. Projection theory is basically a quantum treatment. Thus, the geometrical positions are occupied statistically [12].

This "unconsciously" mind-created space and time refer to a "raw" space–time information and exclusively consists of extensions and changes of extensions [75]. But we usually need numbers for space and time, that is, we need numbers for the variables x, y, z, t. For this purpose, the human being puts a fictitious net over the extensions and the extension changes. Clearly, the mind does not create this fictitious net of space coordinates and time in an unconscious way. The net itself is produced intellectually through the human being.

In addition, human beings have a certain space–time feeling. This space–time feeling is produced unconsciously through the mind without fictitious net of numbers.

5.12.6 Procedure

Let us consider a nanosystem and the corresponding observation procedure. The method itself is of course also applicable to the macroscopic level as, for example, for the observation of a tree.

The wave function of the nanosystem (or tree), formally denoted by $\Psi_{\text{nano}}(x, y, z, t)$, can be calculated on the basis of the system characteristic function $V_{\text{nano}}(x, y, z, t)$ using the fundamental laws of projection theory (Chapter 6). For the representation of $\Psi_{\text{nano}}(x, y, z, t)$ and $V_{\text{nano}}(x, y, z, t)$, we need the variables x, y, z, t as input. How are the elements x, y, z, t determined and how are they applied in practical investigations? Here, three steps are essential:

1. The basic information about the nanosystem is pulled from basic reality and is transferred to the observer's mind. The mind processes this information, restricted through evolutionary processes, and creates a space–time picture before the observer's eyes. The mind produces the characteristics in the space and also those of the time. Space and time appear here in their most basic form, that is, as "extensions" (space) and as "changes of the extensions" (time). All that takes place in an "unconscious" way.
2. The human observer puts a net over the "raw space–time impression" [5] in order to get "numbers" for space and time. This process happens "consciously," that is, in an intellectual way. These are numbers for the elements x, y, z, t, which are primarily defined through the fictitious net.
3. The human being uses exactly these numbers for the calculation of the wave function $\Psi_{\text{nano}}(x, y, z, t)$. Then, he/she can compare the model with the corresponding observation. The parameters of the measuring devices are adapted with respect to the physical characteristics of the nanosystem described by $\Psi_{\text{nano}}(x, y, z, t)$. Clearly, also the functions $\Psi_{\text{nano}}(x, y, z, t)$ and $V_{\text{nano}}(x, y, z, t)$ are units of the mind and have been developed in the human observer's mind.

Human beings are also able to create the raw information about space and time consciously, that is, in an intellectual way within the frame of thinking. It is in the first instance a conscious impression without the fictitious net of numbers.

The space–time picture before the observer's eyes consists basically of statistical events [12]. We cannot dissolve them within the frame of our everyday life experiences, but each macroscopic system, and not only the very small atomic units, is composed of a large number of statistical events.

In this connection, it has to be underlined that the statistical events exist without the x, y, z, t numbers, but not $V_{\text{nano}}(x, y, z, t)$ and $\Psi_{\text{nano}}(x, y, z, t)$. This is a big difference.

5.13 Perception of object-specific properties

For the description of object-specific properties, the system-specific time is of particular relevance. The existence of a system-specific time is a requirement of physics. But this kind of time is missing in traditional physics, in particular with respect to usual quantum theory. In traditional physics, we have only one type of time and this is the external clock time τ.

The projection principle makes the existence of a system-specific time possible that we denoted above by the letter t. From the direct line (Figure 5.10), starting with basic reality, we can only conclude on the "qualitative" existence of a system-specific time t. However, in the mathematical treatment, the second line (based on the variables p_x, p_y, p_z, E), also the "quantitative" aspect of the system-specific time t becomes possible. In this section, we give some principal comments on this topic, just as preliminary to the next chapter so to speak.

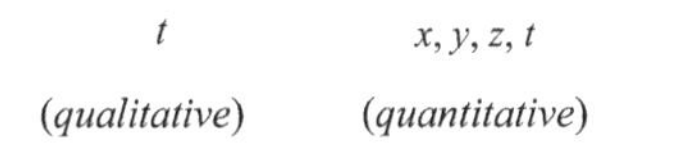

Figure 5.10: In the perception of the system time t, two lines are possible. The "qualitative" perception of t has its origin in the information of a physically real object in basic reality. The object itself is denoted by γ and in reduced form by α. Human beings are caught in space and time and there can be no variables for γ and α. "Caught in space and time" also means that we cannot invent variables without applying space and time. These variables are the momentuns p_x, p_y, p_z and the energy E. In the qualitative description, the letter t reflects a notation and not a variable.

5.13.1 Statistical events

We know from the point of view of the projection principle that there are no material entities in space and time. Here we have exclusively geometrical structures. There are events in space and time, which are characterized by the elements x, y, z, τ. At each clock time τ, the object has a definite geometrical position x_τ, y_τ, z_τ; the object has no mass, no energy and no momentum in space having the coordinates x, y, z.

The object itself is, however, not completely defined by "one" event, but it is described by a statistical ensemble of events, which is expressed by quantum theoretical laws. A specific object is characterized by specific boundary conditions that are used in the application of the quantum laws. The events form a statistical entity, and there is one event at each clock time τ, that is, the events successively occupy space in the course of time, leading to a space–time structure for the event ensemble, which must be in accordance with the quantum laws, and these quantum laws must be in accordance with the projection principle.

5.13.2 Projection theory points on a new type of time

It is important to accentuate that the projection principle points on an essential feature: The description of an object (system) with the coordinates x, y, z is obviously not sufficient for the theoretical description of physically real (material) objects. (Note that the time τ is not a variable for the characterization of the object itself, and this is because τ is an external parameter.) As we already remarked in Section 5.11.2, within projection theory a new type of time becomes necessary. It is the system-specific time t that is needed for the "complete" description of physically real entities.

It must be distinctly underlined that the reality observer principle requires the introduction of the time variable t. This system-specific time t is just the quantum time, which is missing in traditional quantum theory (Chapter 1), that is, instead of x, y, z we have in projection theory at each clock time τ the set x, y, z, t:

$$x, y, z \rightarrow x, y, z, t \tag{5.53}$$

Why does projection theory require the system-specific time t? The answer is straightforward and can be summarized by the arguments given in the following. The details together with a discussion will be given in Chapter 6.

5.13.3 Complete and incomplete facts

Every physically real (material) object is located in basic reality, but it is observed in space and time. They are not observable in basic reality, and this would in fact be superfluous if they are observable in space and time.

Let us consider, for example, one object, say γ. Only the "useful" part α of γ is accessible to human beings. α is pulled by the tools of evolution from basic reality and is located in the observer's mind where α is processed and leads to a picture in space and time of the material object. That is, the information α is given at clock time τ as x, y, z structure.

The picture is what the human being observes, and nothing else. Human beings exclusively work in space and time, but there is no space and no time in basic reality. The observations exclusively take place in space and time. Any other statement in connection with physically real (material) objects is restricted to space and to time as well.

It must be underlined that y and α are given at each time τ in compressed form, i.e., at each time τ the complete information of the object is existent. On the other hand, the observed x, y, z–structures vary in the course of time τ, i.e., they are different from each other. This particularly means that the information at τ cannot be complete and cannot be identified with the compressed information, but can only be a part of it. For such a situation only one solution is possible. Let us briefly discuss this point. We already mentioned this point in Section 5.4.

5.13.4 New Variable for the Time

Some principal statements

We consider again a physically real object, which exists in basic reality and we get its species-dependent form α through the application of the tools of evolution. The quantity α is a raw information in the mind of the human observer and exists here without space and time. Space and time come into play by the processing of α in the mind.

In this way, the human observer gives the raw information *alpha* a „face“, and this face comes exclusively into existence through space and time. The object α is at clock time τ expressed as x, y, z structure. This is the face. α without space and time is merely a raw information and we know nothing more about it. We cannot specify α without space and time.

Let us assume that the envisaged object interacts with its environment in basic reality, and let us assume that this interaction does not change. Then, the object behaves stationary. What does it mean? Since the time is at this stage not defined yet, we cannot couple the object to the clock time τ, that is, α is not expressible in terms of τ and is therefore τ independent. Or we may state that the raw information α is a "constant" for all times τ, i.e., α behaves stationary with respect to time τ. This is in fact compatible with our assumption that the interaction between the object and the environment does not change, that is, α remains a constant in the course of clock time τ.

Then, we may conclude the following: Since α is a constant with respect to τ, it reflects the "complete“ facts of the object, and this complete information of the object is given at each time τ and has therefore to be considered as "compressed" information. This term has already been introduced in Section 5.4.2. It characterizes the fact that the complete information of the physical real object (or even the entire material world) is given all at once at each time τ.

This situation describes the "existence" of the physical real object. But what does a human being observe? He observes at clock time τ a certain space structure of the physically real object, which may be for example a tree. This space structure is expressed in the macroscopic case (everyday life experiences) as three-dimensional x, y, z structure and appears at clock time τ directly before the eyes of the human observer.

Our experiences in daily life, which are the most direct and reliable impressions we can have about the world outside, teach us that these x, y, z structures vary in the course of time τ. That is, this perceived information is "not" a constant with respect to clock time τ. In other words, there seems to be a contradiction between the compressed information α, which "exists" at each clock time τ, and that what we actually "observe" in the course time τ in the form of x, y, z structures. However, there is no contradiction when we work consequently within the frame of the projection principle. Let us briefly analyze the situation.

Analysis

Again, the raw information α behaves stationary and is a constant with respect to clock time τ. The stationary situation is expressed through the following statements:

$$\alpha_1 \to \tau_1,\ \alpha_2 \to \tau_2,\ \alpha_3 \to \tau_3, \cdots$$

with

$$\tau_1 < \tau_2 < \tau_3 < \cdots$$

and

$$\alpha_1 = \alpha_2 = \alpha_3 = \cdots = \alpha.$$

In other words, we have at each time τ exactly the same entity α; it is the information about the object in compressed form.

No doubt, for the complete information α an analytical (mathematical) expression may exist and let us denote this law by $g(a_1, a_2, \cdots)$ with the variables $a_1, a_2, \cdots$, but it is not expressible for human beings without the elements of space and time. This mathematical function $g(a_1, a_2, \cdots)$ is however not dependent on τ and is, like α, exactly the same for all values τ.

Application of the reality-picture conception

Within projection theory, the reality-picture principle is valid and this principle has to be applied on the raw information α. In other words, we want to represent α as picture in space and time. How comes this picture into existence? This question is important, and we have already discussed this point in Chapter 4: The space-time picture is created by the observer's mind. α is pulled from basic reality and is transferred to the mind, where it is processed and the space-time picture (the x, y, z structure at clock time τ) is created, which we feel to be in front of us (Chapter 4).

However, we do not observe the complete information α at time τ, but we observe an x, y, z structure at each time τ, which vary with respect to τ and the x, y, z structures are in general different from each other. That is, the information, observed in the course of time τ, is not the same. In other words, what we actually observe is not α, it is not the complete information in compressed form. What is it what we observe? It can only be a part of the complete information α. This point is essential and must be explained in more detail.

In principle, the information can be considered to be complete if more than one x, y, z structure could be observed simultaneously. Each of these structures, simultaneously observed, would correspond to various clock times $\tau, \tau', \cdots$. Then, there would be a superposition of space structures and a superposition of clock times $\tau, \tau', \cdots$. This situation is however against the experience that all human beings equally have. We observe "one" x, y, z structure at a definite clock time τ. Our clocks do not indicate various time values simultaneously. There is only one clock hand, and we do not observe (perceive) various x, y, z structures simultaneously. These are the facts.

For example, a clock does not show the times τ and τ' simultaneously together with the space structures $(x, y, z\ )$ and $(x, y, z\ )'$ with $(x, y, z\ ) \neq (x, y, z\ )'$, which do also not exist simultaneously and are not overlapped. Such a scenario has never been observed and has therefore to be excluded.

Thus, the problem remains: At clock time τ we "observe" a certain x, y, z structure, which varies in the course of time τ. On the other hand, the complete information of the object, which corresponds to α, is "existent" at each time τ.

This situation seems to reflect a contradiction, but this is not the case. The problem can be solved in a conclusive way. The weak point is here the external time τ. The information α reflects the physically real object. The observed picture of the object, i.e., its space-time structure, is only a representation in restricted form: It is a space structure (x, y, z structure) at time τ, and this x, y, z structure develops in the course of time time τ. This picture has to be considered as incomplete because the clock time τ does not belong to the physically real object; τ is, as we know, an external parameter and is for the characterization of all things in the physically real world.

The external time, displayed by our clocks, reflects a rather naive point of view. The standpoint by Julius Fraser underlines that. He gave an instructive comment in his book "TIME, the familiar Strange", which we already quoted in Chapter 1. Let us repeat the essential idea:

> *"The time is like the cable of the cable railway of San Francisco. The cable is driven by a far and invisible machine, is however hidden. We know that it moves, because the carriages are connected and are carried with it. Completely similarly we usually see the time in everyday life as a universal, cosmic movement of the present, of the now; they are propelled by natural or divine forces and matter, life, humans and society are connected with them and are propelled and along-moved for a while."* [11]

The San Francisco cable railway is of course a metaphor but it exactly reflects the situation how human beings experience the phenomenon "time".

This view is in particular applied in traditional physics (Newton's mechanics, usual quantum theory). Also our physically real object, its x, y, z structure, which we observe at clock time τ, is within the frame of Fraser's view driven by a far and invisible time-machine, which produces the time τ. We obviously use for the representation of the object-picture an incomplete space-time picture because we merely express the object through an x, y, z structure; the time τ characterize the behavior of the object but it does not characterize the material object itself, and this is because τ is an external parameter, which behaves in analogy to the cable railway of San Francisco. That is, the parameter τ does not belong to the physically real object. This is in fact an unsatisfactory situation. However, the projection principle dictates the solution of this problem.

The characterization of the object through an x, y, z structure without time structure has to be considered as an incomplete information. That is, the observed facts, not constant with respect time τ, are due an incomplete space-time picture. This is in fact the key for the solution of our problem analyzed above.

The following step is essential: The raw information α is constant with respect to time τ and is complete. Within the frame of the projection principle, α has to be projected onto space and time. In this connection it is "necessary" to introduce an "object-specific time", which we would like to denote again by the letter t. Then, the picture of the physically real object is characterized by the four variables x, y, z, t and this picture is, like α, complete. In this way, the missing time structure, an effect that we discussed above, is realized. There is no other possibility to solve the problem, and it is a general formulation.

Again, the extension of the scenario through the object-specific time t is necessary, and there is only one way to interpret the situation: This four-dimensional x, y, z, t–picture (block) is equally defined at each clock time τ. Then, the complete raw information α has its complete counterpart in the form of a projection onto space and time leading to the picture of the object. This picture reflects the complete information α if the projection mechanism transfers the complete facts of α. Clearly, in this case the complete object-picture is in fact a constant and is independent on τ (as α itself).

The solution of the problem

Using the system-specific time t we may state the following: We define the physically real object within the time range $t_a \leq t \leq t_b$ where t_a and t_b define the life-time $\Delta t = t_b - t_a$ of the object. Then, the object is not observable for $\tau > t_b$ and $\tau < t_a$. Human beings do not observe the complete facts given by the x, y, z, t–structure for $t_a \leq t \leq t_b$, but the incomplete part $x, y, z, t = \tau$ at clock time τ.

In conclusion, the complete $x, y, z, t-$structure with $t_a \leq t \leq t_b$ (corresponding to α) is "existent" at τ, but only the part $x, y, z, t = \tau$ is "observed" (perceived) at the same time τ. This situation defines a "selection process": The structure $x, y, z, t = \tau$ is selected from the complete $x, y, z, t-$structure with $t_a \leq t \leq t_b$.

Within this scenario, the clock time τ is not created by an obscure external time-machine, but it is created through the human being's mind. The time τ reflects a mind-state. In this way, we obtain a consistent conception and all entities and features of projection theory represent a unified whole.

There is not "one" system existing, singular in character, which produces that what we call time. But time has to be based on the physically real systems (objects) themselves. In other words, not "one" singular system (a hidden time-machine, which works in analogy to the San Francisco cable railway) produces the time, but "each" physically real system (object) does it. This situation defines the term "system-specific time" and it dictates the conception for the solution of the contradiction underlined above: The fact that the "constant" information α is not observed but x, y, z structures, which are "not constant".

A large number of physically real systems (objects) are in the universe, but none of them is preferred and singular, respectively. Each of them must therefore have its own time structure, which is system-specific. This peculiarity is compatible with the individual time feeling of human beings.

Classification

We observe at clock time τ the space structure $g_i(x, y, z)$, where $g_i(x, y, z)$ reflects the material world outside before our eyes. The structure $g_i(x, y, z)$ can be for example a single material object. Whereas $g_i(x, y, z)$ varies in the course of time τ, the quantity α, which is constant with respect to time τ, is the complete information about the object (world outside). The space-structure $g_i(x, y, z)$ cannot be identified with α. That is, we have

$$\begin{aligned} \tau_1 &\rightarrow g_1(x, y, z) \neq \alpha \\ \tau_2 &\rightarrow g_2(x, y, z) \neq \alpha \\ &\vdots \\ \tau_i &\rightarrow g_i(x, y, z) \neq \alpha \\ &\vdots \end{aligned}$$

with

$$g_1(x, y, z) \neq g_2(x, y, z) \neq \cdots \neq g_i(x, y, z) \neq \cdots$$

The facts with respect to α are projected onto space and time, and it is necessary to introduce a system-specific time, which is in the case of a single physically real object "object-specific". We get for the picture an $x, y, z, t-$structure, which exists within the time region $t_a \leq t \leq t_b$ and has to be identified with the complete information α. We have

$$g(x,y,z,t) \triangleq \alpha$$

However, we do not observe the $x, y, z, t-$structure with $t_a \leq t \leq t_b$ all at once but piecewise, i.e., only "one" structure for one t of the complete information $x, y, z, t-$structure with $t_a \leq t \leq t_b$ is observable for human beings. Only the part $x, y, z, t = \tau$ of the complete space-time structure can be „observed" at the time τ, and this part varies with respect to time τ as the observations show. Just this behavior is described by the $x, y, z, t-$structure. The $x, y, z, t = \tau-$structure is selected from the complete $x, y, z, t-$structure with $t_a \leq t \leq t_b$. We have

$$g(x,y,z,t) \xrightarrow{selection} g(x,y,z,t=\tau_1)$$

$$g(x,y,z,t) \xrightarrow{selection} g(x,y,z,t=\tau_2)$$

$$\vdots$$

$$g(x,y,z,t) \xrightarrow{selection} g(x,y,z,t=\tau_i)$$

$$\vdots$$

The various structures $g(x, y, z, t = \tau_i)$ are then identical with observed structures $g_1(x, y, z)$:

$$g(x,y,z,t=\tau_1) = g_1(x,y,z)$$

$$g(x,y,z,t=\tau_2) = g_2(x,y,z)$$

$$\vdots$$

$$g(x,y,z,t=\tau_i) = g_i(x,y,z)$$

$$\vdots$$

The human observer, his mind, scans on the basis of the time τ, which is used as clock time, the $x, y, z, t = \tau-$structure between t_a and t_b and he gets the complete information α after he has observed the object (world outside) within the time interval $\Delta\tau = \Delta t = t_b - t_a$.

It is a piecewise observation of the complete information α, although α is existent at each clock time τ. Why does nature goes this way? The answer is relatively simple and is given by the principles of evolution, in particular the "principle of usefulness" come here into play: It is less burdened for a human being to treat the environment piecewise and not as a compressed block. Such a procedure is more useful for daily life and survival, just in the early phase of evolution (Section 5.4.4).

Since human beings cannot avoid the effects of evolution, they can only perceive a part at time τ of the complete information. From the complete object information $x, y, z, t-$structure with $t_a \leq t \leq t_b$ a human being can only observe the part $x, y, z, t = \tau$ at time τ. Human beings are principally not be able to recognize the "absolute truth". It can only that perceived what the evolution dictates; in this connection not only the $\tau-$selection is effective (see in particular Chapter 4). Human beings are embossed through the effects of evolution.

In this section, we have made qualitative statements. In Chapter 6, we will give a quantitative treatment, and we will recognize that the system-specific time t has its origin at the quantum level and, therefore, t reflects a quantum-aspect of time. Just this type of time is missing in traditional quantum theory.

Essential items

A human being observes one x, y, z structure at time τ, and this x, y, z structure varies in the course of time τ. That is, the x, y, z structure at τ can only be a part of the complete information α. We know however that the complete information α exists at each time τ, but the human being does not observe it at time τ. This is not a contradiction, and the solution is straightforward and basic. Here the system-specific time t is necessary and we cannot resign on it.

The complete information α of the physically real object under investigation is projected onto space and time having the variables x, y, z, t. In contrast to the clock time τ, the letter t refers to the system-specific time since the x, y, z, t structure refers to the physically real object. Instead of α we get a picture of α. Since α is complete, the picture must be complete too. The object is assumed to be existent between the initial time t_a and the final time t_b. Then, the complete x, y, z, t structure, which reflects a static block, is expressed through "all" x, y, z structures in the time interval $t_a \leq t \leq t_b$. However, we observe only one x, y, z structure at time τ, which is identical with the $x, y, z, t = \tau$ structure. The observation of the object starts at $\tau_a = t_a$ and is finished at $\tau_b = t_b$. For $\tau < \tau_a$ and $\tau > \tau_b$ the physically real object cannot be observed.

The time τ goes monotonically from $\tau = t_a$ to $\tau = t_b$. The $t-$spectrum defined through $\tau_a = t_a \leq t \leq \tau_b = t_b$, which is characteristic for the material object, is scanned systematically, that is, the states of the object are not given all at once at time τ, but the states are selected. The τ selection have to be judged as a phenomenon of evolution.

5.13.5 Existence and observation

Let us assume again that the material object exists between the initial time τ_a and the final time τ_b. The object exists for all times τ with $\tau_a \leq \tau \leq \tau_b$, i.e., in the time interval $\Delta\tau = \tau_b - \tau_a \neq 0$. As we have outlined in Section 5.13.4, the facts concerning the object are constant within $\Delta\tau$. This constant information in $\Delta\tau$ defines the information in compressed form and the object "exists" at each time τ in complete form. In other words, the object itself is given at each time τ exactly by the same facts. This situation concerns the "existence" of the object.

But what is "observed" with respect to the object at time τ? At time τ not the complete information is "observable", but only a part of the object is "observable" This is not a contradiction and is explained in Section 5.13.4.

The compressed form (the complete information concerning α) can only be given in terms of a new time-scale with a completely new type of time. We denoted this new type of time by the letter t. There is no other way for the explanation of the facts.

Then, the compressed form of information, i.e., its complete contents, is represented as a function of the four variables x, y, z, t and this complete information is given at each time τ as constant block x, y, z, t with $t_a \leq t \leq t_b$, and this block has to be identified with α. This is however not quite exact. As we will recognize in Chapter 6, this statement is only true for $\tau \pm \varepsilon$, where ε can however be an infinitesimal small time-interval; ε can be arbitrarily small but must be different from zero. Because the quantity ε can in principle be arbitrarily small, τ and $\tau \pm \varepsilon$ are practically undistinguishable: $\tau \pm \varepsilon \cong \tau$ with $\varepsilon \cong 0$.

5.13.6 The Construction for Time-Individuals

One Object

Let us consider one material object, which is observed by a human being. We assign to each clock-time τ one value t of the system-specific time:

$$t = \tau \tag{5.54}$$

The object exists in the time-interval $\Delta\tau = \tau_b - \tau_a$, i.e., the variable t is defined just in this interval and we get

$$\tau_a \leq t \leq \tau_b \tag{5.55}$$

Then, we have with Eq. (5.54)

$$t_a \leq t \leq t_b \tag{5.56}$$

The τ–scale is identical with the life-time of the observer and is not restricted to τ_a and τ_b, which reflect object-properties, i.e. in general we have

$$\begin{aligned} \tau &< t_a \\ \tau &> t_b \end{aligned} \tag{5.57}$$

At each clock time τ, we have a space-time structure given by the variables $x_\tau, y_\tau, z_\tau, t = \tau$ with

$$\tau : x_\tau, y_\tau, z_\tau, t = \tau \tag{5.58}$$

We suppose that the space-positions x_τ, y_τ, z_τ have at each clock-time τ a lower limit $x_{a\tau}, y_{a\tau}, z_{a\tau}$ and an upper limit $x_{b\tau}, y_{b\tau}, z_{b\tau}$:

$$\begin{aligned} x_{a\tau} &\le x_\tau \le x_{b\tau} \\ y_{a\tau} &\le y_\tau \le y_{b\tau} \\ z_{a\tau} &\le z_\tau \le z_{b\tau} \end{aligned} \tag{5.59}$$

with the intervals

$$\begin{aligned} &\Delta x_\tau = x_{b\tau} - x_{a\tau} \\ &\Delta y_\tau = y_{b\tau} - y_{a\tau} \\ &\Delta z_\tau = z_{b\tau} - z_{a\tau} \\ &t = \tau \\ &\tau_a \le t \le \tau_b \end{aligned} \tag{5.60}$$

The complete information with respect to the object under investigation, which is given at each time τ in basic reality in compressed form, is expressed in space and time through the information in the intervals expressed by (5.60).

Then, with Eq. (5.58) the complete information in space and time is given through

$$\begin{aligned} &x_\tau, y_\tau, z_\tau, t = \tau \\ &\tau = \tau_a, \cdots, \tau_b \end{aligned} \tag{5.61}$$

The reference time τ (clock time) moves monotonocally and continuesly from $\tau = \tau_a$ to $\tau = \tau_b$. Equation (5.61) represents the complete information, which is contained in α in compressed form, but is distributed in space and time within the intervals (5.60).

The time t is, as we know, a system-specific time and belongs to the object under investigation. The time-interval Δt is the life-time of the object that will be scanned through the (external) clock time τ. The time τ goes monotonically from $\tau = t_a$ to $\tau = t_b$.

Information Contents

A human being experiences at clock time τ a certain $x, y, z-$ structure. However, the space-structure in terms of the elements $x, y, z, t = \tau$ is only a part of the whole.

The whole reflects the $x, y, z-$ structure of the complete $t-$scale, which is defined through the interval $\tau_a \leq t \leq \tau_b$ and represents the compressed form of the raw information α.

The entire (compressed) information concerning α, which is distributed over the interval $\Delta\tau = \tau_b - \tau_a$, is "existent" at each time τ, but only a part of it is "observed" at time τ.

Nature is organized in this way, where evolutionary processes play a relevant role. In fact, all that is a selection process and has to be seen as a specific process within the frame of biological evolution. It does not reflect an information-reduction but the information is merely elongated.

Remark

The system-specific time t is not observed at clock-time τ. A human being only observes directly $x, y, z-$structures, which they experience spontaneously at each time τ. The relation $t = \tau$ [see Eq. (5.54)] is an assignment and does not mean that we observe t at τ. The space-structure is observed at τ and not t; the existence of the "system-specific" time t is the logical consequence due to the existence of the "compressed" information at each time τ, which follows, on the other hand, from the projection principle. The system-specific time t reflects here a qualitative fact.

Extension to *N* Objects

The Eqs. (5.54) - (5.61) reflect the situation for one material object. It is easy to extend the expressions for the case of N material objects. Instead of (5.54) we obtain

$$\begin{aligned} t_1 &= \tau \\ t_2 &= \tau \\ &\vdots \\ t_N &= \tau \end{aligned} \tag{5.62}$$

if $t_i \neq 0,\ i = 1, 2, \cdots, N$, at clock time τ.

Let us also here assume that the ith material object exists between the initial time τ_{ai} and the final time τ_{bi}, i.e., it exists within the time interval $\Delta\tau_i = \tau_{bi} - \tau_{ai} \neq 0$. All the time intervals $\Delta\tau_i$ are assumed to be within the observation-time interval $\Delta\tau$.

The N objects exist in the time-intervals $\Delta\tau_i = \tau_{bi} - \tau_{ai},\ i = 1, 2, \cdots, N$, i.e., the values t_i are defined just in these intervals and we get

$$\tau_{ai} \leq t_i \leq \tau_{bi},\ i = 1, 2, \cdots, N \tag{5.63}$$

Then, we have with Eq. (5.62)

$$t_{ai} \leq t_i \leq t_{bi},\ i = 1, 2, \cdots, N \tag{5.64}$$

The τ scale is identical with the life-time of the observer and is not restricted to $\tau_{ai}, \tau_{bi},\ i = 1, 2, \cdots, N$, which reflect object-properties, i.e. in general we have

$$\begin{aligned} &\tau < t_{ai}, i = 1, 2, \cdots, N \\ &\tau > t_{bi}, i = 1, 2, \cdots, N \end{aligned} \tag{5.65}$$

In analogy to (5.58), we have N space-time structures given at each clock time τ by the block

$$\begin{aligned} &x_{1\tau}, y_{1\tau}, z_{1\tau}, t_1 = \tau \\ &x_{2\tau}, y_{2\tau}, z_{2\tau}, t_2 = \tau \\ &\vdots \\ &x_{N\tau}, y_{N\tau}, z_{N\tau}, t_N = \tau \end{aligned} \tag{5.66}$$

There are also here lower and upper limits for the space-time positions of all N objects at each clock time τ. We have

$$\begin{aligned} &x_{1a\tau}, y_{1a\tau}, z_{1a\tau}, t_{1a\tau}, x_{1b\tau}, y_{1b\tau}, z_{1b\tau}, t_{1b\tau} \\ &x_{2a\tau}, y_{2a\tau}, z_{2a\tau}, t_{2a\tau}, x_{2b\tau}, y_{2b\tau}, z_{2b\tau}, t_{2b\tau} \\ &\vdots \\ &x_{Na\tau}, y_{Na\tau}, z_{Na\tau}, t_{Na\tau}, x_{Nb\tau}, y_{Nb\tau}, z_{Nb\tau}, t_{Nb\tau} \end{aligned} \tag{5.67}$$

Then, we get in analogy to (5.59) the following conditions:

$$\begin{aligned} &x_{1a\tau} \le x_{1\tau} \le x_{1b\tau}, y_{1a\tau} \le y_{1\tau} \le y_{1b\tau}, z_{1a\tau} \le z_{1\tau} \le z_{1b\tau} \\ &x_{2a\tau} \le x_{2\tau} \le x_{2b\tau}, y_{2a\tau} \le y_{2\tau} \le y_{2b\tau}, z_{2a\tau} \le z_{2\tau} \le z_{2b\tau} \\ &\vdots \\ &x_{Na\tau} \le x_{N\tau} \le x_{Nb\tau}, y_{Na\tau} \le y_{N\tau} \le y_{Nb\tau}, z_{Na\tau} \le z_{N\tau} \le z_{Nb\tau} \end{aligned} \tag{5.68}$$

with the intervals

$$\Delta x_{i\tau}, \Delta y_{i\tau}, \Delta z_{i\tau}, \quad i = 1, 2, \cdots, N \tag{5.69}$$

which are expressed by

$$\begin{aligned} \Delta x_{i\tau} &= x_{ib\tau} - x_{ia\tau} \\ \Delta y_{i\tau} &= y_{ib\tau} - y_{ia\tau} \\ \Delta z_{i\tau} &= z_{ib\tau} - z_{ia\tau} \end{aligned} \tag{5.70}$$

The complete information with respect to the N objects, which is given at each time τ in basic reality in compressed form, is expressed in space and time through the information in the intervals expressed by (5.70).

Then, with Eq. (5.66) the complete information in space and time is given for the N material objects through

$$\begin{aligned} &x_{1\tau}, y_{1\tau}, z_{1\tau}, t_1 = \tau \\ &x_{2\tau}, y_{2\tau}, z_{2\tau}, t_2 = \tau \\ &\vdots \\ &x_{N\tau}, y_{N\tau}, z_{N\tau}, t_N = \tau \\ &\tau = \tau_{ai}, \cdots, \tau_{bi}, \; i = 1, 2, \cdots, N \end{aligned} \tag{5.71}$$

The reference time τ (clock time) moves monotonocally and continuously from $\tau = \tau_a$ to $\tau = \tau_b$. Equation (5.71) represents the complete information, which is contained in the raw information α_i, $i = 1, 2, \cdots, N$ for the N material objects in compressed form, but is distributed in space and time within the intervals (5.60).

5.13.7 Constructions for Space-Individuals

Time-Indivituals (human beings) perceive the compressed object-information as $x, y, z-$structure at time τ. But there is at least a second possibility to perceive exactly the same compressed object-information. In this second case not the $x, y, z-$structure at time τ is observed but the $t-$structure at a certain space-position λ. Since the space has three dimensions, also λ must be given as three-dimensional element having the components $\lambda_x, \lambda_y, \lambda_z$:

$$\lambda = (\lambda_x, \lambda_y, \lambda_z) \tag{5.72}$$

In other words, the compressed object-information is scanned with respect to λ and not with respect to the clock time τ. In this case, the observer is in his biological composition different from human beings. Let us call the type of observer "space-individual". Space-individuals arrange the world space-specific and time-individuals arrange the world time-specific. However, both observer-types experience the same object and have finally exactly the same space-time information about the material object.

One Object

Let us consider also for this case one material object, which is observed by a space-individual (different from a human being). We assign to each position λ one system-specific space-position, described by the three components x, y, z. In analogy to (5.54) we obtain

$$x = \lambda_x, y = \lambda_y, z = \lambda_z \tag{5.73}$$

Let us assume that the material object exists between the initial position $\lambda_a = (\lambda_{xa}, \lambda_{ya}, \lambda_{za})$ and the final position $\lambda_b = (\lambda_{xb}, \lambda_{yb}, \lambda_{zb})$. The object exists for all positions λ with $\lambda_a \leq \lambda \leq \lambda_b$, within the time interval $\Delta t = t_b - t_a$.

The object exists in the following position-intervals

$$\begin{aligned} \Delta\lambda_x &= \lambda_{xb} - \lambda_{xa} \\ \Delta\lambda_y &= \lambda_{yb} - \lambda_{ya} \\ \Delta\lambda_z &= \lambda_{zb} - \lambda_{za} \end{aligned} \tag{5.74}$$

That is, the position x, y, z, described by (5.73), is defined just in these intervals and we get

$$\begin{aligned} \lambda_{xa} \leq x \leq \lambda_{xb} \\ \lambda_{ya} \leq y \leq \lambda_{yb} \\ \lambda_{za} \leq z \leq \lambda_{zb} \end{aligned} \tag{5.75}$$

Then, we get with (5.73)

$$\begin{aligned} x_a \leq x \leq x_b \\ y_a \leq y \leq y_b \\ z_a \leq z \leq z_b \end{aligned} \tag{5.76}$$

The $\lambda-$scale is identical with the life-position (corresponding to the life-time of time-individuals) of the space-individual and is not restricted to $\lambda_a = (\lambda_{xa}, \lambda_{ya}, \lambda_{za})$ and $\lambda_b = (\lambda_{xb}, \lambda_{yb}, \lambda_{zb})$, which reflect object-properties, i.e. in general we have

$$\begin{aligned} \lambda_x < x_a, \lambda_y < y_a, \lambda_z < z_a \\ \lambda_x > x_b, \lambda_y > y_b, \lambda_z > z_b \end{aligned} \tag{5.77}$$

At each space-position λ, we have a space-time structure given by the space position $x_\lambda, y_\lambda, z_\lambda$ and "all" t-values, which we want to summerize by the letter t_λ:

$$\begin{aligned} & x_\lambda = \lambda_x, y_\lambda = \lambda_y, z_\lambda = \lambda_z \\ & t_\lambda \end{aligned} \tag{5.78}$$

For each space-point λ the complete t-spectrum is selected.

We suppose that the space-positions $x_\lambda, y_\lambda, z_\lambda$ have a lower limit x_a, y_a, z_a and an upper limit x_b, y_b, z_b:

$$\begin{aligned} x_a &\le x_\lambda \le x_b \\ y_a &\le y_\lambda \le y_b \\ z_a &\le z_\lambda \le z_b \end{aligned} \tag{5.79}$$

with the space-time intervals

$$\begin{aligned} \Delta x &= x_b - x_a \\ \Delta y &= y_b - y_a \\ \Delta z &= z_b - z_a \end{aligned} \tag{5.80}$$

The complete information with respect to the object under investigation, which is given in basic reality in compressed form, is expressed at each space-position λ through the information in the intervals expressed by (5.80).

Then, with Eq. (5.78) the complete information in space and time is given through

$$\begin{aligned} & x_\lambda, y_\lambda, z_\lambda, t_\lambda \\ & \lambda = \lambda_a, \cdots, \lambda_b \end{aligned} \tag{5.81}$$

where x_λ, y_λ, z_λ represents again one space-position and t_λ the entire t-spectrum.

As in the case of the reference time τ, it is assumed that the reference position λ moves monotonocally and continuesly from $\lambda = \lambda_a$ to $\lambda = \lambda_b$. Equation (5.81) represents the complete information, which is contained in α in compressed form, but is distributed in space and time within the intervals (5.80).

The positions x, y, z are, as we know, system-specific and belong to the object under investigation, observed by the biological individual of the other kind. The space-intervals Δx, Δy, Δz are the life positions (in analogy to the lifetime of time-individuals) of the object that will be scanned through the (external) position λ with $\lambda_x, \lambda_y, \lambda_z$. The position λ goes monotonically from $\lambda = x_a, y_a, z_a$ to $\lambda = x_b, y_b, z_b$.

Information Contents

A space-individual experiences at reference position $\lambda = (\lambda_x.\lambda_y.\lambda_z)$ a certain $t-$structure. However, the time-structure in terms of the elements $x = \lambda_x, y = \lambda_y, z = \lambda_z, t$ is only a part of the whole. The whole reflects the t structures of all x, y, z values on the λ scale, which are defined through the interval $\lambda_a \le x, y, z \le \lambda_b$. This information represents the compressed form of the raw information α.

The entire (compressed) information concerning α, which is distributed over the interval $\Delta\lambda = \lambda_b - \lambda_a$, is "existent" at each reference position λ, but only a part of it is "observed" at position λ.

Nature is organized in this way, where evolutionary processes play a relevant role. In fact, all that is a selection process and has to be seen as a specific process within the frame of biological evolution. It does not reflect an information-reduction but the information is merely elongated.

Remark

The system-specific position x, y, z is not observed at the reference position $\lambda = (\lambda_x, \lambda_y, \lambda_z)$. The biological individuals of the other kind only observe directly t structures, which they experience spontaneously at each position λ. The relations $x = \lambda_x, y = \lambda_y, z = \lambda_z$ are assignments and do not mean that the individual observes x, y, z at $\lambda_x, \lambda_y, \lambda_z$. The time-structure is observed at λ and not x, y, z; the existence of the "system-specific" position x, y, z is the logical consequence due to the existence of the "compressed" information at each position λ, which follows, on the other hand, from the projection principle. As in the case of the system-specific time t, the system-specific positions x, y, z reflect a qualitative fact.

Extension to *N* Objects

The Eqs. (5.73) - (5.81) reflect the situation for one material object. It is easy to extend the expressions for the case of N material objects. Instead of (5.73) we obtain

$$
\begin{aligned}
x_i &= \lambda_x, i = 1, 2, \cdots, N \\
y_i &= \lambda_y, i = 1, 2, \cdots, N \\
z_i &= \lambda_z, i = 1, 2, \cdots, N
\end{aligned}
\tag{5.82}
$$

if $(x_i, y_i, z_i) \neq 0,\ i = 1, 2, \cdots, N$, at space position λ.

Let us also here assume that the ith material object exists between the initial position λ_{ai} and the final position λ_{bi}, i.e., the object exists within the position interval $\Delta\lambda_i = \lambda_{bi} - \lambda_{ai} \neq 0$.

The N objects exist in the position-intervals expressed by eq. (5.75), i.e., the values x_i, y_i, z_i are defined just in these intervals and we get

$$
\begin{aligned}
\lambda_{xa} < x_i \le \lambda_{xb},\ i = 1, 2, \cdots, N \\
\lambda_{ya} \le y_i \le \lambda_{yb},\ i = 1, 2, \cdots, N \\
\lambda_{za} \le z_i \le \lambda_{zb},\ i = 1, 2, \cdots, N
\end{aligned}
\tag{5.83}
$$

Then, we have with Eq. (5.82)

$$
\begin{aligned}
x_{ai} \le x_i \le x_{bi},\ i = 1, 2, \cdots, N \\
y_{ai} \le x_i \le y_{bi},\ i = 1, 2, \cdots, N \\
z_{ai} \le z_i \le z_{bi},\ i = 1, 2, \cdots, N
\end{aligned}
\tag{5.84}
$$

The λ–scale is identical with the life-position of the observer and is not restricted to $\lambda_{ai}, \lambda_{bi},\ i = 1, 2, \cdots, N$, which reflect object-properties, i.e. in general we have

$$\begin{aligned} &\lambda_x < x_{ai}, \lambda_y < y_{ai}, \lambda_z < z_{ai}, i = 1, 2, \cdots, N \\ &\lambda_x > x_{bi}, \lambda_y > y_{bi}, \lambda_z > z_{bi}, i = 1, 2, \cdots, N \end{aligned} \tag{5.85}$$

In analogy to (5.58), we have N space-time structures given at each reference position λ by the block

$$\begin{aligned} &x_{1\lambda}, y_{1\lambda}, z_{1\lambda}, t_{1\lambda} \\ &x_{2\lambda}, y_{2\lambda}, z_{2\lambda}, t_{2\lambda} \\ &\vdots \\ &x_{N\lambda}, y_{N\lambda}, z_{N\lambda}, t_{N\lambda} \end{aligned} \tag{5.86}$$

Also here there lower and upper limits for the time positions of all N objects and we get in analogy to (5.68) and (5.69) corresponding conditions for the λ scale.

The complete information with respect to the N objects, which is given at each reference position λ in basic reality in compressed form, is expressed in space and time through the information in the intervals expressed by (5.70).

Then, with Eq. (5.86) the complete information in space and time is given for the N material objects through

$$\begin{aligned} &x_{1\lambda}, y_{1\lambda}, z_{1\lambda}, t_{1\lambda} \\ &x_{2\lambda}, y_{2\lambda}, z_{2\lambda}, t_{2\lambda} \\ &\vdots \\ &x_{N\lambda}, y_{N\lambda}, z_{N\lambda}, t_{N\lambda} \\ &\lambda = \lambda_{ai}, \cdots, \lambda_{bi},\ i = 1, 2, \cdots, N \end{aligned} \tag{5.87}$$

The values t_λ reflect again the entire t-spectrum of the ith object at position $x_{i\lambda}$, $y_{i\lambda}$, $z_{i\lambda}$,. The space individual perceives at a certain space-position the past, present and future, all together. This is unimagible for time individuals, that is, for human beings. The reference position λ moves, in analogy to the reference time τ, monotonically and continuously from $\lambda = \lambda_a$ to $\lambda = \lambda_b$. Equation (5.87) represents the complete information, which is contained in the raw information $\alpha_i,\ i = 1, 2, \cdots, N$ for the N material objects in compressed form, but is distributed in space and time within the intervals (5.70).

5.13.8 Conclusion

All the statements in connection with time individuals and space individuals can already be concluded from the very basics of projection theory, and the principles are summarized in Figure 5.11. These are exclusively qualitative statements. Qualitative statements are the background for theoretical frames, which allow quantitative descriptions. If, on the other hand, only qualitative statements are possible, the statements are fundamental.

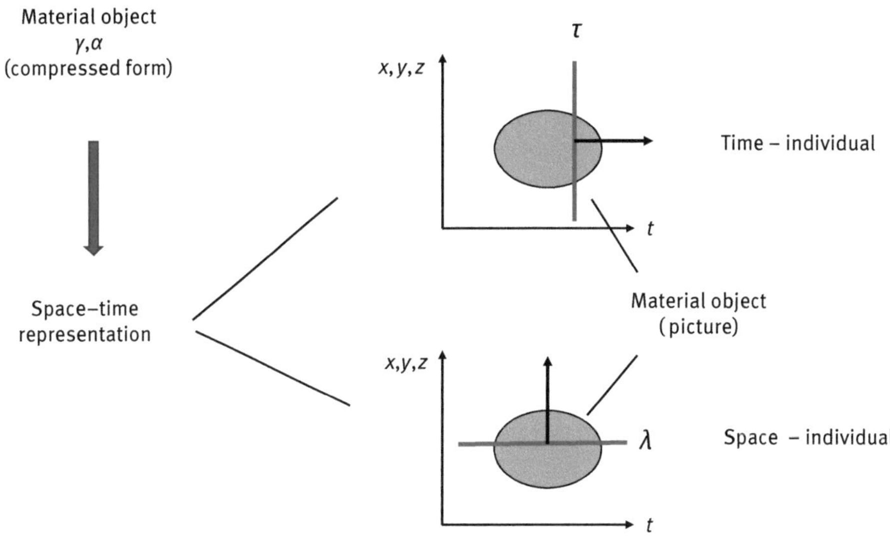

Figure 5.11: Observation of a material object. The object (γ, α) is given in compressed form. Also the picture in space and time is given in compressed form. However, the compressed form is not directly observed but, as we know, exclusively the projected picture in space and time. The picture with the complete object information is scanned, and there are two possibilities for such a scanning process. Here we have to distinguish between time individuals (human beings) and space individuals; how a space individual is biologically structured is not known yet. In the case of time individuals, it is scanned with respect to time and the reference time τ is relevant (vertical line in the figure). In the case of space individuals, it is scanned with respect to space and the reference position λ is relevant (horizontal line in the figure). For time individuals we have: At each clock time τ, the observer experiences a space–time structure given by the variables $x_\tau, y_\tau, z_\tau, t_\tau (t_\tau = \tau)$. For space individuals, the following is: At each reference position λ, we have a space–time structure given by the variables $x_\lambda = \lambda_x, y_\lambda = \lambda_y, z_\lambda = \lambda_z, t_\lambda$.

Concerning time individuals and space individuals, the following has to be mentioned:

Time individuals
Time individuals select the information from basic reality with respect to time τ. In this case, we have a reference time τ. Time individuals experience the world (environment) spontaneously through space structures, which appear successively before them and are expressed through what we call “time.” Human beings belong to the class of time individuals.

Space individuals
Space individuals select with respect to position λ. In this case, we have a reference position λ. At each position the time structure of the object under investigation is different. Space individuals experience the objects in terms of time structures at each selected space position λ.

Time individuals and space individuals experience the same objects in terms of space structures (time individuals) or in terms of time structures (space individuals).

5.13.9 Nanolevel and life functions

We are familiar with time individuals because human beings belong to this class of observers. The space individuals are, however, unimaginable for human beings, but a lot of things are unimaginable for us. For example, the turkey in the experiment by Wolfgang Schleidt responded in a way, which was inscrutable to a human being, that is, it was not comprehensible at all.

The turkey showed a behavior in the experiment, which was completely different from what a human being would do in similar situations. The turkey even killed its own chick, although the environmental situation was changed unessentially. Here, the environment on the earth is concerned. We know, however, that the environments in the cosmos vary strongly and we may assume that individual life is organized differently from what we know in connection with the conditions on the earth. However, we do not know if life is possible at all in the case of extreme situations in the cosmos.

What does the projection principle say? The projection principle does not say something about the details of object manifestations, but it reveals the possibility for various individual life forms. In this connection, it turned out within our analysis that the manifestation of physically real (material) entities is dependent on the observer type. The conditions for space individuality and time individuality can directly be derived within the frames of the projection principle. There are further possibilities for life variations, but we do not want to deepen this point in the book. The entire discussion shows that we can vary the form of individuality if changes with respect to the biological structure are done.

In nanoscience, we work at the ultimate level, where “individuality” emerges. “Ultimate” means that the impact is basic. Nanoscience effectuates therefore a fundamental impact on essential life functions; in particular, brain functions are concerned. For this purpose, we need a detailed knowledge about the physical background, which reflects this field realistically. The projection principle is obviously a suitable frame for such kind of activities.

The possible existence of time individuals and space individuals shows in what direction we think. We have τ individuals (time individual) and we discussed λ individuals (space individuals). In this connection, mixed states, consisting on a superposition of both individual states, become principally possible. Let us denote individuals in mixed states by τ, λ individual.

5.13.10 Interrelations between the basic units

Space and time are influenceable

It is planned to apply the tools of nanoscience in medicine. Here also the brain functions will be in the focus. But we do definitely know what the brain is, in particular its interrelations to the other entities of the human being. In other words, we have to know what the brain is and how it is connected to the mind states and the body functions.

Three units are essential here and these are given by the reality itself, the mind and the brain of the human being. The projection principle goes a step further and declares space and time as the most relevant elements in the assessment of real situations in the world outside (environment). In fact, space and time are created through the mind and are principally influenceable through the mind.

The “interrelations” between reality, mind, brain and the space–time are relevant. These interrelations are obviously more important than the details concerning the functional operations of these modules themselves. This should be the way: first, we have to know the interrelations between the relevant modules forming the perception system. On this basis, we can estimate what functional meaning they have.

Nanotechnological changes

For this purpose, we need a conception for the interrelations between reality, mind, brain and the space–time. The projection principle allows to develop such a conception. In this connection it is important to accentuate that the brain and the mind are coupled (Section 4.9.1) and how this coupling is interlocked with the entire scenario consisting of reality, mind, brain and space and time.

Through the changing of brain functions, which is nanotechnologically planned, new mind states could in principle be created and/or influenced. On the basis of a serious scientific conception, and the projection principle is a serious conception, we can particularly estimate what such interventions effectuate. In other words, we can possibly say something about the impact of such nanotechnological changes on the human being under investigation and on his/her individual behavior.

Misjudgments, as in the case of the experiment with a turkey and its chick (Chapter 1), have to be avoided under all circumstances. Here we deal with creatures and not only with usual matter.

5.13.11 Changing the brain functions at the nanolevel

The actions of a human being essentially indicate in what mind state he/she is. A human being recognizes the physically real objects as space–time picture in the form of geometrical structures.

His/her activities are exclusively based on these experiences. In the case of τ individual (human beings), the x, y, z structure of the observed material object at clock time τ is in fact a creation of the observer's mind.

It belongs to the characteristics of the projection principle that the mind is located outside of space and time. Nevertheless, the space–time structures, on which everything is based, are influenced by thc mind; moreover, they are created by the mind. It belongs to the characteristics of the projection principle too that only a part of the brain is material in character, but it has, in addition, a mindlike state as well (see in particular Section 5.2.2).

Because of the coupling between mind and brain, changes with respect to the brain (brain functions) influence the mind and mind states, respectively. Influences with respect to the mind effectuates, on the other hand, the space–time structure before the eyes of the human being. This can in principle very far-reaching: a τ individual (human being) or a λ individual can become, for example, a τ, λ individual. That is, through nanotechnological brain changes, drastic alterations about the perception of the material world outside can occur. Again, the form of appearance is changed, not the objective existence of the object. The objective existence of the material objects remains unaffected when we experience them within the τ state, the λ state or the τ, λ state. Here the mind and the brain are in the center. As we know, the mind and the brain are necessarily coupled.

5.13.12 Space and time individuals

Timescales

If there is one observer (a human being in the case of time individuals) and only one object, the object is defined on the following timescale:

$$\tau \text{ scale: } \tau_a \leq \tau \leq \tau_b \tag{5.88}$$

The τ scale itself refers to the observer. There are also τ states for

$$\begin{aligned} \tau &< \tau_a \\ \tau &> \tau_b \end{aligned} \tag{5.89}$$

That is, we suppose that the lifetime of the human being exceeds the time range of the object.

If there is one human being, we get in the case of N objects N time ranges for the reference timescale, that is, we get:

$$\tau \text{ scale: } \tau_{a1}, \tau_{a2}, \ldots, \tau_{aN}, \tau_{b1}, \tau_{b2}, \ldots, \tau_{bN} \tag{5.90}$$

and

$$\tau < \tau_{a1}, \tau_{a2}, \ldots, \tau_{aN}, \tau_{b1}, \tau_{b2}, \ldots, \tau_{bN} \tag{5.91}$$

$$\tau > \tau_{a1}, \tau_{a2}, \ldots, \tau_{aN}, \tau_{b1}, \tau_{b2}, \ldots, \tau_{bN} \tag{5.92}$$

That is, we suppose that the lifetime of the human being exceeds the time ranges of the N physically real (material) objects. The reference timescale is defined through the lifetime of the observer, which vary of course from observer to observer.

For the N objects, we have N timescales depending on the lifetimes of the physically real (material) objects:

$$\begin{aligned} &(t \text{ scale})_1\text{: } t_{a1}, t_{b1} \\ &(t \text{ scale})_2\text{: } t_{a2}, t_{b2} \\ &\quad\vdots \\ &(t \text{ scale})_N\text{: } t_{aN}, t_{bN} \end{aligned} \tag{5.93}$$

In conclusion, we have one reference timescale, but N system-specific timescales. The observer is able to observe N material objects simultaneously.

Position scales

In the case of space individuals, the observers are biological systems of another kind, that is, they are different from human beings. If there is one observer (space individual) and only one physically real (material) object, the object is defined on the following reference position scale:

$$\lambda \text{ scale: } \lambda_{xa}, \lambda_{ya}, \lambda_{za}, \lambda_{xb}, \lambda_{yb}, \lambda_{zb} \tag{5.94}$$

with

$$\lambda_a = (\lambda_{xa}, \lambda_{ya}, \lambda_{za}) \tag{5.95}$$

$$\lambda_b = (\lambda_{xb}, \lambda_{yb}, \lambda_{zb}) \tag{5.96}$$

The λ scale itself refers to the observer. There are also λ states for

$$\begin{aligned}&\lambda_x < \lambda_{xa}, \lambda_y < \lambda_{ya}, \lambda_z < \lambda_{za}\\&\lambda_x > \lambda_{xb}, \lambda_y > \lambda_{yb}, \lambda_z > \lambda_{zb}\end{aligned} \tag{5.97}$$

That is, we suppose that the life position of the space individual exceeds the position range of the object.

If there is one space individual but N objects, we get N position ranges for the reference position scale, that is, we get:

$$\lambda \text{ scale: } \lambda_{xa1}, \lambda_{ya1}, \lambda_{za1}, \ldots, \lambda_{xaN}, \lambda_{yaN}, \lambda_{zaN} \tag{5.98}$$

and

$$\lambda < \lambda_{xa1}, \lambda_{ya1}, \lambda_{za1}, \ldots, \lambda_{xaN}, \lambda_{yaN}, \lambda_{zaN} \tag{5.99}$$

$$\lambda > \lambda_{xb1}, \lambda_{yb1}, \lambda_{zb1}, \ldots, \lambda_{xbN}, \lambda_{ybN}, \lambda_{zbN} \tag{5.100}$$

That is, we suppose that the life position of the space individual exceeds the position ranges of the N physically real (material) objects. The reference position scale is defined through the life position of the observer, which vary of course also in this case from observer to observer.

For the N objects, we have N position scales depending on the life positions of the objects:

$$\begin{aligned}&(x, y, z \text{ scale})_1\text{: } x_{a1}, y_{a1}, z_{a1}, x_{b1}, y_{b1}, z_{b1}\\&(x, y, z \text{ scale})_2\text{: } x_{a2}, y_{a2}, z_{a2}, x_{b2}, y_{b2}, z_{b2}\\&\vdots\\&(x, y, z \text{ scale})_N\text{: } x_{aN}, y_{aN}, z_{aN}, x_{bN}, y_{bN}, z_{bN}\end{aligned} \tag{5.101}$$

In conclusion, we have one reference position scale, but N system-specific position scales. The observer is able to observe N material objects simultaneously.

5.13.13 Interchanging the roles of space and time

For human beings, which are time individuals after our classification above, at clock time τ the positions $\mathbf{r}_i = (x_i, y_i, z_i)$, $i = 1, 2, \ldots, N$, of N objects are directly be recognizable. The existence of the system-specific time t is the logical consequence, and this is due to the existence of the compressed information with respect to the entities y and α. Again, here human beings are concerned; a human being has at clock time τ the x, y, z structure directly before his/her eyes.

For space individuals (biological systems, which are different from human beings), we suppose that merely space and time are interchanged without changing the general argumentation. That is, we transfer the conception with respect to time individuals to the case of space individuals.

Then, we may argue as follows: in this case, the system-specific time t is directly recognizable and the existence of the space information $\mathbf{r}_i = (x_i, y_i, z_i), i = 1, 2, \ldots, N$, becomes the logical consequence, also here due to the existence of the compressed information concerning γ and α.

Here, we do not deal with a human being but, as we remarked earlier, with a biological system of another kind. For this kind of observer, the complete t structure would directly be recognizable at a certain space position, say λ. That is, in the case of space individuals, the functions of space and time are interchanged, when we compare the situation with that of time individuals.

When we go from time individuals to space individuals, there is a change in perspective. The material object is perceived in various ways, and each way is tailor-made to the idiosyncrasies of the observing biological system. It is not the objectively existing (true) object, which is observed, but a picture of it, depending on the type of the biological system.

The transition from τ to λ indicates a transition to another perception. Instead of the reference time τ we have now the reference position $\lambda = (\lambda_x, \lambda_y, \lambda_z)$, which is three-dimensional because the space is three-dimensional. In connection with time τ, there is a one-dimensional scanning, and the scanning process is three-dimensional with respect to λ.

Instead of time τ, indicated by a clock (clock time), we have a reference position λ indicated by a device, which is principally different from a clock and indicates a standard position λ in analogy to τ. Both the τ observer and the λ observer experience the same object but create different pictures of it, and each picture is useful for the respective life.

5.14 Time: qualitative and quantitative formulations

The system-specific time t, introduced in the way of Section 5.13, is independent on any physical theory. The existence of the element t follows from the compressed form of physically real (material) objects, which we denoted by γ and α. In fact, the compressed block concerning a material object is not expressible in terms of variables (e.g., see Section 4.9.6), which is needed for giving entities a physical face.

5.14.1 Principal statements

Once again, the letter t is a notation and is not a variable in this basic interpretation. This is of course also the case for the space elements x, y, z. This is the interpretation of the most basic state of a physically real (material) object occupying basic reality. In this state, the elements x, y, z, t can only be notations and have no numbers but, nevertheless, the notions "time individuals" and "space individuals" can be moved into, as we did in Section 5.13.

The entity α, describing the material object in its basic form, is not accessible for human beings and is therefore not directly ascertainable in terms of a mathematical expression. A human being expresses the effects in basic reality as space–time phenomenon outside of space and time. Since the entity α cannot be ascertained as mathematical expression, its projection onto space and time can only be described through "qualitative" arguments. This qualitative formulation is, on the other hand, the most basic articulation of a physical phenomenon at the basic level.

The system-specific time t can only be formulated in "quantitative" form through a physical theory, as is usual way in physics. Such a theory is, however, not assignable within the frame of basic reality. But a mathematical description for system-specific time t is possible in another way, also here using the projection principle.

We will recognize in Chapter 6 that this is obviously possible by means of a "quantum theoretical" treatment. The quantum aspect of the world is the starting point for the formulation of t and not the classical approach. The system-specific time t is basically a quantum time.

Thus, the system-specific time t reflects the quantum aspect of time that is missing in traditional quantum theory. It can in fact be moved on the basis of the projection principle. In other words, for the quantitative treatment we need a mathematical formulation of the projection principle. In Chapter 6 such a way is described.

5.14.2 Time-selection processes

The world behaves stationary in basic reality, that is, outside of space and time. Stationary means that the world in its basic form does not develop. The "entire" (complete) world or any object in basic reality is given at each time τ, that is, all at once.

What is existent and what is observed?

That is, the characteristics of an object or the complete material world outside are existent all at once and are given in "compressed" form. Then, the information in the picture must also be given in compressed form, that is, the complete information with respect to the space–time picture is "existent" at each time τ all at once, but only a part of it is "observed."

At clock time τ, a certain information is selected, but it cannot be the complete information. At time τ only a part of the compressed data is observed. Then, τ is applied to a timescale, which cannot be identical with the τ scale. The reason is simple: The complete τ scale cannot be defined at a certain time τ. As we know, the $x, y, z, t = \tau$ structures at various times τ are different from each other, that is, a certain $x, y, z, t = \tau$ structure at a certain clock time τ does not contain the complete information about the material object under investigation.

In other words, the compressed form of the data set requires a new type of time with a specific timescale. It is a system-specific timescale, that is, each object (system) has its own timescale, which we denoted above by the letter t. The complete data set of an object or even the entire world must be distributed on this t scale; the information within the entire t scale is then identical with the compressed information, which is defined at each time τ as stationary block (Section 5.4.4).

The entire $t$scale is successively and systematically scanned by the clock time τ, and the information at time $\tau = t$ is directly observed in daily life and appears before the eyes of human observers. It is a x, y, z, t structure (trees, houses, cars, etc.) at clock time $\tau = t$.

Again, human beings are caught in space and time and there can be no variables for γ and α since these entities are not connected to space and time. γ is located in basic reality where we have no space and time, and the entity α belongs to the observer's mind where also no space and no time exist; the mind creates space and time for making a picture of reality. Once again, human beings are caught in space and time. "Caught in space and time" also means that we cannot invent variables without applying space and time. Human beings are not able to say something about material objects (the physically real world) outside of space and time. The projection principle requires a world outside of space and time, but human beings cannot perceive these things directly.

The origin of the system-specific time

Therefore, there can be no variable assigned to α and γ; they merely represent raw information and remain "qualitative" in character and no numerical data can be produced for them. The letter t for the system-specific time here reflects a notation and not a variable. Since α describes a physically real (material) object, also the time describes the same object (it is a projection of α) and is therefore object-specific (system-specific).

It is important to accentuate that already on these qualitative statements the existence of a new type of time necessarily emerges, which is the sytem-specific time t in contrast to the clock time τ.

5.14.3 On the mathematical formulation

We have outlined several times that we have within the projection principle the following situation: at each clock time τ, an object has a definite space-time position $x_\tau, y_\tau, z_\tau, t=\tau$, but the object has, at each time τ, no mass, no energy and no momentum in space but the coordinates x, y, z. These qualitative statements are nevertheless the most basic information, which has been underlined in the earlier discussions.

But an object without physically real features cannot exist. The physical real features are associated with basic reality. Only "qualitative" statements are possible when we treat the features with respect to basic reality (Section 5.9). However, we must have the possibility for analytical treatments of the physically real features in connection with theoretical physics. In such cases, we need so-called fictitious realities as a replacement for basic reality.

Within projection theory, fictitious realities are described within the so-called $(\mathbf{p}, E)$ space where $\mathbf{p}$ is again the momentum with $\mathbf{p} = (p_x, p_y, p_z)$ and E is the energy, and both variables refer to the object under investigation. Thus, the physically real features are expressed by the variables $\mathbf{p}$ and E. The object itself is projected from $(\mathbf{p}, E)$ space onto space and time. Under consideration of the system-specific time t, it is a projection onto $(\mathbf{r}, t)$ space with $\mathbf{r} = (x, y, z)$. This line ("fictitious reality-picture line") leads to "quantitative" statements about the material object. The other line, that is, the projection from basic reality onto space and time leads to "qualitative" statements, but the system-specific time has its origin in the "basic reality-picture line."

The interaction processes take place in $(\mathbf{p}, E)$ space, that is, in fictitious reality, and there is an exchange of momentum and energy in the form of $\Delta\mathbf{p}$ and ΔE chunks, that is, we do not need further description elements for the understanding of interactions. The interaction problem is here solved elegantly without inventing the new notion as, for example, in the case of Newton's gravitational law. This point has already been discussed in Section 5.11.

The reality-observer principle with the five units consisting of basic reality, the tools of evolution, the mind, the processing characteristic η and the space–time picture enables to introduce a quantum aspect of time that is definitely missing in traditional physics.

5.15 Summary and final remarks

In this chapter, we gave detailed statements about what we call "matter" and we discussed qualitatively the existence of a system-specific time, which is an outcome of projection theory. In this section, we will summarize the main facts. But let us first underline the relevance of the projection principle for physics and in particular with respect to nanoscience.

5.15.1 Nanoscience: one theory for all phenomena

It is generally accepted that nanoscience will influence our future considerably. In particular, nanobiology and nanomedicine will change our life sustainable. For the understanding of all these phenomena, theoretical and computational nanoscience will play an important role. What mathematical tools do we need here and what are the most important description elements?

The impact of nanotechnology will be tremendous. The manipulation of atoms and molecules will allow to construct new technological devices and will bring fundamentally new possibilities in the field of medicine. Within the frame of nanobiotechnology, big changes are planned. It has been speculated that through nanoscience, the human bodies will become undecaying systems of "infinite life." In order to increase man's intelligence, it is particularly planned to change the brain functions. Do we really know how brain functions work and how they are connected to the other parts of the individuals? We have to be careful!

No doubt, the treats in nanotechnological manipulation are large, just in connection with self-organizing processes, and such processes can lead to uncontrolled situations that may appear. It has been speculated that even the entire earth could be transformed destructively and could in principle be even hostile toward life. Then, biological individuality would no longer be possible on the earth.

In nanoscience, we move between two poles, and these poles are given by "total destruction" and "infinite life." This is in fact a challenge for theoretical physics. For all nanodisciplines, we will have to develop specific theoretical modules (architectures) on the basis of theoretical physics,

Here we work at the ultimate level where the properties of materials, biological systems and others come into existence and, therefore, we have to apply the basic physical laws. Do we really have a comprehensive knowledge about the basic physical laws? No, we have not. Just through nanoscience, open questions in this field that come into play are depicted.

This, in particular, means that we have one theory for all phenomena at the nanolevel. This theory is expressed by the laws of theoretical physics. We need those basic laws of theoretical physics, which are formulated for the nanolevel. We do not need of course the theoretical structures of elementary particle physics, string theory or cosmology.

Within the frame of traditional technologies, for example, in micro- and macroengineering, engineers normally do not work at the ultimate level. Here, phenomenological descriptions are in the foreground which, in general, cannot be derived from the basic physical laws. Each (traditional) technological discipline has its own theoretical conception.

Nanoscience opens the door for fantastic new developments. A large variety of effects are thinkable leading to the invention of a lot of nanofacets and disciplines, respectively. For these disciplines, we will have specific theoretical modules (architectures), developed by the laws of theoretical physics. Only theoretical physics is here the adequate tool.

In nanoscience, we have various systems and topics, for example, functional nanomaterials, food chemistry, medicine with brain research, quantum and molecular computing, bioinformatics and nanoelectronics.

Although we have here a lot of directions and disciplines, respectively, which are partly very different in character, we are working in nanoscience at the same theoretical footing – in contrast to the traditional technologies. Again, in nanoscience we have one theory for all these disciplines and phenomena, and this is given by the laws of theoretical physics.

Such a situation requires a reliable theoretical frame for the understanding of the various nanophenomena. Just the corresponding basic elements, in particular, the notion "quantum time" belongs to them and has to be available in realistic form.

5.15.2 The direct impressions

The most direct impression about the world outside is what we have in everyday life immediately before us, before our eyes. These are objects in space, and we experience them in the course of time. In other words, it is a sequence of space structures. The space structures vary with clock time τ.

These everyday life experiences are the basis for the developments of physical theories. Every physical theory, consciously developed, must be compatible with these unconsciously experienced facts. There are sophisticated physical developments from cosmology to elementary particle physics (strings, etc., are included); they delivered world views that broaden considerably our unconsciously perceived everyday life experiences.

5.15.3 The notion "time"

It is astonishing that everything developed in all disciplines conceptually further, but not the "time." For time, the clock time τ is used without modification. On the question "What is time?" Albert Einstein answered: "Time is that what we measure with a clock." He analyzed the notion "time" not further.

Again, physics proceeded and the notions used in a theory had to be adapted to the respective conception, but the notion "time" remained. For example, scientists developed the quantum theory but the time is also here given by the clock time τ, which is classical in character, that is, we have no quantum aspect of time. We are missing a quantum time. Erwin Schrödinger tried to construct an operator for the time, but without success.

What is the reason for that? The interpretation of the fundamental input, that is, what we have in everyday life before our eyes, is obviously an approximation and is not complete. These are object structures in space at each time τ and are obviously not the complete description for the understanding of the phenomenon "time."

To believe that the material objects are embedded in space is not compatible with the basic peculiarity of space and time (clock time). Already Ernst Mach pointed distinctly on the particularity of space. Mach's principle motivated Albert Einstein to think about the true nature of space and time, but his theory of relativity does not fulfill Mach's principle and, moreover, it does not touch quantum phenomena and a quantum aspect of time was not a topic within the theory of relativity. Once again, are the things, which we have directly before us, entities made of matter?

We answered this question in this chapter and came to the conclusion that the things before us can merely be geometrical structures. We touch and feel these things as physically real (material) entities, but this is not a contradiction, as we have underlined in this chapter.

Everything follows from the fact that space and time are not physically real elements because they are not accessible to empirical tests, that is, space and time cannot be observed as "isolated" entities but always in connection with objects. Then, it is necessary to conclude that space and time are the frame for the representation of the physically real world. The physically real (material) world outside is projected onto space and time and it appears here at each time τ as global geometrical structure. The material world itself is embedded in basic reality, which does not contain space and time; they do reflect mind states.

We are caught in space and time. No statement about the real world can be done without space and time. There is obviously no exception in physics, and this fact makes this behavior to a principle.

The classical treatment of nanosystem is an approximation and is in many cases not possible and not sufficient. That is, in the behavior of nanosystems quantum effects are important and are distinctly reflected in the properties. This is particularly relevant when we want to investigate "self-organizing processes" within the nanorealm.

A self-organizing process can be characterized as follows: a certain quantum state develops in the course of time from one quantum state to another and reaches, after a certain time, a stationary state.

Self-organizing processes are relevant in nanoscience and they belong to the heart of it. We cannot resign on them. Self-organizing processes are considered as a sequence of quantum states. In Chapter 6, we will give more details in connection with the treatment of such systems on the basis of the quantum time.

Is a full quantum theoretical treatment of self-organizing processes possible at all in traditional quantum theory? No, this is not possible because the "time" is treated classically in traditional quantum theory, and the time is a relevant factor in connection with self-organizing processes.

In other words, the treatment of self-organizing processes within traditional quantum theory is problematic, and this is because the time, which is in the center of self-organizing processes, is treated here as a "classical" entity, even when the process itself is classified as a quantum mechanical process. We have to be careful and consequent.

Let us repeat: This transition occurs in the course of time and behaves quantum mechanically. However, the time itself, the relevant factor in the transition, does not reflect a quantum aspect when we work within traditional quantum theory. The time is here a classical element and is measured with our clocks used in everyday life, indicated in this book by the Greek letter τ.

In other words, when we go from classical physics to traditional quantum theory the character of time is not changed and remains classically in character, although the realm of quantum theory is entered. In both cases, we have merely the external clock time τ. But just the time is a relevant element in the case of self-organizing processes and, as we already remarked, such kind of processes belong to the heart of nanoscience.

5.15.4 Levels of description

Within the frame of the projection principle, observation means that reality, mind and brain cooperate systematically and the created information is projected onto space and time leading to a picture of reality. The observation itself appears exclusively as picture before the eyes of the human being.

How are the mind and the brain ordered when the physical reality is related with what a human being actually perceives? This question has been answered in detail within the reality-observation principle. We order the appearances in terms of levels of reality. The notion "levels of reality" has already been introduced in Chapter 2.

The observations in everyday life are positioned at the "macroscopic level" where the five senses are effective. Above the macroscopic level is the "atomic level," which is the ultimate level where the properties of usual matter and biological systems emerge. It is in particular the level of nanoscience. Usual matter is located at the macroscopic level and is that what we have directly before our eyes in everyday life.

We need at the ultimate level "one" theory with one description complex. The notion "ultimate level" demands that! Usual quantum theory represents a closed complex, but is this complex comprehensive enough for the understanding of the phenomena in nanoscience? In this chapter and also in the previous chapters, we discussed this point and came to the conclusion that this is not the case. A typical point is here the problem we have with the phenomenon "time." Nanoscience suggests and even requires the expansion of the present laws of quantum theory.

Matter is what a human being unconsciously experiences with his five senses. It is the macroscopic level. There are other levels above the macroscopic level. In Chapter 4, we introduced the atomic level and we saw that there is a connection between these two levels. The atomic level is the ultimate level at which the properties of usual matter and biological systems (e.g., living systems) come into existence. This is the level of nanoscience. We introduced other forms of life, which we summarized under the term "species."

5.15.5 The level of matter

In Chaps. 4 and 5, we defined the notion "matter." What is matter? When we base our considerations on a realistic standpoint, matter is basically an "experienced" feature and has to be based in the theoretical analysis on certain superordinated facts, for example, the principles of evolution.

Matter is what an observer (human being) pulls from an objectively existing world outside (basic reality), and this consists of physically real objects that we call "matter."

Existence-conserving characteristic

The basic information α is the entity on which everything has to be based when we deal with the physically real world. We experience the objects as physically real units. α is an existing entity with an "existence-conserving" property. Concerning the notion "existence," we discuss this feature in more detail in Chapter 6 when we treat the raw information α on an analytical footing within the frame of fictitious realities.

What we call "matter" can however only be the observer's view of matter, that is, matter is an observer-dependent notion. It is not the objectively existing world, which we observe, but it reflects a "useful" part of it and has been selected through evolutionary processes. This selected part depends in particular on the observer type. The observer-independent physically real (material) objects are located in basic reality and are not accessible to human beings and possibly also not for other types of biological systems.

Physical reality, mind and brain

This means that the human being produces the physically real objects himself. In this connection it is important to specify the notion "physically real": an object can be considered as physically real if it has constant properties in a certain environment, that is, if it behaves stationary. The adjectives "stationary" and "existence conserving" belong together.

Evolution moves toward a stationary state. In this connection, sensory apparatuses are important and are widely different for different observer types. They developed evolutionary. The sensory apparatus is in the case of human beings given by the five senses.

Besides the sensory apparatus, within projection theory the observer's mind and his/her brain are needed for the perception of the world outside. What exactly is the mind and what is the task of the brain? At this stage, we do not need the details of the units mind and brain, but their existence is required from the point of view of projection theory. The projection principle makes definite statements about the functionalities between mind and brain; the actions of both units relative to each other are relevant. In this connection, projection theory makes clear statements, which are discussed in this chapter but also in Chapter 4.

5.15.6 Time: more details

The effects in nanoscience are essentially quantum phenomena and are time sensitive. This in particular means that a "quantum time" cannot be factored out. However, the quantum aspect of time is missing in usual quantum theory, and this is without any doubt a challenge for theoretical physics.

We may state quite generally that novel conditions in science require a deeper penetration into the nature of the world. In connection with the notion "quantum time," we have in fact such a situation: We have obviously penetrating deeper into the nature of time. The traditional view is not sophisticated enough.

The external time

In traditional physics, we only know one type of time. It is the time τ, which we called "clock time." We already quoted the famous statement by Albert Einstein: "Time is that what we measure with a clock." However, time is obviously more than that: A human not only looks at the clock, but he/she has a certain "time feeling." The time-feeling effect is obviously not merely an illusion. However, this clock time τ is only an "external" time and has nothing to do with the objects under investigation and can in particular not explain the time feeling.

This external time cannot be rationally explained by the conceptions of physics. In physics, we deal with individual objects, which can be small and large. But there is no "global unit" known for the inoculation of what we call "time" and which equally influences everything. In other words, there is no time machine anywhere in the universe that produces the externally defined clock time τ. Such conceptions do not exist in traditional physics but, nevertheless, the time effect cannot be denied, and the individual human beings perceive the time effect comparably. The often-used San Francisco metaphor (Chapter 1) is a typical example for explaining the external clock time τ, but it reflects merely a metaphysical standpoint.

Nanoscience requires a realistic conception for the time

All that indicates that the time is not adequately treated in traditional physics. This is particularly accentuated by the fact that there is no operator for the time in traditional quantum theory. Erwin Schrödinger tried to introduce such an operator within the formalism of traditional quantum theory but without success [4]. Mario Bunge underlined [77] that the role of time in the usual form of quantum theory is not acceptable. We already discussed this point in Chapter 1 and will continue this topic in Chapter 7.

In conclusion, a realistic consideration of time on the basis of scientific arguments is necessary. For further developments in this field, Mach's principle should be seminal, that is, we have to go beyond the traditional form of quantum theory.

Just in connection with nanoscience, where self-organizing processes are important and which belong to the heart of this new technology, a realistic conception for the time is required.

No doubt, a time operator is missing in traditional quantum theory and its introduction on the basis of the "same" theoretical frame did not lead to the goal as Schrödinger demonstrated. His efforts just demonstrated that drastic steps are necessary. This assessment has been distinctly underlined by Marion Bunge [77]. In fact, we cannot resign on a qualified solution of this problem when we enter the realm of nanoscience. We obviously have to expand the theoretical frame of traditional quantum theory.

Mach's principle

Once again, in this connection Mach's principle should be seminal. This principle reschedules the space–time complex at the fundamental level of physics. When we take Mach's principle seriously, we come to the following conception: Space and time are not physically real entities and can therefore not responsible for physically real effects. In accordance with the philosopher Immanuel Kant, we came to the conclusion that space and time are elements of the human being's mind.

Is the mind made through matter or vice versa? This and other variations with respect to the mind-matter problem are extensively discussed in literature without coming to a convincing result. When we start from the basic properties of space and time, we come to a conception, which is based on only one characteristic: Matter is projected onto space and time via the observer's mind. All that reflects basic physics and has primarily nothing to do with philosophy.

5.15.7 What is the level of intervention?

As we already mentioned, Erwin Schrödinger tried to introduce the quantum time, but without leaving the principal frame of traditional quantum theory. Thus, we have to go a step further. We have obviously go deeper into the scientific background.

But at which level have we to intervene? In the search for the level at which the quantum time exists, we touch the most basic features of physics. That is, we have to recourse to the most basic form of space and time.

As we already remarked, here Mach's principle is such a basic feature and is therefore seminal. Mach himself did not formulate his basic ideas within a mathematical frame, as is the standard way in physics. Nevertheless, his qualitative arguments are easy to formulate and the following characteristics supplement our earlier statement: Space points may not be used in isolated form and, furthermore, they may not be used in theories without material objects.

These statements reflect fundamental truths. Space and time only fulfill Mach's principle if they are not "physically real." There is in fact no possibility to "measure" isolated space points, that is, isolated space points are not provable. That what is principally not observable does not exist as physically real element.

That was the reason why we worked with space–time structures that are "not" physically real (Chapter 1). This is basically a drastic step and has consequences on the entire observation procedure. But all that is a consequence of Mach's principle, which is underlined through the statements of Immanuel Kant.

What is the consequence? Schrödinger wanted to expand quantum theory on the basis of the already existing frame. But this way does not work. The already existing frame is the problem. This frame is burdened with the container model in which space and time play the role of physically real entities and this against Mach's principle.

However, we have to be careful. The more we penetrate into the details of nature, the more we need realistic conceptions for their description, that is, we need conceptions with increasing sophistication. That is, in such cases the basic laws will be more and more externalized. This often means that the complete theoretical frame has to be renewed from the very first. The use of a modified version of the space–time is such step.

With nanoscience, we have reached the ultimate level where the properties of usual matter and those of biological systems emerge. Thus, we need here "one" conception for the description of "all" phenomena. In other words, instead of a variety of phenomenological frames as in the realm of engineering, we need one theory in nanoscience for the comprehensive coverage of all phenomena in this field and it is obvious that at this level an adequate theory is needed.

The problem of time, recognized approximately more than 90 years ago, did not disturb so far in practical calculations, but cannot be ignored when we enter the realm of nanoscience. In this situation, ideas will be underlined and new aspects are created.

Concerning the notion "time," projection theory goes a step further. This theory requires in fact the existence of a system-specific time. What is the background of that? Let us summarize the main features: A human being has at clock time τ the positions $\mathbf{r}_i = (x_i, y_i, z_i)$, $i = 1, 2, \ldots, N$, of N objects directly before his/her eyes. The existence of the system-specific time t is the logical consequence, which directly follows from the existence of the compressed information with respect to the entities γ and α. Again, here human beings are concerned; a human being has at clock time τ the x, y, z structure directly before his/her eyes. In this way, the existence of the system-specific time t has to be defined necessarily, but this connection is given only in qualitative form. A quantitative formulation is outlined in Chapter 6.

The reality-observer principle

For the observation of physically real entities in basic reality the entire block consisting of "observer, tools of evolution, processing characteristic and the picture of reality" is involved and must be applicable at "each" time τ. Therefore, procedure (5.34) from basic reality (observer $\rightarrow$ object, process) to the picture occurs instantaneously, that is, without any delay. Scheme (5.34) reflects therefore "one" state.

In fact, block (5.34) reflects a mind state, and physically real (material) mechanisms do not take place. In particular, space and time are not involved in the steps within block (5.34). Clearly, the picture contains the elements x, y, z, τ but it does not refer to the creation of the picture itself but is for the representation of the facts in basic reality.

The entire block consisting of the four entities "observer, tools of evolution, processing characteristic and the picture of reality" belong together and act simultaneously. This block with these specific features defines the notion "observation." It is "one" unit and is not separable. None of the entities can exist without the others.

This is the reality-observation principle, which reflects a "compressed state."

Chapter 6
Fictitious realities

6.1 Quantum states

In the preceding chapters we treated space and time, that is, the elements x, y, z and τ, as auxiliary entities for the representation of physically real systems (objects). The coordinates x, y, z and the time τ, which is indicated by a usual clock, are not physically real.

6.1.1 Other spaces

Fourier space

Within projection theory, the space–time is defined by the variables $\mathbf{r} = (x, y, z)$ and τ. Let us therefore denote space–time "$(\mathbf{r}, \tau)$ space." Physical reality is projected onto $(\mathbf{r}, \tau)$ space. Because space and also time are auxiliary elements, it should be possible to "invent" other spaces on which *reality* can be projected. Such a space is, for example, the so-called Fourier space (the so-called reciprocal space) having the variables $\mathbf{k}$, ω with the characteristics of a wave:

$$\begin{gathered}\mathbf{k} = (k_x, k_y, k_z) \\ |\mathbf{k}| = \frac{2\pi}{\lambda}, \quad \omega = 2\pi\nu\end{gathered} \tag{6.1}$$

where λ is the wavelength and ν the frequency of the wave.

Let $\rho(x, y, z, \tau)$ be the information of a classical system, which is a projection onto $(\mathbf{r}, \tau)$ space. Furthermore, let $F(k_x, k_y, k_z, \omega)$ be the projection of the *same* physically real process on $(\mathbf{k}, \omega)$ space. Then, $\rho(x, y, z, \tau)$ can be obtained from $F(k_x, k_y, k_z, \omega)$ by the Fourier transform

$$\rho(x, y, z, \tau) = \frac{1}{(2\pi)^3} \int_{-\infty}^{\infty} F(k_x, k_y, k_z, \omega) \exp[i(\mathbf{k} \cdot \mathbf{r} - \omega\tau)] \, dk_x dk_y dk_z \, d\omega \tag{6.2}$$

Using the inverse Fourier transformation, the determination of $F(k_x, k_y, k_z, \omega)$ from $\rho(x, y, z, \tau)$ is possible.

For example, in the case of classical many-particle systems (e.g., see Refs. [78, 79]), the function $\rho(x, y, z, \tau)$ is a correlation function, and the function $F(k_x, k_y, k_z, \omega)$ is in principle accessible to measurements within the frame of scattering experiments. The information concerning the many-particle system is completely transformed from $(\mathbf{k}, \omega)$ space to $(\mathbf{r}, \tau)$ space, and vice versa. Either the $(\mathbf{r}, \tau)$ space or the $(\mathbf{k}, \omega)$ space can be used for the description the many-particle system. Both representations are equivalent.

https://doi.org/10.1515/9783110525595-006

In conclusion, the variables x, y, z, τ can in principle be replaced by another set of variables. As an example we used k_x, k_y, k_z, and ω in the description of reality; there is a coexistence of several spaces. The measurements are performed in $(\mathbf{k}, \omega)$ space and the intuition of the observer is adapted to the $(\mathbf{r}, \tau)$ space. More details are given in Refs. [78, 79].

6.1.2 The influence of Planck's constant

We discussed an important point: Besides the elements x, y, z, τ other variables can be used for the description of reality. There is a coexistence of various spaces. The following question is relevant: To what space does a given set of variables belong? What role are played, for example, by the momentum $\mathbf{p}$ with the components p_x, p_y, p_z and the energy E? Here, the $(\mathbf{k}, \omega)$ space, introduced earlier, becomes an essential factor, just then when we go from classical physics to the quantum realm. The following way is straightforward.

We introduce Planck's constant $\hbar$ by the well-known basic relations

$$\begin{aligned} &\mathbf{p} = \hbar\mathbf{k} \\ &p_x = \hbar k_x, p_y = \hbar k_y, p_z = \hbar k_z \\ &E = \hbar\omega \end{aligned} \tag{6.3}$$

Because the momentum $\mathbf{p}$ is proportional to $\mathbf{k}$ and the energy E to ω, we have to consider the momentum $\mathbf{p}$ and the energy E as the elements of a new space, which we would like to denote $(\mathbf{p}, E)$ space. This $(\mathbf{p}, E)$ space has the characteristics of the $(\mathbf{k}, \omega)$ space, discussed earlier.

Consequently, the $(\mathbf{r}, \tau)$ space and the $(\mathbf{p}, E)$ space are equivalent, and are connected to each other by a Fourier transform. However, we have to be careful. In connection with $\mathbf{p} = \hbar\mathbf{k}$, $E = \hbar\omega$ [eq. (6.3)], the time τ can no longer be identified with the external clock time τ of classical mechanics and traditional quantum theory. It will turn out that τ becomes a real quantum variable in connection with $\mathbf{p} = \hbar\mathbf{k}$, $E = \hbar\omega$. Therefore, we want to use the letter t instead of τ:

$$\tau \to t \tag{6.4}$$

Then, instead of "$(\mathbf{r}, \tau)$ space," we get the marking $(\mathbf{r}, t)$ space:

$$(\mathbf{r}, \tau)\ \text{space} \to (\mathbf{r}, t)\ \text{space} \tag{6.5}$$

Nevertheless, the time τ will also be used in the novel quantum-theoretical description but is applied here as reference time and remains of course an external time measured by our clocks. The time t is a real quantum time and describes the quantum aspect of time, which is missing in traditional quantum theory. Thus, the quantum time t is a completely new variable. It should already mentioned here that t has to be identified with the system-specific time introduced in Section 5.13.5.

Then, we have to replace in eq. (6.2) the variables $\mathbf{k}$ and ω by the momentum $\mathbf{p}$ and the energy E using eq. (6.3); The variable τ has to be replaced by t:

$$\phi\,(x,y,z,t) = \frac{1}{\hbar^4 (2\pi\hbar)^3} \int_{-\infty}^{\infty} \phi\,(p_x,p_y,p_z,E) \exp\left\{ i \left[\frac{\mathbf{p}}{\hbar} \cdot \mathbf{r} - \frac{E}{\hbar} t \right] \right\} dp_x dp_y dp_z dE \tag{6.6}$$

$$\phi\,(p_x,p_y,p_z,E) = \frac{1}{2\pi} \int_{-\infty}^{\infty} \phi\,(x,y,z,t) \exp\left\{ -i \left[\frac{\mathbf{p}}{\hbar} \cdot \mathbf{r} - \frac{E}{\hbar} t \right] \right\} dt dx dy dz \tag{6.7}$$

The two spaces, $(\mathbf{r},\ t)$ and $(\mathbf{p}, E)$space, are equivalent concerning their information. Both spaces are connected to each other by a Fourier transform. That is, due to $\hbar$ the variables $\mathbf{r}, t, \mathbf{p}, E$ are arranged in a quite different way than in classical mechanics, where $\hbar$ is not defined.

The function $\phi\,(x,y,z,t)$ reflects a certain "information" in $(\mathbf{r},\ t)$ space, and the function $\phi\,(p_x,p_y,p_z,E)$ is the equivalent information in $(\mathbf{p}, E)$ space. As argued in Ref. [4], the determination of $\phi\,(x,y,z,t)$ and $\phi\,(p_x,p_y,p_z,E)$, respectively, cannot be done with the arguments of classical mechanics as in the case of $\rho(x,y,z,\tau)$.

Remark

Since the momentum $\mathbf{p}$ and the energy E characterize physically real objects in ($\mathbf{p}$, E) space and are system-specific, also the equivalent information in ($\mathbf{r}$, t) space, i.e., the projection, must be system-specific as well. Thus, the variables $\mathbf{r}$, t must be system-specific like the variables $\mathbf{p}$, E; in particular, the time t is a system-specific quantity and has nothing to do with the external clock time τ. Again, ($\mathbf{p}$, E) space and ($\mathbf{r}$, t) space contain the same information about the physically real object.

6.1.3 Reality and its picture

Instead of $\phi\,(x,y,z,t)$ and $\phi\,(p_x,p_y,p_z,E)$ in the following we want to use the symmetrical forms for the wave function, which are expressed by

$$\Psi(x,y,z,t) = \frac{1}{(2\pi\hbar)^2} \int_{-\infty}^{\infty} \Psi(p_x,p_y,p_z,E) \exp\left\{ i \left[\frac{\mathbf{p}}{\hbar} \cdot \mathbf{r} - \frac{E}{\hbar} t \right] \right\} dp_x dp_y dp_z dE \tag{6.8}$$

and

$$\Psi(p_x,p_y,p_z,E) = \frac{1}{(2\pi\hbar)^2} \int_{-\infty}^{\infty} \Psi(x,y,z,t) \exp\left\{ -i \left[\frac{\mathbf{p}}{\hbar} \cdot \mathbf{r} - \frac{E}{\hbar} t \right] \right\} dt dx dy dz \tag{6.9}$$

Since both spaces, $(\mathbf{r},\ t)$ and $(\mathbf{p}, E)$space, are equivalent, the wave functions $\Psi(x, y, z, t)$ and $\Psi(p_x, p_y, p_z, E)$ are equivalent too. They contain the same information. Some essential properties of $\Psi(x, y, z, t)$ and $\Psi(p_x, p_y, p_z, E)$ will be discussed later. Furthermore, we will formulate for $\Psi(x, y, z, t)$ and $\Psi(p_x, p_y, p_z, E)$ equations for their determination on the basis of interaction laws.

The following question arises: Why in nature is the *same* information repeated? According to projection theory and Mach's principle, physically real processes are projected onto space and time; we have the following situation (see also Figure 6.1): The $(\mathbf{p}, E)$ space reflects the physical reality, and this reality (i.e., its information content) is completely projected onto $(\mathbf{r},\ t)$ space. This is consistent with the fact that we do not measure the elements x, y, z, t, but we measure the quantities p_x, p_y, p_z, E.

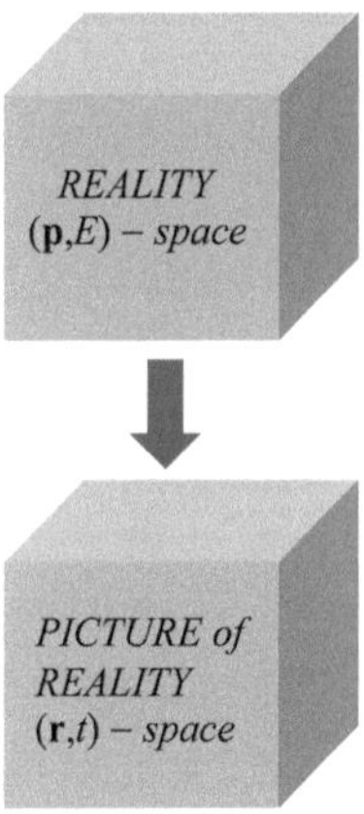

Figure 6.1: We measure the quantities $\mathbf{p}$ and E, but never space–time points $\mathbf{r} = (x, y, z)$ and t, that is, the space–time elements are not accessible to empirical tests; we can only say something about distances in connection with physically real objects (masses) and time intervals in connection with real physical processes (see also Chapter 1). The variables $\mathbf{p}$ and E have to be assigned to reality, whereas the variables x, y, z and t are the elements of the picture. In other words, reality is not embedded in space–time, as in traditional physics, but it is projected onto the space–time. Space and time are exclusively elements of the picture, that is, they do not appear in reality with the variables $\mathbf{p}$ and E. On the other hand, this kind of reality with the variables $\mathbf{p}$ and E cannot be identified with basic reality, since we cannot make statements about basic reality (see also Ref. [4]). A picture-independent point of view is not conceivable. We have to consider reality with the variables $\mathbf{p}$ and E as "fictitious reality," which is constructed from the human observer's point of view on the basis of space and time. Fictitious reality describes the following situation: The actual physically real world is embedded in basic reality, which is however not directly accessible to human beings. Therefore, a human being needs what we named "fictitious realty." Fictitious realities do not say what actually "is" in basic reality, but "as if" it were so.

Measurement means that an interaction process takes place and, on the other hand, interaction means an exchange of energy and momentum. On the basis of this measured information, pictures are formed in $(\mathbf{r},\ t)$ space, that is, we obtain the following scheme:

$$\begin{array}{c}\text{reality}\\ \downarrow\\ \text{picture of reality}\end{array} \tag{6.10}$$

This projection scheme is close to what we have analyzed in connection with basic reality γ and its picture, which human beings experience in everyday life (see in particular Chapters 4 and 5).

6.1.4 What is energy?

In the preceding chapters we have stated that we cannot make any statement about basic (absolute) reality. This is true even with respect to the variables of it. On the other hand, the variables $\mathbf{p}, E$ represent what we called "fictitious reality." The variables $\mathbf{p}, E$ have to be considered as abstract ideas, a product of the human's mind, so to speak. Thus, fictitious reality, which is embedded in $(\mathbf{p}, E)$ space, has to be considered as an abstract idea, that is, it cannot be identified with basic reality.

The following has to be kept in mind: The actual physically real world is embedded in basic reality, which is however not directly accessible to human beings. Therefore, a human being needs what we named "fictitious realty." Fictitious realities do not say what actually "is" in basic reality, but "as if" it were so.

The energy E is introduced on the basis of $(\mathbf{r},\ t)$ space to which our intuition is adapted. In fact, in the unit of energy (g × length^2/time^2), the units of space and time explicitly appear. In other words, E does not occur in nature. This point has often been discussed in literature (e.g., see Ref. [12]). In connection with energy E, a good statement is given by Christian von Baeyer [80]:

> The gradual crystallization of the concept of information during the last hundred years contrasts sharply with the birth of the equally abstract quantity called energy in the middle of the nineteenth century. Then, in the brief span of 20 years, energy was invented, defined and established as a cornerstone, first in physics, then of all science. We don't know what energy is, any more than we know what information is, but as a now robust scientific concept we can describe it in precise mathematical terms, and as a commodity we can measure, market, regulate and tax it.

Concerning energy, Edgar Lüscher remarked the following (see Ref. [4] and the references therein):

> Energy is not a quantity that actually occurs in nature, but it is an abstract idea, a product of the human mind who tries to understand nature within his/her capabilities.

Reality is characterized by the variables $\mathbf{p}$ and E and, on the other hand, the pictures of reality are expressed in terms of the variables $\mathbf{r}$ and t. Both spaces are products of mind. We will deepen this point in Section 6.2.

6.1.5 Existence of objects and processes between objects

Within the frame of the projection principle, there are two spaces, the $(\mathbf{r}, t)$ space and the $(\mathbf{p}, E)$ space. We stated earlier that the reality is embedded in $(\mathbf{p}, E)$ space, but it is not the basic reality and, therefore, we named this kind of reality, which is located in $(\mathbf{p}, E)$ space, "fictitious reality." Within $(\mathbf{r}, t)$ space, there are no physically real objects, but merely "pictures" of them in the form of geometrical positions.

There is a principal difference between the variables x, y, z, t of $(\mathbf{r}, t)$ space and the variables p_x, p_y, p_z, E of $(\mathbf{p}, E)$ space. This point is essential and let us discuss it for the momentums and the energy conservation laws that exist, but not for the co-ordinates and the time.

The importance of momentum and energy

Let us underline once again that the quantities $\mathbf{p} = (p_x, p_y, p_z)$ and E are basic entities. Modern physics started approximately 300 years ago with Newton's point of view. Within Newton's mechanics, the variables p_x, p_y, p_z and E for the momentum and the energy are well-established quantities. In fact, the variables p_x, p_y, p_z and E belong to the most relevant basic entities in physics up to the present day.

In projection theory, p_x, p_y, p_z and E reflect description elements because they are constructed on the basis of space and time, that is, on the elements x, y, z, t, which are elements of the mind. Also p_x, p_y, p_z, E and the wave function $\Psi(p_x, p_y, p_z, E)$ are elements of the mind.

Again, the variables p_x, p_y, p_z and E are introduced via the space–time elements x, y, z, t and this shows that the entire construction is a creation of the mind. Nevertheless, p_x, p_y, p_z and E may be treated as physically real elements and x, y, z, t as auxiliary elements. We may treat the unit with p_x, p_y, p_z and E "as if" they were physically real entities, but they are entities of a fictitious reality, which is projected onto space and time with x, y, z, t. The actual reality, which we named "basic reality," is principally not accessible to human beings; this point has been discussed in the earlier chapters.

Let us repeat: The actual physically real world is embedded in basic reality, which is, however, not directly accessible to human beings. Therefore, a human being needs what we named "fictitious realty." Fictitious realities do not say what actually "is" in basic reality and what actually takes place in basic reality, but "as if" it were so. This is a realistic standpoint and considers the fact that only a human-being-dependent reality is perceivable (Chapters 4 and 5).

In conclusion, we have a (fictitious) reality, where the physically real processes are simulated. This information is projected onto space and time and we get a picture of this (fictitious) reality. This is the highest level, which human beings can take in the concrete description of the world. This is a clear and realistic point of view and we can distinguish between what is describable and observable and what not.

The traditional view does not clearly distinguish between all that, that is, between physically real things and those that behave metaphysical. Just the never-ending debate with respect to Mach's principle underlines that distinctly. No doubt, most of the laws work excellently but, from the epistemological point of view, they are more or less recipes.

The quantities $\mathbf{p}$ and E substitute the raw information α that we introduced in Chapter 4. Only α reflects directly physical reality but not $\mathbf{p}$ and E. Nevertheless, $\mathbf{p}$ and E are reality-describing quantities and form that what we named above fictitious reality.

6.1.6 Conservation laws

For the momentum $\mathbf{p}$ and the energy E, "conservation laws" exist, which is of essential relevance for the assessment of the role of $(\mathbf{p}, E)$ space and justifies the designation fictitious reality.

The variables $\mathbf{p} = (p_x, p_y, p_z)$ and E are not unrelated points in $(\mathbf{p}, E)$ space. If the values of $\mathbf{p}$ and E of an object (say i) are changed, the values $\mathbf{p}$ and E of another object (say k) are changed simultaneously. This correlated exchange of momentum and energy, expressed by $\Delta\mathbf{p}$ and ΔE, occurs so that the sums of $\mathbf{p}$ and E remain constant. In other words, we have

$$\begin{aligned} \mathbf{p}_k &= \mathbf{p}_i \pm \Delta\mathbf{p} \\ E_k &= E_i \pm \Delta E \\ \mathbf{p}_i &= \mathbf{p}_k \mp \Delta\mathbf{p} \\ E_i &= E_k \mp \Delta E \end{aligned} \tag{6.11}$$

Equations (6.11) reflect conservation laws for the momentum and the energy and they are valid simultaneously, that is, at each time τ. This defines an "interaction process" and, therefore, $(\mathbf{p}, E)$ space has to be assessed as reality (fictitious reality).

It should be emphasized that an object (system) can only exist if there are $(\mathbf{p}, E)$ fluctuations [12], that is, an object, having properties $\mathbf{p}$ and E, is only able to exist if $\mathbf{p}$ and E fluctuate without interruption. That is, the quantities fluctuate incessantly with respect to $\mathbf{p}$ and E. That is, the objects i and k "interact."

The conservation laws for the momentum and the energy are not only responsible for interaction processes between objects, but they are existence preserving as well. Existence through interaction! Within projection theory, objects do not exist without interaction [12]. All that refers of course to $(\mathbf{p}, E)$ space.

No conservation laws in space and time

There are no conservation laws in $(\mathbf{r}, t)$ space, that is, with respect to the coordinates x, y, z and the time t. The space–time positions themselves do not fluctuate in the above sense. Existence-preserving effects and interactions between different space–time points are not known in $(\mathbf{r}, t)$ space and also not thinkable. There is no exchange of space and time in the form of $\Delta\mathbf{r}$ and Δt pieces (see also Chapter 1). There are no physically real processes in space and time.

Thus, $(\mathbf{r}, t)$ space cannot be assessed as reality, but justifies the designation "picture," that is, we get a "picture of reality." There is in fact no other possibility. In this way, we come to the "reality-picture conception."

6.1.7 Counterparts

Within projection theory, the systems are described by the wave functions $\Psi(x, y, z, t)$ and $\Psi(p_x, p_y, p_z, E)$. We have two equivalent descriptions, one in $(\mathbf{r}, t)$ space and another in $(\mathbf{p}, E)$ space. The $(\mathbf{p}, E)$ space reflects (fictitious) reality and the $(\mathbf{r}, t)$ space is the frame where the picture of reality is located.

The momentum and the energy of each object (system) fluctuate incessantly because the systems are only defined (existent) if their momentum and their energy fluctuate at each time τ [12].

In other words, due to the conservation laws for the momentum and the energy, the $(\mathbf{p}, E)$ values fluctuate for each quantum object. In principle, such fluctuations take place in the range $-\infty < p_x, p_y, p_z, E < \infty$.

Because of these $(\mathbf{p}, E)$ preserving fluctuations, another system is required, which has to be considered as a "counterpart" of the system under investigation. Without such a counterpart, the conservation laws for the momentum and the energy could not be fulfilled. In connection with eq. (6.11), object i and object k exchange momentum and energy in accordance with the conservation laws.

Laws (6.11) are independent on the positions of both objects; the conservation laws must be fulfilled independent on the relative positions of object i and object k. Object i cannot exist without the existence of object k and vice versa.

In conclusion, we need at least two systems; the conservation laws do demand that. The $(\mathbf{p}, E)$ fluctuations (interactions) between object i and object k lead to distance-independent correlations in $(\mathbf{r}, t)$ space. This kind of interaction produces the objects themselves, and a certain space–time structure for an object is created. This kind of interaction can therefore be named "form interaction." We showed in [12] that also distance-dependent correlations can be introduced within the framework of the projection principle.

6.1.8 Description of properties

Important properties can be deduced if we require certain conditions with respect to the wave function $\Psi(x, y, z, t)$ and $\Psi(p_x, p_y, p_z, E)$. Let us study four-dimensional shifts within $(\mathbf{r}, t)$ space as well as within $(\mathbf{p}, E)$ space.

Shifts within $(\mathbf{r}, t)$ space

The wave function $\Psi(x, y, z, t)$ describes the object (system) under investigation, which can be an atom, an electron or even a macroscopic object like a stone. We know that the properties of an object (stone, etc.) are not changed when we move it from one position x, y, z to another $x + x_c, y + y_c, z + z_c$. The same is the case when we consider the system-specific time t: The properties are not changed when we go from t to $t + t_c$. Therefore, let us require the following: a function $D_{\mathbf{p},E}$ describes the observable reality in $(\mathbf{p}, E)$ space if it remains unchanged when we shift the position and the time by the constant values x_c, y_c, z_c and t_c

$$\Psi(x, y, z, t) \rightarrow \Psi(x + x_c, y + y_c, z + z_c, t + t_c) \tag{6.12}$$

The function $D_{\mathbf{p},E}$ characterizes the appearance of the system under investigation in $(\mathbf{p}, E)$ space. With eq. (6.8)

$$\Psi(x, y, z, t) = \frac{1}{(2\pi\hbar)^2} \int\limits_{-\infty}^{\infty} \Psi(p_x, p_y, p_z, E) \exp\left\{ i \left[\frac{\mathbf{p}}{\hbar} \cdot \mathbf{r} - \frac{E}{\hbar} t \right] \right\} dp_x dp_y dp_z dE$$

we obtain immediately

$$\begin{aligned} \Psi(x + x_c, y + y_c, z + z_c, t + t_c) &= \frac{1}{(2\pi\hbar)^2} \times \\ \int\limits_{-\infty}^{\infty} \phi(p_x, p_y, p_z, E) \, \exp &\left\{ \frac{i}{\hbar} (\mathbf{p} \cdot \mathbf{r} - Et) \right\} dp_x dp_y dp_z \, dE \end{aligned} \tag{6.13}$$

where

$$\phi(p_x, p_y, p_z, E) = \Psi(p_x, p_y, p_z, E) \, \exp\left\{ \frac{i}{\hbar} (\mathbf{p} \cdot \mathbf{r}_c - Et_c) \right\} \tag{6.14}$$

Then the function $D_{\mathbf{p},E}$ fulfills the above required feature and is expressed by

$$\begin{aligned} D_{\mathbf{p},E} = \phi^*(p_x, p_y, p_z, E) \, \phi(p_x, p_y, p_z, E) = \\ \Psi^*(p_x, p_y, p_z, E) \Psi(p_x, p_y, p_z, E) \end{aligned} \tag{6.15}$$

The wave functions $\Psi(x, y, z, t)$ and $\Psi(p_x, p_y, p_z, E)$ are not of relevance for observational physical process but the quantities $\Psi^*(x, y, z, t)\Psi(x, y, z, t)$ and $\Psi^*(p_x, p_y, p_z, E)$ $\Psi(p_x, p_y, p_z, E)$.

In the case of eq. (6.13) we obtain the following scheme with shifts of the form $x + x_c, y + y_c, z + z_c, t + t_c$:

$$\begin{array}{ccc} \Psi(\mathbf{r} + \mathbf{r}_c, t + t_c), t + t_c) & \leftrightarrow & \phi(p_x, p_y, p_z, E) \\ \downarrow & & \downarrow \\ \Psi^*(\mathbf{r} + \mathbf{r}_c, t + t_c)\Psi(\mathbf{r} + \mathbf{r}_c, t + t_c), t + t_c) & \leftrightarrow & \Psi^*(p_x, p_y, p_z, E)\Psi(p_x, p_y, p_z, E) \end{array} \tag{6.16}$$

where

$$\mathbf{r} + \mathbf{r}_c = (x+x_c, y+y_c, z+z_c) \tag{6.17}$$

Only "real numbers" are correlated within observations, and these real numbers are not defined by functions like Ψ but through quantities like $\Psi^*\Psi$. This has to be considered in connection with eq. (6.15).

On the other hand, we have the following scheme in the case of eq. (6.8) "without" the shifts $x+x_c, y+y_c, z+z_c, t+t_c$:

$$\begin{array}{ccc} \Psi(x,y,z,t) & \leftrightarrow & \Psi(p_x,p_y,p_z,E) \\ \downarrow & & \downarrow \\ \Psi^*(x,y,z,t)\Psi(x,y,z,t) & \leftrightarrow & \Psi^*(p_x,p_y,p_z,E)\Psi(p_x,p_y,p_z,E) \end{array} \tag{6.18}$$

Also here, only real numbers are relevant when we consider observational processes, which are expressed through quantities like $\Psi^*\Psi$. This has also to be considered in connection with eq. (6.16).

Note that human beings perceive in everyday life a stone or other physically real things around them as $\Psi^*(x,y,z,t)\Psi(x,y,z,t)$ structure and not as wave function $\Psi(x,y,z,t)$.

When we compare eq. (6.16) with eq. (6.18), we recognize that the shift of the functions $\Psi^*\Psi$ by constant values x_c, y_c, z_c, t_c in $(\mathbf{r},t)$ space does not influence the corresponding functions $\Psi^*\Psi$ in $(\mathbf{p},E)$ space. That is, the $(\mathbf{p},E)$ fluctuations, which reflect physically real processes, are not changed when the object $\Psi^*\Psi$ in the picture is shifted.

Shifts within $(\mathbf{p}, E)$ space

Furthermore, we may require the following: a function $D_{\mathbf{r},t}$ describes reality in $(\mathbf{r},t)$ space if it remains unchanged when both the momentum and the energy are shifted by the constant values p_{xc}, p_{yc}, p_{zc} and E_c:

$$\Psi(p_x,p_y,p_z,E) \to \Psi(p_x+p_{xc}, p_y+p_{yc}, p_z+p_{zc}, E+E_c) \tag{6.19}$$

The function $D_{\mathbf{r},t}$ characterizes the appearance of a physically real object system in $(\mathbf{r},t)$ space.

This property is due to the fact that only fluctuations of p_x, p_y, p_z and E are relevant (Section 6.1.6) and not their absolute values. The wave function $\Psi(x,y,z,t)$ is zero for constant values of p_x, p_y, p_z and E (see also Ref. [12]).

With eq. (6.9)

$$\Psi(p_x,p_y,p_z,E) = \frac{1}{(2\pi\hbar)^2}\int_{-\infty}^{\infty} \Psi(x,y,z,t)\exp\left\{-i\left[\frac{\mathbf{p}}{\hbar}\cdot\mathbf{r} - \frac{E}{\hbar}t\right]\right\}dtdxdydz$$

we get

$$\Psi(p_x + p_{xc}, p_y + p_{yc}, p_z + p_{zc}, E + E_c) = \frac{1}{(2\pi\hbar)^2} \times \int_{-\infty}^{\infty} \phi(x, y, z, t) \exp\left\{-\tfrac{i}{\hbar}(\mathbf{p}\cdot\mathbf{r} - Et)\right\} dxdydzdt \tag{6.20}$$

with

$$\phi(x, y, z, t) = \Psi(x, y, z, t) \exp\left\{-\frac{i}{\hbar}(\mathbf{p}_c \cdot \mathbf{r} - E_c t)\right\}. \tag{6.21}$$

Then, the function $D_{\mathbf{r},t}$ with the feature required above is expressed by

$$D_{\mathbf{r},t} = \phi^*(x, y, z, t)\phi(x, y, z, t) = \Psi^*(x, y, z, t)\Psi(x, y, z, t). \tag{6.22}$$

that is, also here not the functions $\Psi(x, y, z, t)$ and $\Psi(p_x, p_y, p_z, E)$ are of relevance for the physically real process but the quantities $\Psi^*(x, y, z, t)\Psi(x, y, z, t)$ and $\Psi^*(p_x, p_y, p_z, E)\Psi(p_x, p_y, p_z, E)$. In analogy to eq. (6.16), we obtain the following scheme:

$$\begin{array}{ccc} \phi(x, y, z, t) & \leftrightarrow & \Psi(\mathbf{p} + \mathbf{p}_c, E + E_c) \\ \downarrow & & \downarrow \\ \Psi^*(x, y, z, t)\Psi(x, y, z, t) & \leftrightarrow & \Psi^*(\mathbf{p} + \mathbf{p}_c, E + E_c)\Psi(\mathbf{p} + \mathbf{p}_c, E + E_c) \end{array} \tag{6.23}$$

where

$$\mathbf{p} + \mathbf{p}_c = (p_x + p_{xc}, p_y + p_{yc}, p_z + p_{zc}) \tag{6.24}$$

When we compare eq. (6.23) with eq. (6.18), we recognize that the shift of the functions $\Psi^*\Psi$ through constant values $p_{xc}, p_{yc}, p_{zc}, E_c$ in $(\mathbf{p}, E)$ space does not influence the corresponding functions $\Psi^*\Psi$ in $(\mathbf{r}, t)$ space. Only the relative $(\mathbf{p}, E)$ values are of relevance and not the absolute quantities p_x, p_y, p_z, E. As we have stated in Section 6.1.6, the $(\mathbf{p}, E)$ values fluctuate in the course of time τ and this means "interaction" with the environment. In this connection only $(\mathbf{p}, E)$ changes occur, that is, there are only $\Delta\mathbf{p}, \Delta E$ effects. These types of processes define the form (shape) of the objects in $(\mathbf{r}, t)$ space, but the form itself remains preserved. These features are reflected by scheme (6.23).

6.1.9 Instantaneous transition from reality to the picture

Because space and time are not physically real, they are entities of the mind. Since we are caught in space and time, the starting point for the mathematical description of fictitious reality is space and time, which are characterized by the elements x, y, z, t. The variables, on which fictitious reality is based, have to be constructed by means of x, y, z, t. These variables of reality are given by the momentum $\mathbf{p} = (p_x, p_y, p_z)$ and the energy E, that is, reality is located in $(\mathbf{p}, E)$ space.

In fact, **p** and E are introduced through space and time; the unit of momentum is $\text{g} \times \text{length}/\text{time}$ the unit of energy is $\text{g} \times \text{length}^2/\text{time}^2$.

Reality is described by $\Psi(p_x, p_y, p_z, E)$ and the picture through $\Psi(x, y, z, t)$. Both functions are creations of the mind (a more detailed analysis is given in Section 6.2). There exist definite connections between $\Psi(p_x, p_y, p_z, E)$ and $\Psi(x, y, z, t)$, which are given by eqs. (6.8) and (6.9):

$$\Psi(x, y, z, t) = \frac{1}{(2\pi\hbar)^2} \int_{-\infty}^{\infty} \Psi(p_x, p_y, p_z, E) \exp\left\{ i \left[\frac{\mathbf{p}}{\hbar} \cdot \mathbf{r} - \frac{E}{\hbar} t \right] \right\} dp_x dp_y dp_z dE \tag{6.25}$$
$$\Psi(p_x, p_y, p_z, E) \leftarrow \Psi(x, y, z, t)$$

and

$$\Psi(p_x, p_y, p_z, E) = \frac{1}{(2\pi\hbar)^2} \int_{-\infty}^{\infty} \Psi(x, y, z, t) \exp\left\{ -i \left[\frac{\mathbf{p}}{\hbar} \cdot \mathbf{r} - \frac{E}{\hbar} t \right] \right\} dt dx dy dz \tag{6.26}$$
$$\Psi(p_x, p_y, p_z, E) \rightarrow \Psi(x, y, z, t)$$

These connections (equations) are basic and necessary. Otherwise, there would be no reality-picture principle definable.

The thinking process with respect to reality and the picture, which should not be confused with a physically real situation, can be done on an equal footing in two directions, even when the actual information is projected from $(\mathbf{p}, E)$ space to $(\mathbf{r}, t)$ space.

In connection with eq. (6.26), the transition can be thought from $\Psi(p_x, p_y, p_z, E)$ to $\Psi(x, y, z, t)$ and also from $\Psi(x, y, z, t)$ to $\Psi(p_x, p_y, p_z, E)$ when we consider eq. (6.25). Again, these are thinking processes, which take place in the human being's mind in a "conscious" way. We model $\Psi(p_x, p_y, p_z, E)$ or $\Psi(x, y, z, t)$ and determine the functions $\Psi(x, y, z, t)$ or $\Psi(x, y, z, t)$ with eq. (6.25) or (6.26). In both cases, a cause–effect procedure is applied, that is, cause and effect must take place simultaneously. Otherwise, there would be a contradiction.

The actual transition takes place "unconsciously" and corresponds to the transition from $\Psi(p_x, p_y, p_z, E)$ to $\Psi(x, y, z, t)$ and is of course also a process in mind. Therefore, this unconscious process from reality to the pictures is also a cause–effect procedure in the above sense, and cause and effect take place simultaneously.

Remark

The following has to be noted: In the traditional way, the cause–effect procedure is defined with respect to two physical events: first one is the cause, the other is the effect and both events occur successively. If the cause takes place at clock time τ_C and the effect at τ_E, we usually have $\tau_C < \tau_E$. The effect is after the cause. In connection with eqs. (6.25) and (6.26), the situation has to be assessed differently: it is a cause–effect procedure, where cause and effect take place without time delay but simultaneously.

That is, it is not causality in the classical sense: effect "after" the cause. The procedure is in fact different from the traditional view. Within the traditional view, there are two physically real processes, which are involved in a cause–effect procedure. But within the frame of the projection principle, there is only one physically real process which is projected onto space and time and leads to a picture, which describes a physically real process. The picture itself does not reflect a physically real process. Space and time, the frame of the picture, are not permanently installed in the observer's mind, but they are created if there is something to represent.

6.1.10 Reality-picture connection

Equations (6.25) and (6.26) are basic relations. They connect the wave functions $\Psi(x, y, z, t)$ and $\Psi(p_x, p_y, p_z, E)$, and this connection is irresolvable because the relations are basic. This situation reflects a law.

Reality and its picture are tightly linked. No reality without picture and this reflects the purpose of the whole procedure given through eqs. (6.25) and (6.26). Let us repeat: physically real objects are pulled from basic reality γ by the tools of evolution and transfer an observer-relevant part α to the human observer's mind, and the mind creates the picture of reality (see Chapters 4 and 5).

Again, the purpose of this procedure is to observe reality and, therefore, reality is not perceivable without picture. We come to the "reality-picture conception," which we already introduced, but not in connection with the wave functions $\Psi(x, y, z, t)$ and $\Psi(p_x, p_y, p_z, E)$ and their couplings by eqs. (6.25) and (6.26).

We may model $\Psi(x, y, z, t)$ in accordance with definite principles but without $\Psi(p_x, p_y, p_z, E)$ and vice versa, but such a modeling does not eliminate the basic connections (6.25) and (6.26).

The discussion in Section 6.1.9 particularly showed that the reality and its picture exist simultaneously at time τ. In other words, the projection process takes place instantaneously.

6.1.11 Probabilities

Equations (6.25) and (6.26) are not τ dependent and are valid at each time τ. There are no τ effects in connection with these equations. There is no coupling to time τ. This means that $\Psi(x, y, z, t)$ and $\Psi(p_x, p_y, p_z, E)$ exist simultaneously. This property can be analyzed further. The variables of both functions are defined in the range where $\Psi(x, y, z, t)$ and $\Psi(p_x, p_y, p_z, E)$ are different from zero. This can be expressed in general form by

$$\Psi(x, y, z, t), \quad -\infty < x, y, z, t < \infty \tag{6.27}$$

$$\Psi(p_x, p_y, p_z, E), \quad -\infty < p_x, p_y, p_z, E < \infty \tag{6.28}$$

The sets of variables x, y, z, t and p_x, p_y, p_z, E are "defined" at each time τ, all at once. However, they do not "exist" all at once. The variables "can" exist at each time τ. However, it is not imaginable that all possible values x, y, z, t, expressed by $\Psi(x, y, z, t)$, are simultaneously existent at clock time τ. It is also not imaginable that all possible values p_x, p_y, p_z, E, expressed by $\Psi(p_x, p_y, p_z, E)$, are simultaneously existent at τ. The x, y, z, t details and the p_x, p_y, p_z, E details would get lost in the case of an all-at-once existence. The description in terms of the variables x, y, z, t and p_x, p_y, p_z, E would be eliminated.

In other words, the variables "can" exist at each time τ, but they do not exist at each time τ. This situation defines "probability." We get back to this point in the next section.

6.1.12 The interpretation of the wave functions

The basic information about a system are the wave functions $\Psi(x, y, z, t)$ and $\Psi(p_x, p_y, p_z, E)$, which are representations in $(\mathbf{r}, t)$ space and $(\mathbf{p}, E)$ space. The basic quantities $\Psi(x, y, z, t)$ and $\Psi(p_x, p_y, p_z, E)$ are completely equivalent concerning their physical content.

No deterministic behavior

Both functions are not independent of each other, which is expressed by the Fourier transforms (6.8) and (6.9). However, the quantities $\Psi^*(x, y, z, t)\Psi(x, y, z, t)$ and $\Psi^*(p_x, p_y, p_z, E)\Psi(p_x, p_y, p_z, E)$ are more suitable to interpret the role of the variables x, y, z, t and p_x, p_y, p_z, E, and this is because these quantities fulfill certain natural conditions (Section 6.1.8), which are obviously also valid at the macroscopic level.

Deterministic laws for the variables

$$\begin{gathered} x = x(\tau),\ y = y(\tau),\ z = z(\tau),\ t = t(\tau), \\ p_x = p_x(\tau),\ p_y - p_y(\tau),\ p_z = p_z(\tau),\ E = E(\tau) \end{gathered} \tag{6.29}$$

cannot be formulated; trajectories are not defined. It is relatively easy to verify that if eqs. (6.8) and (6.9) are valid simultaneously. The clock time τ is not concerned and goes also here strictly from the past to the future; the time τ has primarily nothing to do with the physically real objects.

In other words, there is no longer a physical law that tells us *when* and how and *where* a "particle" moves. This behavior means that the variables x, y, z, t and p_x, p_y, p_z, E must behave "statistically." There is no alternative. In other words, the elements x, y, z, t and p_x, p_y, p_z, E are statistical variables and we have to find a description for the corresponding probabilitics.

What quantities are relevant here? We have outlined in Section 6.1.8 that the functions $\Psi^*(p_x, p_y, p_z, E)\Psi(p_x, p_y, p_z, E)$ and $\Psi^*(x, y, z, t)\Psi(x, y, z, t)$ are responsible for the appearances (observations) in both spaces [$(\mathbf{p}, E)$ space and $(\mathbf{r}, t)$ space]. These quantities fulfill certain natural conditions. On the other hand, the variables x, y, z, t and p_x, p_y, p_z, E behave statistically.

In conclusion, for the description of physically real objects (systems) two points are obviously in the center:

1. There is a statistical behavior of the variables x, y, z, t in $(\mathbf{r}, t)$ space, and there is a statistical behavior of the variables p_x, p_y, p_z, E in $(\mathbf{p}, E)$ space as well.
2. The description should be done in both spaces on the basis of the functions $\Psi^*(p_x, p_y, p_z, E)\Psi(p_x, p_y, p_z, E)$ and $\Psi^*(x, y, z, t)\Psi(x, y, z, t)$.

The statistical behavior refers to the variables x, y, z, t for which the wave function $\Psi(x, y, z, t)$ is not zero. Furthermore, the statistical behavior also refers to the variables p_x, p_y, p_z, E for which the wave function $\Psi(p_x, p_y, p_z, E)$ is not zero. However, statistical behavior means that there are frequency rates for each of the variables. Frequency rates are represented by real numbers and the functions $\Psi(x, y, z, t)$ and $\Psi(p_x, p_y, p_z, E)$ are not real functions in the mathematical sense but consist of a real and an imaginary part.

We recognized in Section 6.1.8 that the functions $\Psi^*(p_x, p_y, p_z, E)\Psi(p_x, p_y, p_z, E)$ and $\Psi^*(x, y, z, t)\Psi(x, y, z, t)$ fulfill relevant observational conditions, and statistical frequency rates belong to the category of observational quantities. Moreover, the functions $\Psi^*(p_x, p_y, p_z, E)\Psi(p_x, p_y, p_z, E)$ and $\Psi^*(x, y, z, t)\Psi(x, y, z, t)$ are real functions in the mathematical sense. This is important in this connection.

6.1.13 Probability distributions

At clock time τ we have one value for each of the coordinates x, y, z and one value for the system-specific time t:

$$\tau : x, y, z, t \tag{6.30}$$

At time τ we have also one value for each of the momentums p_x, p_y, p_z and one value for the energy E:

$$\tau : p_x, p_y, p_z, E \tag{6.31}$$

Once again, all the variables behave statistically, that is, we can principally nothing say about definite values x', y', z', t' and p'_x, p'_y, p'_z, E' at time τ'. However, due to eqs. (6.8) and (6.9), the variables x, y, z, t of $(\mathbf{r}, t)$ space are completely equivalent to the variables p_x, p_y, p_z of $(\mathbf{p}, E)$ space.

Thus, if there is a probability distribution for x, y, z, t there must also be a probability distribution for p_x, p_y, p_z, E. In other words, at time τ we measure one of the possible sets x, y, z, t of the x, y, z, t structure and, simultaneously, one of the possible sets p_x, p_y, p_z, E of the p_x, p_y, p_z, E structure. But how are the probability distributions in $(\mathbf{r}, t)$ space and in $(\mathbf{p}, E)$ space determined? What are the physical laws? Here the functions $\Psi^*(p_x, p_y, p_z, E)\Psi(p_x, p_y, p_z, E)$ and $\Psi^*(x, y, z, t)\Psi(x, y, z, t)$ with their suitable properties come into play.

If the functions $\Psi^*(x, y, z, t)\Psi(x, y, z, t)$ and $\Psi^*(p_x, p_y, p_z, E)\Psi(p_x, p_y, p_z, E)$ are probability densities for the determination of the variables, that is, for x, y, z, t and p_x, p_y, p_z, E, the following arguments should be valid: The sum (integral) over the functions $\Psi^*(x, y, z, t)\Psi(x, y, z, t)$ expresses the peculiarity that one set x, y, z, t of the possible x, y, z, t sets is given with "certainty," which is reflected by the value $S_{\mathbf{r},t}$ defined by

$$\int_{-\infty}^{\infty} \Psi^*(x, y, z, t)\Psi(x, y, z, t)dxdydzdt = S_{\mathbf{r},t} \tag{6.32}$$

This refers to the situation of $(\mathbf{r}, t)$ space.

Exactly, the same arguments must hold for $(\mathbf{p}, E)$ space: the sum (integral) over the functions $\Psi^*(p_x, p_y, p_z, E)\Psi(p_x, p_y, p_z, E)$ expresses the peculiarity that one set p_x, p_y, p_z, E of the possible p_x, p_y, p_z, E sets is given with "certainty," which is reflected by the value $S_{\mathbf{p},E}$ defined by

$$\int_{-\infty}^{\infty} \Psi^*(p_x, p_y, p_z, E)\Psi(p_x, p_y, p_z, E)dp_x dp_y dp_z dE = S_{\mathbf{p},E} \tag{6.33}$$

At time τ there is one event in $(\mathbf{r}, t)$ space and, simultaneously, there is one event in $(\mathbf{p}, E)$ space. This is in accordance with the reality-picture conception. Thus, we must have

$$S_{\mathbf{r},t} = S_{\mathbf{p},E} \tag{6.34}$$

If the functions $\Psi^*(x, y, z, t)\Psi(x, y, z, t)$ and $\Psi^*(p_x, p_y, p_z, E)\Psi(p_x, p_y, p_z, E)$ fulfill eq. (6.34), we may state that these functions are probability densities.

In fact, we get with eqs. (6.8) and (6.9) the following relationship:

$$\begin{aligned} &\int_{-\infty}^{\infty} \Psi^*(x, y, z, t)\,\Psi(x, y, z, t)\,dxdydzdt = \\ &\qquad \int_{-\infty}^{\infty} \Psi^*(p_x, p_y, p_z, E)\Psi(p_x, p_y, p_z, E)\,dp_x dp_y dp_z dE \end{aligned} \tag{6.35}$$

Furthermore, if we set

$$\int_{-\infty}^{\infty} \Psi^{*}(x,y,z,t)\Psi(x,y,z,t)dxdydzdt = 1 \tag{6.36}$$

we immediately get

$$\int_{-\infty}^{\infty} \Psi^{*}(p_x,p_y,p_z,E)\Psi(p_x,p_y,p_z,E)dp_xdp_ydp_zdE = 1 \tag{6.37}$$

The probability $dW_{\mathbf{r},t}$ of an event that occurs in the space–time region $dxdydzdt$ of $(\mathbf{r},t)$ space is given by

$$dW_{\mathbf{r},t} = \Psi^{*}(x,y,z,t)\Psi(x,y,z,t)dxdydzdt \tag{6.38}$$

On the other hand, the probability $dW_{\mathbf{p},E}$ of an event that occurs in the momentum–energy region $dp_xdp_ydp_zdE$ of $(\mathbf{p},E)$ space is given by

$$dW_{\mathbf{p},\mathbf{E}} = \Psi^{*}(p_x,p_y,p_z,E)\Psi(p_x,p_y,p_z,E)dp_xdp_ydp_zdE \tag{6.39}$$

From our considerations, in the preceding sections a statistical event in $(\mathbf{p},E)$ space and a statistical event in $(\mathbf{r},t)$ space are described by the probability densities $\Psi^{*}(x,y,z,t)\Psi(x,y,z,t)$ and $\Psi^{*}(p_x,p_y,p_z,E)\Psi(p_x,p_y,p_z,E)$ and occur simultaneously.

6.1.14 Connections

The functions $\Psi^{*}(x,y,z,t)\Psi(x,y,z,t)$ and $\Psi^{*}(p_x,p_y,p_z,E)\Psi(p_x,p_y,p_z,E)$ fulfill specific properties and have therefore to be identified with probability densities for the variables x,y,z,t of $(\mathbf{r},t)$ space and for the variables p_x,p_y,p_z,E of $(\mathbf{p},E)$ space.

Within the frame of the projection principle theoretical picture given here, physically real objects are never located in $(\mathbf{r},t)$ space. Thus, it is not possible to interpret the quantity $\Psi^{*}(x,y,z,t)\Psi(x,y,z,t)$ as a probability distribution of finding a physically real object in the intervals

$$\begin{gathered} x, x+dx \\ y, y+dy \\ z, z+dz \\ t, t+dt \end{gathered} \tag{6.40}$$

This contradicts, of course, Born's statistical interpretation of the wave function [12]. Clearly, in those cases, where $\Psi^{*}(x,y,z,t)\Psi(x,y,z,t)$ has sharp peaks, we can define objects in $(\mathbf{r},t)$ space although the material object is not embedded here, but exclusively in $(\mathbf{p}, E)$ space. However, such an object definition is based on $\Psi(x,y,z,t)$ and is not used for the interpretation of $\Psi(x,y,z,t)$ as within Born's statistical interpretation of the wave function.

When there are no physically real objects in $(\mathbf{r}, t)$ space, the following question arises: of what is $\Psi^*(x,y,z,t)\Psi(x,y,z,t)$ the probability density? Of course, it is the probability distribution for the variables x,y,z,t.

The measurement of the variables x,y,z,t is only possible in connection with physically real objects and processes. However, within projection theory such objects and processes are embedded in $(\mathbf{p}, E)$ space, that is, real objects and physically real processes are described by the variables p_x, p_y, p_z, E.

Thus, in order to answer the above question (of what is $\Psi^*(x,y,z,t)\Psi(x,y,z,t)$ the probability density?) we have to use eq. (6.8). Accordingly, $\Psi(x,y,z,t)$ is determined at location x,y,z,t in space–time by all possible values p_x, p_y, p_z and E $(-\infty < p_x, p_y, p_z, E < \infty)$, which are given with the probability density $\Psi^*(p_x, p_y, p_z, E)$ $\Psi(p_x, p_y, p_z, E)$. Therefore, $\Psi^*(x,y,z,t)\Psi(x,y,z,t)$ can only be interpreted in connection with the variables p_x, p_y, p_z and E. There is in fact no other way. We may analyze the situation with respect to $(\mathbf{r}, t)$ space and with respect to $(\mathbf{p}, E)$ space. Both views are equivalent with the frame of the reality-picture conception, which is based on eqs. (6.8) and (6.9).

1. (r, *t*) space

"One of the possible values for p_x, p_y, p_z and for E is present in the intervals $x, x+dx$; $y, y+dy$; $z, z+dz$ and $t, t+dt$ with a probability, which is proportional to the probability density $\Psi^*(x,y,z,t)\Psi(x,y,z,t)$."

But also eq. (6.9), which is equivalent to eq. (6.8), can be read in a similar way Then, $\Psi(p_x, p_y, p_z, E)$ is determined at location p_x, p_y, p_z, E of $(\mathbf{p}, E)$ space by all possible values x,y,z and t $(-\infty < x,y,z,t < \infty)$ which are given with the probability density $\Psi^*(x,y,z,t)\Psi(x,y,z,t)$. In this way, the probability density $\Psi^*(p_x, p_y, p_z, E)\Psi(p_x, p_y, p_z, E)$ in $(\mathbf{p}, E)$ space is interpreted in terms of the variables x,y,z,t. This equivalent way, based on (6.9), is a direct consequence that follows from the reality-picture conception. We get the following connection:

2. (p, *t*) space

"One of the possible values for x,y,z and for t is present in the momentum-energy intervals $p_x, p_x + dp_x; p_y, p_y + dp_y; p_z, p_z + dp_z$ and $E, E + dE$ with a probability, which is proportional to the probability density $\Psi^*(p_x, p_y, p_z, E)\Psi(p_x, p_y, p_z, E)$."

These combined views in $(\mathbf{r}, t)$ space and $(\mathbf{p}, E)$ space confirm that there is one x,y,z,t event at clock time τ and, simultaneously, one p_x, p_y, p_z, E event. Both events are directly based on the fundament eqs (6.8) and (6.9) and are also based on the fact that both spaces, $(\mathbf{r}, t)$ and $(\mathbf{p}, E)$ –, are equivalent.

Human beings are caught in space and time and, therefore, measurements exclusively take place in $(\mathbf{r}, t)$ space. Only the momentums p_x, p_y, p_z and the energy E, which are the variables of (fictitious) reality, are accessible to physically real effects but not the space–time elements x, y, z, t. This situation defines a measurement, which are exclusively done in $(\mathbf{r}, t)$ space:

"The measurement of one of the possible values for p_x, p_y, p_z and for E is done in the space–time intervals $x + dx, y + dy, z + dz$ and $t, t + dt$ with the probability density $\Psi^*(x, y, z, t)\Psi(x, y, z, t)$."

This in fact reflects the reality-picture conception. Reality [located in $(\mathbf{p}, E)$ space] is projected onto $(\mathbf{r}, t)$ space and the human being gets the picture of reality. The following should be underlined: in reality, there are $(\mathbf{p}, E)$ fluctuations and there is an exchange of $\Delta\mathbf{p}, \Delta E$ pieces (Section 6.1.6). There are also $\mathbf{r}, t$ fluctuations, but there is no exchange of $\Delta\mathbf{r}, \Delta t$ pieces.

6.1.15 Detections

Assignments

The Fourier integral (6.8) can be read as follows: each single x, y, z, t state of $(\mathbf{r}, t)$ space is created through all p_x, p_y, p_z, E states of $(\mathbf{p}, E)$ space. This feature can be expressed by

$$\Psi(x, y, z, t) \Leftarrow \{\Psi(p_x, p_y, p_z, E), \ -\infty < p_x, p_y, p_z, E < \infty\} \tag{6.41}$$

One x, y, z, t event in $(\mathbf{r}, t)$ space takes place with a definite probability defined by the probability density $\Psi^*(x, y, z, t)\Psi(x, y, z, t)$. But there is simultaneously one of the possible p_x, p_y, p_z, E events in $(\mathbf{p}, E)$ space that is realized with certainty. This statement can be read from the Fourier transform (6.8).

On the other hand, the inverse Fourier integral (6.9) can be read in a similar way: each single p_x, p_y, p_z, E state of $(\mathbf{p}, E)$ space is created through all x, y, z, t states of $(\mathbf{r}, t)$ space. In analogy to eq. (6.41), this feature is formulated by

$$\Psi(p_x, p_y, p_z, E) \Leftarrow \{\Psi(x, y, z, t), \ -\infty < x, y, z, t < \infty\} \tag{6.42}$$

The inverse case, based on eq. (6.9), is defined as follows: one p_x, p_y, p_z, E event in $(\mathbf{p}, E)$ space takes place with a definite probability defined by the probability density $\Psi^*(p_x, p_y, p_z, E)\Psi(p_x, p_y, p_z, E)$. But there is simultaneously one of the possible x, y, z, t events that is realized with certainty in $(\mathbf{r}, t)$ space.

These combined views in $(\mathbf{r}, t)$ space and $(\mathbf{p}, E)$ space underline that there is "one" x, y, z, t event and, simultaneously, "one" p_x, p_y, p_z, E event. Both events are directly based on the fundament eqs (6.8) and (6.9) [schemes (6.41) and (6.42)] and are essentially based on the fact that both spaces, $(\mathbf{r}, t)$ space and $(\mathbf{p}, E)$ space, are equivalent.

In conclusion, there is an ensemble of x, y, z, t probabilities and there is simultaneously an ensemble of p_x, p_y, p_z, E probabilities. A detector for the registration of quantum states must be adapted and constructed with respect to such a situation.

Simultaneous registrations

At clock time τ, all physically real objects (systems) exist in $(\mathbf{r}, t)$ space and, simultaneously in $(\mathbf{p}, E)$ space. These objects are characterized in $(\mathbf{r}, t)$ space by the wave function $\Psi(x, y, z, t)$ and the probability density $\Psi^*(x, y, z, t)\Psi(x, y, z, t)$. In $(\mathbf{p}, E)$ space they are described by the corresponding quantities, expressed by $\Psi(p_x, p_y, p_z, E)$ and $\Psi^*(p_x, p_y, p_z, E)\Psi(p_x, p_y, p_z, E)$.

In other words, the quantum states exist in $(\mathbf{r}, t)$ space and simultaneously in $(\mathbf{p}, E)$ space. For the detection of these quantum states, we need measuring instruments with specific properties. We need detectors, which also exist simultaneously in $(\mathbf{r}, t)$ space and $(\mathbf{p}, E)$ space. Both detector versions are connected through a Fourier transform, as in the case of the physically real objects (systems).

The detectors must have specific properties: They must have the ability to detect in $(\mathbf{r}, t)$ space all quantum states x, y, z, t of the system under investigation for which the probability density $\Psi^*(x, y, z, t)\Psi(x, y, z, t)$ is not zero. They must also have, simultaneously, the ability to detect in $(\mathbf{p}, E)$ space all quantum states p_x, p_y, p_z, E of the same physically real system for which the probability density $\Psi^*(p_x, p_y, p_z, E)\Psi(p_x, p_y, p_z, E)$ is not zero.

How such detectors can be constructed is at first an open question, but they must fulfill the properties just quoted. These properties are dictated by the basic principles of the projection principle.

At clock time τ, the detector measures in $(\mathbf{r}, t)$ space one of the x, y, z, t states with a certain probability, which is expressed by the probability density $\Psi^*(x, y, z, t)$ $\Psi(x, y, z, t)$, and the detector measures, simultaneously, one of the possible p_x, p_y, p_z, E states also with a certain probability, which is proportional to the probability density $\Psi^*(p_x, p_y, p_z, E)\Psi(p_x, p_y, p_z, E)$.

Conclusion

Human beings are caught in space and time and do the measurements exclusively in $(\mathbf{r}, t)$ space. In $(\mathbf{r}, t)$ space, the detector indicates at the space–time position x, y, z, t changes of its geometrical structure caused by p_x, p_y, p_z and E in $(\mathbf{p}, E)$ space. The geometrical structure changes are the result of an interaction of the detector with the object in $(\mathbf{p}, E)$ space, which comes into existence through an exchange of momentum and energy $\Delta\mathbf{p}$ and ΔE. The entire scenario is projected onto $(\mathbf{r}, t)$ space by means of a Fourier transform.

6.2 Dialogue with nature

How can we connect the wave functions $\Psi(x, y, z, t)$ and $\Psi(p_x, p_y, p_z, E)$, analyzed in Section 6.1, with what we defined as basic reality γ and its pulled part α? The entities γ and α have been introduced and discussed in Chapters 4 and 5. In particular, for human beings the species-dependent part α is relevant. Only the quantity α reflect that what is accessible for human beings with their sensory apparatus and the developed measuring instruments. Human beings know only a little about the tools of evolution, which pull the relevant part α from γ.

6.2.1 Caught in space and time

In order to be able to answer the above question with respect to the relation between the functions $\Psi(x, y, z, t)$ and $\Psi(p_x, p_y, p_z, E)$ and the basic entities γ and α, we have to recourse to the very basics of the projection principle. However, we have to underline already here that this connection, which we want to express by

$$\Psi(x, y, z, t), \Psi(p_x, p_y, p_z, E) \quad \leftrightarrow \quad \gamma, \alpha. \tag{6.43}$$

can in principle only be explained in a qualitative way. Relation (6.43) reflects a "dialogue with nature." Let us discuss the situation in this section.

What does the notion "nature" mean within projection theory? Nature is what a human being experiences without individual conceptions. These are the most direct impressions about the world outside. Within the projection theory, these direct impressions are expressed through the picture of physically real objects, which are located in basic reality γ (see in particular Chapters 4 and 5). Then the picture itself can be defined as nature. The picture is unconsciously created by the human being's mind. Why unconsciously? What does it mean?

All what we experience and do on the basis of our five senses and what is registered by measuring instruments are events in space and time and are exclusively represented in the picture. That is, the picture of reality (nature) is exclusively a pattern represented in space and time. Human beings are caught in space and time.

Human beings have the impression that everything is embedded in space and time, but this is a fallacy as we have outlined in the earlier chapters. Nevertheless, the actions and also the thinking with respect to physically real entities are in fact restricted to space and time. This justifies to use the notion "nature" for the picture of reality.

6.2.2 The unconscious way

But how come all these impressions into existence and why unconsciously? As we have stated several times, space and time do not belong to the class of physically real objects. Thus, space and time are exclusively entities of the picture and do not belong to basic reality γ and of course also the information α. Once again, α is pulled from γ with the tools of evolution and is transferred to the human being's mind.

γ and α are located outside space and time. More details are given in Chapters 4 and 5.

The following is here essential: since human beings are caught in space and time, human beings have no direct access to γ and α, and this is because γ and α are located outside space and time. Thus, the theoretical description of γ and α requires representations "without space and time." For the mathematical description of γ and α "without space and time," we therefore need other characteristic variables, but we do not know them and even do not know a way to find them, which is because human beings are caught in space and time. Such variables simply do not exist.

The steps from basic reality to the picture cannot be described theoretically and, therefore, we cannot understand the steps from γ to the space–time picture. We know that all these steps exist within the frame of the projection principle, that is, we know the steps from γ to the space–time picture but not the analytical details.

The way from α, which is the human-relevant part of γ, to the picture is done through the human being's mind but, as we outlined, this way is theoretically not accessible. However, the real world goes this way. Both points are only in accordance with each other if the mind processes "unconsciously." In other words, we recognize nature through a space–time picture. The picture itself is an unconscious creation of the human being's mind. The situation is summarized in the following scheme:

$$\alpha \quad \xrightarrow[\text{(unconscious transiton)}]{\text{mind}} \quad \text{picture in space and time} \tag{6.44}$$

Scheme (6.44) reflects an important feature, which we would like to discuss in the next section.

6.2.3 The automatic execution of information

We know that γ exists but we cannot express its features with the means of theoretical physics. The features of the entity γ can only be formulated in terms of a picture in space and time. However, from the picture it is not directly possible to draw conclusions on the objects in basic reality γ.

The intuition of human beings is not sensitive to physically real entities without the presence of space and time. Thus, a human being cannot develop conceptions (ideas) for what is outside space and time. However, without ideas no plans for conscious actions are possible.

From all that follows that the functionalities of γ and α can merely be expressed within a "qualitative frame." The detailed functionality of the tools of evolution is unknown as well. The mind processes the physically real information α but it can only do that in an "unconscious" way because a conscious way would need a certain kind of thinking that is usually based on an information, which is given, in the case of physically real facts, in analytical form.

A conscious processing of α has to be excluded. Such a conscious processing would take place in the human being's mind, and we would need a theoretical conception for the creation of space and time in the mind; space and time are needed for the representation of the physically real facts in the picture of reality.

The reason why the transition from the entities γ and α to the picture is performed in an unconscious way is the fact that a human is principally not be able to analyze the situation. We simply do not know the structures (details) of γ and α; therefore, we cannot understand the transition from the entities γ and α to the picture of reality. Therefore, that what we do not understand cannot be theoretically described. In other words, the situation is not analyzable for human beings. We have in particular no theoretical conception for the creation of space and time within the mind; we do not know the mechanism how space and time come into existence.

However, the transition exists, that is, the real world actually goes this way. The real world is not caught in space and time. The conscious way is obviously excluded on principle. This situation obviously reflects a principle. But it does allow to do that in an unconscious way. The automatic execution is more reliable and is not dependent on individual intentions. Thus, we may state that the mind has the ability to act in an unconscious way.

For an optimal living, the members of a species need the same preconditions that are required for all. Then, human beings are able to relate their actions optimal to each other. Optimal actions are obviously necessary for an effective living and, in particular, for survival, at least in the early phase of evolution. The same preconditions, that is, in this case the same impression of the physically real world before the eyes, are an essential point. This is guaranteed through an automatic (unconscious) execution.

In Chapter 1, we discussed this topic in connection with members of another species. We recognized that the picture of reality of a turkey is obviously quite different from that of a human being. This situation led to a disaster.

6.2.4 The conscious way

We recognize the world outside as a space–time picture before our eyes, which comes into existence in an "unconscious way." All human beings equally experience the physically real objects around us in this unconscious way. Such space–time pictures are the basis for actions. However, we merely recognize the objects and their relative distances to each other, but not their properties and features. What physical laws are behind the space–time structures of these pictures? Why are they not revealed by the picture? Such kind of statements are not directly possible on the basis of the observed picture alone. It is not directly possible to draw any conclusion on the objects in basic reality γ. The "unconscious" way is not open for such kind of analysis. We have to go the "conscious" way with adequate methods. How is such a conscious way organized?

We are caught in space and time. Therefore, the starting point for the "conscious way" must be given through the elements of space and time. Before we analyze the situation, let us still summarize the main items on which the conscious (intellectual) way is based.

Summary of the main items

Space and time have to be considered as the basis of physics. However, the coordinates and the time elements are not accessible to measurements. Space and time have not the ability to interact, also not with the sensory apparatus of human beings. Thus, space and time are not observable, but we experience them. Space and time do not exist in physically real form in the world outside and, simultaneously, in the picture as entities that are not physically real. The evolution would not have chosen such space–time elements independently from each other, but with exactly the same features. This is in fact a rather artificial conception and has to be excluded.

Thus, space and time cannot be considered as physically real entities like matter. This must have consequences: in the physically real world outside there can be no entities that are not physically real. This means there can be no space in the world outside and also no time.

The projection principle incorporates this feature. We have a reality (the world outside) without space and time, but the reality is projected onto space and time. Since human beings are caught in space and time, they are not able to describe the material reality outside. We simply do not know the variables that we need for the formulation of the physical laws in its most basic form. Furthermore, it is not imaginable to perform experiments outside of space and time. This means that we are not able to experience and to describe what we call "basic reality." The details are simply not accessible to human beings

On the other hand, we experience the picture of reality, that is, the projection of reality onto space and time. That is, we know that there is a physically real world outside and we also know that there must be a way, a procedure, to go from basic reality to the picture. But this way remains hidden, and the entire procedure from reality to the picture can only be described "qualitatively" in an unconscious way.

We analyzed the situation in Chapter 4 and argued quite generally that the information about the physically real objects is pulled from basic reality γ by the tools of evolution and is transferred to the observer's mind, which we called α. The entities γ and α do not exist in mathematical form but in the form of raw data, which are processed by the mind, and this procedure leads to a picture in space and time, where space and time themselves are creations of the mind (in accordance with the view of Immanuel Kant). Since we are not able to formulate all that by means of mathematical tools, the procedure reflects a strict "qualitative view" and is achieved in an unconscious way.

The pictures of reality come into existence without knowing how they come into existence. We do not know the creation laws. Therefore, we cannot directly read the properties of physically real systems from the space–time picture. They are located in basic reality, but are represented in the space–time picture. We have principally no idea (possibility) how the information about the systems, and also the human being's specific part α is expressed in the real world outside.

6.2.5 Fictitious realities and basic reality

We cannot describe the basic information α and a projection of the physically real properties onto space and time is therefore not possible in a conscious way, and this is because a conscious formulation for α is not possible. We have therefore introduce the so-called fictitious realities having the feature to be ascertainable mathematically and can be projected onto space and time.

Let us briefly underline what we have stated in Section 6.1. The actual physically real world is embedded in "basic reality," which is however not directly accessible to human beings. Therefore, a human being needs what we named "fictitious realty." Fictitious realities do not say what actually "is" in basic reality, but "as if" it were so.

Development of physical realities and the dialogue with nature

Only via fictitious realities, a human being is able to find out what the "unconsciously" created space–time picture (nature) contains with respect to the physically real situation in the world outside? There is basically only one possibility: the human being develops "consciously" a theoretical conception (a fictitious reality) for the analysis of the unconsciously produced space–time picture. It is the aim to reproduce the observed space–time pictures, which we called nature, by means of theoretical space–time pictures. Both versions of the picture are then compared, and the conception, on which the theoretical picture is based, reflects the physics of the observed picture. It is a "dialogue with nature," so to speak.

This way (dialogue with nature) is throughout applied in traditional physics [5]. Also in traditional physics, there is basically only one possibility: the human being creates a physical conception of the system under investigation (or the world outside) by thinking and this gives rise to questions, that is, questions that are put to nature itself: that is to say the human being carries out specific experiments, and the deflection of the pointer on the measuring device is the answer to our questions. If the theory describes the experiments sufficiently well, we may conclude that the elements of this theoretical conception reflect the properties of the system or even the structure of the world outside.

Exactly the same is done within projection theory. The only difference here is the notion "nature." It is defined more specifically within the frame of the projection principle.

The general principle

What is the general principle? The way is to develop first fictitious realities on the basis of the general principles of the space–time elements x, y, z, t. These are the basic variables on which human beings are adapted. The elements $\mathbf{r} = (x, y, z)$, t form the so-called $(\mathbf{r}, t)$ space.

Like the information α, fictitious reality is assumed to be able to create the physically real objects in basic reality. The situation in connection with α and fictitious reality is shown in Figure 6.2. In the next step, the structure of fictitious reality has to be formulated mathematically in order to make a statement about the information α, which reflects the system (object) under investigation. This is the conscious way; the transition from α to the observed space–time picture is the unconscious way and reflect what we defined above as nature.

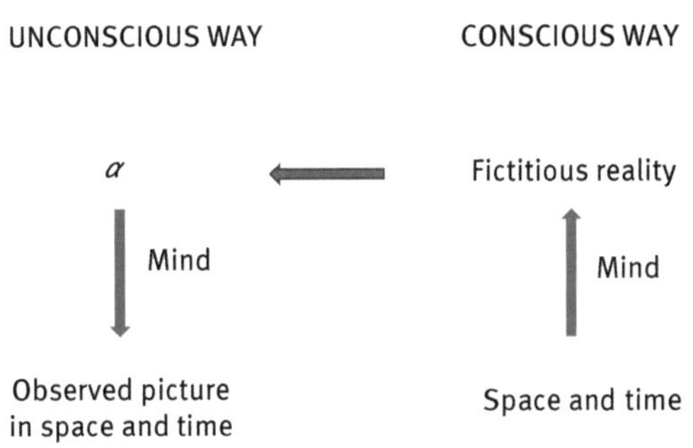

Figure 6.2: A human being needs physical information in order to learn something about the world outside. Here the species-dependent entity α is relevant, which is located in the mind of the human being. The transition of physically real information from basic reality to the mind is mastered through the tools of evolution. The entity α is "unconsciously" processed within the mind of the observing human being. In this way, a picture in space and time is unconsciously produced, whereby space and time are produced unconsciously as well and belong to the mind-processing procedure. As we outlined several times, the physically real information α may not contain space and time. Since human beings are caught in space and time, it is therefore not possible to describe α. Thus, we have to recourse to another way for the understanding of the unconsciously produced space–time picture. This is the "conscious" way. The starting point is here the space–time itself, and the elements of space and time are used to create fictitious realities, which describe the physically real world in close connection to α.

The variables of fictitious reality

In order to understand the relationships in the observed space–time picture, we need a mathematical formulation of the raw data α, but this is not possible in a direct way. As we have outlined earlier and in Figure 6.2, this is only possible on the basis of fictitious realities. The variables, which belong to a fictitious reality, are the momentum $\mathbf{p} = (p_x, p_y, p_z)$ and the energy E. Human beings are caught in space and time and, therefore, the variables x, y, z, t are the starting point for the construction of the variables for fictitious reality, that is, for p_x, p_y, p_z and E.

Already in Newton's mechanics, the momentum p_x, p_y, p_z and the energy E are essential quantities for the characterization of physically real objects. We would like to assume that also within the frame of the projection principle, the quantities p_x, p_y, p_z and E reflect adequately the properties of physically real objects (systems). We particularly assume that such a description of physically real objects is close to the corresponding characterization through the basic information α. This is of course an analogy consideration. Since α reflects physically real features, also the definition for the mathematical expression must reflect physically real features. We have of course criteria to check that; here the "dialogue with nature" is seminal.

Direct comparisons: Possibilities for checking

For the understanding of the structures in the space–time pictures, we need mathematical formulations for the adequate description of the systems and the processes, which are represented through the picture. The mathematical descriptions, developed by physicists, contain specific principles that, in the opinion of the physicist, actually take place.

The mathematical formulation is necessary to base the considerations on a reality, which has however to be seen as a "fictitious realty" since the actual reality (basic reality) remains hidden. That is, the fictitious reality creates also a picture of reality in space and time [$(\mathbf{r}, t)$ space]. The structures within this "theoretical" space–time picture, indicated here by $[x, y, z, \tau]_{\text{theory}}$, are then compared with the "observed" space–time picture $[x, y, z, \tau]_{\text{observation}}$, produced in an unconscious way discussed earlier. If the structures within both pictures are close together, that is, if

$$[x, y, z, \tau]_{observation} = [x, y, z, \tau]_{theory} \tag{6.45}$$

we may conclude that the model for the description of basic reality, expressed within the frame of a fictitious reality, is characteristic for the system under investigation.

Since we are caught in space and time, the starting point for the mathematical description of what we called "fictitious reality" is space and time. Using the elements x, y, z, t, the variables, on which fictitious reality is based, have to be constructed. As we already stated, these variables are given by the momentum $\mathbf{p} = (p_x, p_y, p_z)$ and the energy E, that is, fictitious reality is characterized by $(\mathbf{p}, E)$ space. In fact, p_x, p_y, p_z, E

and E are introduced through space and time; the unit of momentum is $g \times length/$ *time* the unit of energy is $g \times length^2/time^2$.

The physical properties of the system are expressed by the wave function $\Psi(p_x, p_y, p_z, E)$. This information is projected onto $(\mathbf{r}, t)$ space and we obtain the wave function $\Psi(x, y, z, t)$. The function $\Psi(p_x, p_y, p_z, E)$ describes (fictitious) reality and $\Psi(x, y, z, t)$ its picture. $\Psi(p_x, p_y, p_z, E)$ corresponds to α and is a substitute for α:

$$\alpha \leftarrow \Psi(p_x, p_y, p_z, E) \tag{6.46}$$

On the other hand, the wave function $\Psi(x, y, z, t)$ describes the theoretical picture $[x, y, z, t = \tau]_{\text{theory}}$. The situation is illustrated in Figure 6.3.

UNCONSCIOUS WAY		CONSCIOUS WAY				
γ						
$\downarrow$						
α				$\Psi(p_x, p_y, p_z, E)$	$\leftarrow$	p_x, p_y, p_z, E
$\downarrow$ mind				$\downarrow$ mind		$\uparrow$ mind
$[(x, y, z, \tau)\ \text{picture}]_{\text{OBS}}$	$\Leftrightarrow$	$[(x, y, z, \tau)\ \text{picture}]_{\text{TH}}$	$\Leftarrow$	$\Psi(x, y, z, t)$	$\leftarrow$	x, y, z, t

Figure 6.3: The situation in connection with fictitious realities (Figure 6.2) is described in more detail. The unconscious creation of the observed space–time structure, expressed by $[(x, y, z, \tau)\ \text{picture}]_{\text{OBS}}$, is explained by the consciously created theoretical space–time structure of the physically real system under investigation; it is indicated by $[(x, y, z, \tau)\ \text{picture}]_{\text{TH}}$. This consciously created picture is constructed by means of a fictitious reality described by the wave function $\Psi(p_x, p_y, p_z, E)$ in $(\mathbf{p}, E)$ space. The information is projected onto $(\mathbf{r}, t)$ space and we obtain the function $\Psi(x, y, z, t)$ through a Fourier transform. $\Psi(x, y, z, t)$ directly leads to the theoretical picture of reality expressed as $[(x, y, z, \tau)\ \text{picture}]_{\text{TH}}$. If the observed picture is close to the theoretical picture, the model for $\Psi(p_x, p_y, p_z, E)$ can be considered as realistic and describes the physically real information α adequately. The starting point for this development is the conscious construction of the variables p_x, p_y, p_z, E for the description of fictitious reality through the set x, y, z, t of space and time to which human beings are adapted.

We face the unconscious way, that is, the transition from basic reality γ to the observed space–time picture, with the conscious way, which is based on theoretical considerations in connection with the wave functions $\Psi(p_x, p_y, p_z, E)$ and $\Psi(x, y, z, t)$.

The entire conception, developed in Section 6.1, contains the projection principle in analytical form. The principles behind this conception can also be formulated as follows: on the basis of space and time, that is, on the elements x, y, z, t, we have found a new set of variables $[\mathbf{p}, E$ with $\mathbf{p} = (p_x, p_y, p_z)]$ for the description of the features of fictitious reality, and fictitious reality is a substitute for the raw information α. In other words, within the conscious way we used in Section 6.1 as a starting point of our consideration the space–time elements x, y, z, t for the "conscious" formulation of the variables p_x, p_y, p_z, E, which allow to express the properties of fictitious reality. We have

$$p_x, p_y, p_z, E \xleftarrow[\text{of the mind}]{\text{conscious action}} x, y, z, t \tag{6.47}$$

In contrast to eq. (6.44), scheme (6.47) reflects a conscious action of the human being's mind.

6.3 Operators

The "picture of reality" is connected to "reality." We experience the world in $(\mathbf{r}, t)$ space and exclusively perform our measurements in $(\mathbf{r}, t)$ space. In other words, due to the connection between the "picture of reality" and "reality," we are able to observe the properties of the world in $(\mathbf{r}, t)$ space. This mutual dependence of the variables $\mathbf{r}, t, \mathbf{p}$ and E can be expressed by the relation $h(\mathbf{r}, t, \mathbf{p}, E) = 0$.

However, due to the equivalence of $(\mathbf{r}, t)$ space and $(\mathbf{p}, E)$ space, it makes only sense to express the properties of physically real objects in one of the both spaces, either we work in $(\mathbf{r}, t)$ space or in $(\mathbf{p}, E)$ space. Then, due to $h(\mathbf{r}, t, \mathbf{p}, E) = 0$, we have to know how the variables p_x, p_y, p_z, E are expressed in $(\mathbf{r}, t)$ space and, on the other hand, if we would like to work in $(\mathbf{p}, E)$ space we have to know how the variables x, y, z, t are expressed in $(\mathbf{p}, E)$ space.

6.3.1 The variables are not always simple numbers

It is not possible to give definite values for the coordinates x, y, z and the time t if p_x, p_y, p_z, E take definite values. This is due to the structure of the Fourier transform. And, on the other hand, it is not possible to give definite values for p_x, p_y, p_z and E if the coordinates x, y, z and the time t take definite values. For example, for the determination of $\Psi(x, y, z, t)$ for definite values x, y, z and t, say x_1, y_1, z_1 and t_1, we need all possible values for p_x, p_y, p_z and E, in principle from $-\infty$ to ∞. This is a property of the Fourier transformation (6.8).

Thus, in the analysis of quantum phenomena, given within projection theory, the variables p_x, p_y, p_z and E, expressed in $(\mathbf{r}, t)$ space, cannot be simple numbers. On the other hand, the variables x, y, z and t, expressed in $(\mathbf{p}, E)$ space, can also not be simple numbers.

But how can we express the variables p_x, p_y, p_z and E in $(\mathbf{r}, t)$ space and, on the other hand, x, y, z and t in $(\mathbf{p}, E)$ space? In order to answer this question, we have to apply eq. (6.8). Let us consider the following identities:

$$-i\hbar\frac{\partial}{\partial x}\Psi(x,y,z,t)=\frac{1}{(2\pi\hbar)^2}\times$$
$$\int_{-\infty}^{\infty} p_x\Psi(p_x,p_y,p_z,E)\exp\left\{\frac{i}{\hbar}\left[p_xx+p_yy+p_zz-Et\right]\right\}dp_xdp_ydp_z\,dE \tag{6.48}$$

$$-i\hbar\frac{\partial}{\partial y}\Psi(x,y,z,t)=\frac{1}{(2\pi\hbar)^2}\times$$
$$\int_{-\infty}^{\infty} p_y\Psi(p_x,p_y,p_z,E)\exp\left\{\frac{i}{\hbar}\left[p_xx+p_yy+p_zz-Et\right]\right\}dp_xdp_ydp_z\,dE \tag{6.49}$$

$$-i\hbar\frac{\partial}{\partial z}\Psi(x,y,z,t)=\frac{1}{(2\pi\hbar)^2}\times$$
$$\int_{-\infty}^{\infty} p_z\Psi(p_x,p_y,p_z,E)\exp\left\{\frac{i}{\hbar}\left[p_xx+p_yy+p_zz-Et\right]\right\}dp_xdp_ydp_z\,dE \tag{6.50}$$

Interpretation of eqs. (6.48) – (6.50): any information given in $(\mathbf{r},t)$ space can be completely transformed into $(\mathbf{p},E)$ space, and vice versa. Both information must be physically equivalent. We have two representations of the same thing. $\Psi(p_x,p_y,p_z,E)$ in eqs. (6.48) – (6.50)) is equivalent to $\Psi(x,y,z,t)$, and vice versa [see eq. (6.9)]. Also, $-i\hbar\partial/\partial x\Psi(x,y,z,t)$ and $p_x\Psi(p_x,p_y,p_z,E)$ in eq. (6.48) must be equivalent. Thus, the operators

$$\hat{p}_x=-i\hbar\frac{\partial}{\partial x},\hat{p}_y=-i\hbar\frac{\partial}{\partial y},\hat{p}_z=-i\hbar\frac{\partial}{\partial z} \tag{6.51}$$

must be equivalent to the momentum p_x,p_y,p_z, that is, the momentums take the form of operators in $(\mathbf{r},t)$ space. We have the transitions

$$p_x\rightarrow-i\hbar\frac{\partial}{\partial x},p_y\rightarrow-i\hbar\frac{\partial}{\partial y},p_z\rightarrow-i\hbar\frac{\partial}{\partial z} \tag{6.52}$$

when we go from $(\mathbf{p},E)$ space to $(\mathbf{r},t)$ space.

6.3.2 The operator for the time coordinate

In the same way, we can find on the basis of eqs. (6.8) and (6.9) operators for E, $\mathbf{r}$ and t. For example, with eq. (6.9) we have

$$-i\hbar\frac{\partial}{\partial E}\Psi(p_x,p_y,p_z,E)=\frac{1}{(2\pi\hbar)^2}\times \int_{-\infty}^{\infty} t\,\Psi(x,y,z,t)\exp\left\{-i\left[\frac{\mathbf{p}}{\hbar}\cdot\mathbf{r}-\frac{E}{\hbar}t\right]\right\}dxdydzdt \tag{6.53}$$

Thus, the operator

$$\hat{t}=-i\hbar\frac{\partial}{\partial E} \tag{6.54}$$

must be equivalent to the system-specific time t, that is, the time t takes the form of an operator in $(\mathbf{p},E)$ space. There is the transition

$$t\rightarrow -i\hbar\frac{\partial}{\partial E}$$

when we go from $(\mathbf{r},t)$ space to $(\mathbf{p},E)$ space.

Furthermore, it is straightforward to show by means of eq. (6.9) that the coordinates x,y,z take in $(\mathbf{p},E)$ space the following operator form:

$$\hat{x}=i\hbar\frac{\partial}{\partial p_x},\ \hat{y}=i\hbar\frac{\partial}{\partial p_y},\ \hat{z}=i\hbar\frac{\partial}{\partial p_z} \tag{6.55}$$

That is, we have the transitions

$$x\rightarrow i\hbar\frac{\partial}{\partial p_x},y\rightarrow i\hbar\frac{\partial}{\partial p_y},z\rightarrow i\hbar\frac{\partial}{\partial p_z} \tag{6.56}$$

when we go from $(\mathbf{r},t)$ space to $(\mathbf{p},E)$ space.

Moreover, the energy takes in $(\mathbf{r},t)$ space and operator form as well. We get the operator

$$\hat{E}\rightarrow i\hbar\frac{\partial}{\partial t} \tag{6.57}$$

Here we have the transitions

$$E\rightarrow i\hbar\frac{\partial}{\partial t} \tag{6.58}$$

when we go from $(\mathbf{p},E)$ space to $(\mathbf{r},t)$ space.

6.3.3 Conclusion and summary

Within the frame of the projection principle we have the following situation: both spaces, $(\mathbf{r},t)$ space, $(\mathbf{p},E)$ space, are equivalent concerning their information. Each physically real object contains in both spaces exactly the same physical information. In particular, we have the following peculiarities:

1. In $(\mathbf{r}, t)$ space, the coordinates x, y, z and the time t are numbers and the momentums and the energy are operators that are given in the form $-i\hbar\partial/\partial x$, $-i\hbar\partial/\partial y$, $-i\hbar\partial/\partial z, i\hbar\partial/\partial t$.
2. In $(\mathbf{p}, E)$ space, the coordinates and the time are expressed by the operators $i\hbar\partial/\partial p_x$, $i\hbar\partial/\partial p_y$, $i\hbar\partial/\partial p_z$, $-i\hbar\partial/\partial E$. The momentums p_x, p_y, p_z and the energy E are numbers.

There are two interesting points:

1. We have shown that either the coordinates or the momentums are numbers. If the coordinates are numbers, the momentums must be operators in $(\mathbf{r}, t)$ space. If the momentums are numbers, the coordinates must be operators in $(\mathbf{p}, E)$ space.

These rules can only be obtained in traditional quantum theory in connection with Schrödinger's equation, and this equation is an assumption and cannot be derived in usual quantum theory; this point has been discussed in Ref. [12]. Therefore, also the rules concerning numbers and operators have to be considered as assumed within the traditional view of quantum theory. The source of these rules is unknown within usual quantum theory and become therefore postulates within this view.

In contrast to projection theory (physical reality is projected onto space–time), within conventional quantum theory (physical reality is embedded in space–time) we cannot recognize what physical reality hide behind these laws.

2. The time coordinate is not always a simple number within projection theory, but can also be an operator $(-i\hbar\,\partial/\partial E)$; this is of fundamental importance as we already discussed in the preceding chapters. Within traditional quantum theory, the time is always a simple external parameter and is exclusively given by the clock time τ.

The laws, which we have found in this section, are general in character and do not reflect specific properties of physically real objects (systems).

6.4 Operator equations

Let us start with the well-known classical equation for a system (particle) in an external field $U(x, y, z)$:

$$E = \frac{p_x^2 + p_y^2 + p_z^2}{2m_0} + U(x, y, z) \tag{6.59}$$

where m_0 is the mass of the object, p_x, p_y, p_z are the components of the momentum and E is its energy. Here the particle is still point-like and it is embedded in space. The effect of $\hbar$ (see Section 6.1) is that we have no longer "one" space, but "two" coexisting spaces [$(\mathbf{p}, E)$ space and $(\mathbf{r}, t)$ space] representing reality and its picture.

Both spaces are equivalent and we may describe the physical object (system) in $(\mathbf{p}, E)$ space but in $(\mathbf{r}, t)$ space as well. Using eq. (6.59) and the rules derived in Section 6.3, we can formulate the corresponding quantum-mechanical equations for $\Psi(p_x, p_y, p_z, E)$ and $\Psi(x, y, z, t)$.

The operators, which we derived for the variables x, y, z, t and p_x, p_y, p_z, E, are generally valid. In this section and in Section 6.6 we will formulate operator equations that refer to specific objects and systems, respectively. First, let us give the following general remark concerning the notion "equivalency".

6.4.1 Equivalencies

In the derivation of these equations we have to notice the following: the function f of a variable, say x, becomes in those cases, where the variable becomes an operator $i\hbar\,\partial/\partial p_x$ (Section 6.3), a function of this operator, that is, we have the following equivalency:

$$f(x)\Psi(x,y,z,t) \leftrightarrow f\left(i\hbar\frac{\partial}{\partial p_x}\right)\Psi(p_x,p_y,p_z,E), \tag{6.60}$$

where rule (6.55) has been used.

Equation (6.60) can easily be verified. Expanding the function $f(x)$ in a Taylor series, we have

$$f(x) = \sum_{n=0}^{\infty} a_n x^n. \tag{6.61}$$

With eq. (6.9) we obtain immediately

$$\begin{aligned} i^n\hbar^n a_n \frac{\partial^n}{\partial p_x^n}\Psi(p_x,p_y,p_z,E) = \frac{1}{(2\pi\hbar)^2} \times \\ \int_{-\infty}^{\infty} a_n x^n \Psi(x,y,z,t)\, \exp\left\{-\frac{i}{\hbar}(\mathbf{p}\cdot\mathbf{r}-Et)\right\} dxdydzdt \end{aligned} \tag{6.62}$$

and we recognize that the nth term of eq. (6.61) takes in $(\mathbf{p}, E)$ space the form

$$i^n\hbar^n a_n \frac{\partial^n}{\partial p_x^n}, \tag{6.63}$$

and it is equivalent to

$$a_n x^n. \tag{6.64}$$

This argumentation is valid for all terms in eq. (6.61), and we obtain

$$f\left(i\hbar\frac{\partial}{\partial p_x}\right)\Psi(p_x,p_y,p_z,E)=\frac{1}{(2\pi\hbar)^2}\times \int\limits_{-\infty}^{\infty} f(x)\,\Psi(x,y,z,t)\,\exp\left\{-\frac{i}{\hbar}(\mathbf{p}\cdot\mathbf{r}-Et)\right\}dxdydzdt \tag{6.65}$$

That is, we get the equivalency given by eq. (6.60).

6.4.2 Space-specific formulations

In order to formulate the equations for the wave functions $\Psi(p_x,p_y,p_z,E)$ and $\Psi(x,y,z,t)$, we have to consider the situations in the two spaces.

(p, *t*) space

Only the function $U(x,y,z)$ in eq. (6.59) has to be transformed since the variable $r=(x,y,z)$ does not belong to $(\mathbf{p},E)$ space. In analogy to eq. (6.65) we have the equivalency

$$U(x,y,z)\leftrightarrow U\left(i\hbar\frac{\partial}{\partial p_x},i\hbar\frac{\partial}{\partial p_y},i\hbar\frac{\partial}{\partial p_z}\right) \tag{6.66}$$

That is, $U(x,y,z)$ is given in ($\mathbf{p}$, E) space by $U\left(i\hbar\frac{\partial}{\partial p_x},i\hbar\frac{\partial}{\partial p_y},i\hbar\frac{\partial}{\partial p_z}\right)$

(r, *t*) space

The quantities E and $(p_x^2+p_y^2+p_z^2)/2m_0$ in eq. (6.59) belong to $(\mathbf{p},E)$ space, and we have to use the corresponding relations in $(\mathbf{r},t)$ space [see also eqs. (6.52) and (6.57)]:

$$E\,\Psi(p_x,p_y,p_z,E)\leftrightarrow i\hbar\frac{\partial}{\partial t}\Psi(x,y,z,t) \tag{6.67}$$

$$\frac{p_x^2+p_y^2+p_z^2}{2m_0}\Psi(p_x,p_y,p_z,E)\leftrightarrow -\frac{\hbar^2}{2m_0}\Delta\Psi(x,y,z,t) \tag{6.68}$$

Then, we can formulate the following quantum-mechanical equations for the determination of $\Psi(p_x,p_y,p_z,E)$ and $\Psi(x,y,z,t)$. For the classical system, characterized by (6.59), we can formulate the following quantum-mechanical equivalency by means of (6.8)

$$\Psi(x,y,z,t)=\frac{1}{(2\pi\hbar)^2}\int\limits_{-\infty}^{\infty}\Psi(p_x,p_y,p_z,E)\exp\left\{i\left[\frac{\mathbf{p}}{\hbar}\cdot\mathbf{r}-\frac{E}{\hbar}t\right]\right\}dp_xdp_ydp_zdE$$

as follows

$$\left\{i\hbar\frac{\partial}{\partial t}+\frac{\hbar^2}{2m_0}\Delta-U(x,y,z)\right\}\Psi(x,y,z,t) = \frac{1}{(2\pi\hbar)^2}\times$$
$$\int_{-\infty}^{\infty}\left\{E-\frac{p_x^2+p_y^2+p_z^2}{2m_0}-U\left(i\hbar\frac{\partial}{\partial p_x},i\hbar\frac{\partial}{\partial p_y},i\hbar\frac{\partial}{\partial p_z}\right)\right\}\times \tag{6.69}$$
$$\Psi(p_x,p_y,p_z,E)\,\exp\left[\frac{i}{\hbar}(\mathbf{p}\cdot\mathbf{r}-Et)\right]\,dp_xdp_ydp_zdE$$

Equation (6.69) leads to

$$\left\{i\hbar\frac{\partial}{\partial t}+\frac{\hbar^2}{2m_0}\Delta-U(x,y,z)\right\}\Psi(x,y,z,t)=f(x,y,z,t) \tag{6.70}$$

and

$$\left\{E-\frac{p_x^2+p_y^2+p_z^2}{2m_0}-U\left(i\hbar\frac{\partial}{\partial p_x},i\hbar\frac{\partial}{\partial p_y},i\hbar\frac{\partial}{\partial p_z}\right)\right\}\times$$
$$\Psi(p_x,p_y,p_z,E)=f(p_x,p_y,p_z,E) \tag{6.71}$$

with

$$f(x,y,z,t)=\frac{1}{(2\pi\hbar)^2}\int_{-\infty}^{\infty}\left\{E-\frac{p_x^2+p_y^2+p_z^2}{2m_0}-U\left(i\hbar\frac{\partial}{\partial p_x},i\hbar\frac{\partial}{\partial p_y},i\hbar\frac{\partial}{\partial p_z}\right)\right\}\times \tag{6.72}$$
$$\Psi(p_x,p_y,p_z,E)\,\exp\left[\frac{i}{\hbar}(\mathbf{p}\cdot\mathbf{r}-Et)\right]\,dp_xdp_ydp_zdE$$

and

$$f(x,y,z,t)=\frac{1}{(2\pi\hbar)^2}\int_{-\infty}^{\infty}f(p_x,p_y,p_z,E)\,\exp\left[\frac{i}{\hbar}(\mathbf{p}\cdot\mathbf{r}-Et)\right]dp_xdp_ydp_zdE \tag{6.73}$$

The classical potential $U(x,y,z)$ is unknown but the function $f(x,y,z,t)$ as well. Therefore, we want to combine the functions $U(x,y,z)$ and $f(x,y,z,t)$ to one term, which we want to name $V(x,y,z,t)$.

This re-formulation leads to the final quantum-mechanical equation of the classical case (6.69):

$$i\hbar\frac{\partial}{\partial t}\Psi(x,y,z,t)=-\frac{\hbar^2}{2m_0}\Delta\Psi(x,y,z,t)+V(x,y,z,t)\Psi(x,y,z,t) \tag{6.74}$$

Equation (6.74) is a formulation given for $(\mathbf{r},t)$ space, where

$$V(x,y,z,t) = U(x,y,z) + g(x,y,z,t) \tag{6.75}$$

with

$$g(x,y,z,t) = \frac{f(x,y,z,t)}{\Psi(x,y,z,t)} \tag{6.76}$$

$V(x,y,z,t)$ should be considered as a generalized interaction potential. In the quantum-mechanical case we have no longer the classical static potential $U(x,y,z)$ but the time-dependent potential $V(x,y,z,t)$, which is now four-dimensional.

In other words, we obtain from the classical stationary case (6.59) a nonstationary quantum-mechanical equation [eq. (6.74)].

Using the procedure applied earlier, $V(x,y,z,t)$ takes in $(\mathbf{p},E)$ space the form

$$V\left(i\hbar\frac{\partial}{\partial p_x}, i\hbar\frac{\partial}{\partial p_y}, i\hbar\frac{\partial}{\partial p_z}, -i\hbar\frac{\partial}{\partial E}\right) \tag{6.77}$$

resulting from the equivalence [using eq. (6.9)]

$$\begin{aligned} &V\left(i\hbar\frac{\partial}{\partial p_x}, i\hbar\frac{\partial}{\partial p_y}, i\hbar\frac{\partial}{\partial p_z}, -i\hbar\frac{\partial}{\partial E}\right)\Psi(p_x,p_y,p_z,E) \\ &= \frac{1}{(2\pi\hbar)^2}\int_{-\infty}^{\infty} V(x,y,z,t)\,\Psi(x,y,z,t)\exp\left\{-i\left[\frac{\mathbf{p}}{\hbar}\cdot\mathbf{r}-\frac{E}{\hbar}t\right]\right\}dtdxdydz \end{aligned} \tag{6.78}$$

Note that the time t takes in $(\mathbf{p},E)$ space the form $-i\hbar\,\partial/\partial E$ [see eq. (6.54)]. Using the equivalencies (6.68) and (6.78), the quantum-mechanical case of eq. (6.59) is given in $(\mathbf{p},E)$ space by

$$\begin{aligned} E\Psi(p_x,p_y,p_z,E) = {} & \frac{p_x^2+p_y^2+p_z^2}{2m_0}\Psi(p_x,p_y,p_z,E) + \\ & V\left(i\hbar\frac{\partial}{\partial p_x}, i\hbar\frac{\partial}{\partial p_y}, i\hbar\frac{\partial}{\partial p_z}, -i\hbar\frac{\partial}{\partial E}\right)\Psi(p_x,p_y,p_z,E) \end{aligned} \tag{6.79}$$

with [see eq. (6.71)]

$$\begin{aligned} &V\left(i\hbar\frac{\partial}{\partial p_x}, i\hbar\frac{\partial}{\partial p_y}, i\hbar\frac{\partial}{\partial p_z}, -i\hbar\frac{\partial}{\partial E}\right)\Psi(p_x,p_y,p_z,E) \\ &= U\left(i\hbar\frac{\partial}{\partial p_x}, i\hbar\frac{\partial}{\partial p_y}, i\hbar\frac{\partial}{\partial p_z}\right)\Psi(p_x,p_y,p_z,E) + f(p_x,p_y,p_z,E) \end{aligned} \tag{6.80}$$

Equation (6.79) is completely equivalent to eq. (6.74): $\Psi(x,y,z,t)$ is determined by eq. (6.74) and $\Psi(p_x,p_y,p_z,E)$ by eq. (6.79), and both solutions [$\Psi(x,y,z,t)$ and $\Psi(p_x,p_y,p_z,E)$] are connected by eqs. (6.8) and (6.9), respectively.

6.4.3 Comparison with Schrödinger's equation

Equation (6.79) does not exist in usual quantum theory and this is because the operator $-i\hbar\,\partial/\partial E$ for the time coordinate does not exist in traditional quantum theory. Since eqs. (6.79) and (6.74) are completely equivalent, also eq. (6.74) cannot exist in traditional quantum theory. If we know, for example $\Psi(p_x, p_y, p_z, E)$, we can determine $\Psi(x, y, z, t)$ without explicitly to use eq. (6.74); $\Psi(p_x, p_y, p_z, E)$ and $\Psi(x, y, z, t)$ are connected to each other by a Fourier transform [see eqs. (6.8) and (6.9)].

The time t is system specific in character. The time within traditional quantum theory is exclusively given by the clock time τ which we use in everyday life. The system-specific time t is not defined in traditional quantum theory. That is, the time used in traditional quantum theory must be identified with $\tau(t = \tau)$, and when we use $V(x, y, z, t) = U(x, y, z)$ we obtain in the case of eq. (6.59) Schrödinger's equation of traditional quantum theory:

$$i\hbar\frac{\partial}{\partial\tau}\Psi(x, y, z, \tau) = -\frac{\hbar^2}{2m_0}\Delta\Psi(x, y, z, \tau) + U(x, y, z)\Psi(x, y, z, \tau) \tag{6.81}$$

The system-specific time t does not appear in eq. (6.81). Instead of eq. (6.74) (projection theory] we get Schrödinger's equation of usual quantum theory. Both equations are based on eq. (6.59). Concerning Schrödinger's equations, more details are given in Ref. [12].

6.5 Time operator

For the realistic treatment of nanosystems we need a quantum aspect of time, which is not known in traditional quantum theory. Self-organizing processes belong to the heart of nanoscience and such processes have to be based on a quantum-theoretical background. Self-organizing processes are transitions of quantum states within a certain time interval. The realistic description of time sequences is relevant here and it influences the process itself and the outcome as well. In other words, the notion "time" is an essential factor.

6.5.1 Time in traditional quantum theory

As we underlined in Chapter 1, the physics of time is underdeveloped in traditional quantum theory. This is a serious problem. Here we have merely an external time, which is strictly classical in character. It is the clock time τ that we often used in the earlier chapters. No other type of time is known here. A quantum aspect of time is missing in traditional quantum theory. That is, the phenomenon time appears in usual quantum theory only in restricted form.

Erwin Schrödinger tried to find a quantum operator for the time, but without success; he tried to construct such an operator within the given frame of traditional quantum theory itself, that is, without expanding the basic frame, but this is obviously inescapable.

A system-specific quantum aspect of time can be developed within the frame of the projection principle. The system-specific time is given in $(\mathbf{r}, t)$ space by a real number but in $(\mathbf{p}, E)$ space as operator of the form $-i\hbar\,\partial/\partial E$. This operator $-i\hbar\,\partial/\partial E$ reflects the time coordinate in general form and it is valid for each kind of physically real object and system, respectively. In this section, we formulate a system-specific time operator, which reflects the quantum properties of the system under investigation.

Such kind of formulation is in fact needed when we want to treat nanosystems in a realistic way. However, not only nanosystems are concerned, but the problem with time is a general problem in physics.

6.5.2 Uncertainties with respect to the variables

We have statistical fluctuations in both spaces: $(\mathbf{r}, t)$ space and $(\mathbf{p}, E)$ space. It is easy to recognize that there are uncertainties with respect to all variables x, y, z, t and p_x, p_y, p_z, E. These uncertainties are expressed through the following relations (e.g., see Ref. [12]):

$$\begin{aligned} \delta p_x \delta x &\geq \frac{\hbar}{2} \\ \delta p_y \delta y &\geq \frac{\hbar}{2} \\ \delta p_z \delta z &\geq \frac{\hbar}{2} \\ \delta E \delta t &\geq \frac{\hbar}{2} \end{aligned} \tag{6.82}$$

In connection with these uncertainty relations, the appearance of $\delta E\delta t \geq \hbar/2$ is remarkable: the energy E and the system-specific time t are inherently uncertain within projection theory. This is in fact a general requirement, but this is not fulfilled in traditional quantum theory.

In fact, such an equation does not exist in the usual form of quantum theory. Since time is still a classical parameter within usual quantum theory, there is no uncertainty relation of type $\delta E\delta t \geq \hbar/2$ in traditional quantum theory. However, here we have the energy–time relation of the form

$$\Delta E \Delta \tau \geq \frac{\hbar}{2} \tag{6.83}$$

But the quantities ΔE and $\Delta\tau$ are differences and they have nothing to do with uncertainties. Relation (6.83) is a quasiclassical equation. The value of such an equation is, however, questionable. In Mario Bunge's opinion, the energy–time relation $\Delta E \Delta\tau \geq \hbar/2$ "is a total stranger to quantum theory." We will still discuss this point in Chapter 7.

If the uncertainties of the variables of $(\mathbf{r}, t)$ space become zero [5], that is, if

$$\delta x \rightarrow 0, \delta y \rightarrow 0, \delta z \rightarrow 0, \delta t \rightarrow 0 \tag{6.84}$$

the reality-picture conception requires that also the variables of $(\mathbf{p}, E)$ space become zero:

$$\partial p_x \rightarrow 0, \partial p_y \rightarrow 0, \partial p_z \rightarrow 0, \partial E \rightarrow 0 \tag{6.85}$$

Then, instead of the wave functions $\Psi(x, y, z, t)$ and $\Psi(p_x, p_y, p_z, E)$, which are connected through relations (6.8) and (6.99, at each clock time τ there are definite values for the variables x, y, z, t and p_x, p_y, p_z, E and there are no fluctuations:

$$\begin{aligned} &\tau\colon x, y, z, t, \\ &p_x, p_y, p_z, E \end{aligned} \tag{6.86}$$

In other words, there are no probability densities $\Psi^*(x, y, z, t)\Psi(x, y, z, t)$ and $\Psi^*(p_x, p_y, p_z, E)\Psi(p_x, p_y, p_z, E)$ effective. In other words, the variables x, y, z, t and p_x, p_y, p_z, E behave classically, that is, without any uncertainty.

However, also in this classical case the reality-picture conception must be valid. This means that the variables must be connected and this is reflected through relations of the form

$$t = t\,(x, y, z, p_x, p_y, p_z, E) \tag{6.87}$$

and also in the form of

$$E = E(x, y, z, t, p_x, p_y, p_z) \tag{6.88}$$

Both relations, eqs. (6.87) and (6.88), can be used as a starting point for quantum-theoretical formulations. We would like to derive first the quantum laws for eq. (6.87), which reflects the time representation of the problem.

6.5.3 Time representation

With the basic eq. (6.9), we immediately obtain

$$\left\{-i\hbar\frac{\partial}{\partial E}\right\}\Psi(p_x,p_y,p_z,E)=\frac{1}{(2\pi\hbar)^2}\int_{-\infty}^{\infty} t\,\Psi(x,y,z,t)\exp\left\{-i\left[\frac{\mathbf{p}}{\hbar}\cdot\mathbf{r}-\frac{E}{\hbar}t\right]\right\}dxdydzdt \tag{6.89}$$

The inverse transformation leads to

$$t\Psi(x,y,z,t)=\frac{1}{(2\pi\hbar)^2}\int_{-\infty}^{\infty}\left\{-i\hbar\frac{\partial}{\partial E}\right\}\Psi(p_x,p_y,p_z,E)\exp\left\{i\left[\frac{\mathbf{p}}{\hbar}\cdot\mathbf{r}-\frac{E}{\hbar}t\right]\right\}dp_xdp_ydp_zdE \tag{6.90}$$

Furthermore, in order to get a quantum-mechanical formulation for the quantity $t\,(x,y,z,p_x,p_y,p_z,E)$ [see eq. (6.87)] in $(\mathbf{r},t)$ space, we have to apply the operator rules derived in Section 6.3. Then instead of $t\,(x,y,z,p_x,p_y,p_z,E)$ we get an operator, which we want to indicate by $\hat{C}$, and $\hat{C}$ is the time operator. We have the transition

$$t\,(x,y,z,p_x,p_y,p_z,E)\ \rightarrow\ \hat{C}\left(x,y,z,-i\hbar\frac{\partial}{\partial x},-i\hbar\frac{\partial}{\partial y},-i\hbar\frac{\partial}{\partial z},i\hbar\frac{\partial}{\partial t}\right) \tag{6.91}$$

With (6.8) we get

$$\hat{C}\left(x,y,z,-i\hbar\frac{\partial}{\partial x},-i\hbar\frac{\partial}{\partial y},-i\hbar\frac{\partial}{\partial z},i\hbar\frac{\partial}{\partial t}\right)\Psi(x,y,z,t)=\frac{1}{(2\pi\hbar)^2}\times\int_{-\infty}^{\infty}\hat{C}(x,y,z,p_x,p_y,p_z,E)\ \Psi(p_x,p_y,p_z,E)\exp\left\{i\left[\frac{\mathbf{p}}{\hbar}\cdot\mathbf{r}-\frac{E}{\hbar}t\right]\right\}dp_xdp_ydp_zdE \tag{6.92}$$

Here the procedure used in Section 6.4 has been applied piecewise; we may of course also use the so-called rational functions.

On the other hand, when we apply the inverse Fourier transform (6.9), we get the following relation:

$$\hat{C}\left(i\hbar\frac{\partial}{\partial p_x},i\hbar\frac{\partial}{\partial p_y},i\hbar\frac{\partial}{\partial p_z},p_x,p_y,p_z,E\right)\Psi(p_x,p_y,p_z,E)=\frac{1}{(2\pi\hbar)^2}\times\int_{-\infty}^{\infty}\hat{C}\left(x,y,z,-i\hbar\frac{\partial}{\partial x},-i\hbar\frac{\partial}{\partial y},-i\hbar\frac{\partial}{\partial z},i\hbar\frac{\partial}{\partial t}\right)\Psi(x,y,z,t)\times\exp\left\{-i\left[\frac{\mathbf{p}}{\hbar}\cdot\mathbf{r}-\frac{E}{\hbar}t\right]\right\}dtdxdydz \tag{6.93}$$

From this formulation, we get with transformation (6.8) the expression, which we need for the further treatment

$$\hat{C}\left(x, y, z, -i\hbar\frac{\partial}{\partial x}, -i\hbar\frac{\partial}{\partial y}, -i\hbar\frac{\partial}{\partial z}, i\hbar\frac{\partial}{\partial t}\right)\Psi(x, y, z, t) = \frac{1}{(2\pi\hbar)^2} \times \int_{-\infty}^{\infty} \hat{C}\left(i\hbar\frac{\partial}{\partial p_x}, i\hbar\frac{\partial}{\partial p_y}, i\hbar\frac{\partial}{\partial p_z}, p_x, p_y, p_z, E\right)\Psi(p_x, p_y, p_z, E) \times \exp\left\{i\left[\frac{\mathbf{p}}{\hbar}\cdot\mathbf{r} - \frac{E}{\hbar}t\right]\right\}dp_x dp_y dp_z dE \tag{6.94}$$

Combining eq. (6.90) with eq. (6.94), we obtain a formula that is suitable to formulate basic equations in both spaces

$$\left\{t - \hat{C}\left(x, y, z, -i\hbar\frac{\partial}{\partial x}, -i\hbar\frac{\partial}{\partial y}, -i\hbar\frac{\partial}{\partial z}, i\hbar\frac{\partial}{\partial t}\right)\right\}\Psi(x, y, z, t) = \frac{1}{(2\pi\hbar)^2} \times \int_{-\infty}^{\infty} \left\{-i\hbar\frac{\partial}{\partial E} - \hat{C}\left(i\hbar\frac{\partial}{\partial p_x}, i\hbar\frac{\partial}{\partial p_y}, i\hbar\frac{\partial}{\partial p_z}, p_x, p_y, p_z, E\right)\right\}\Psi(p_x, p_y, p_z, E) \times \exp\left\{i\left[\frac{\mathbf{p}}{\hbar}\cdot\mathbf{r} - \frac{E}{\hbar}t\right]\right\}dp_x dp_y dp_z dE \tag{6.95}$$

Then, the solution of eq. (6.95) is straightforward and allows to formulate two equations within the frame of the “time representation,” one for $(\mathbf{r}, t)$ space and another for $(\mathbf{p}, E)$ space.

Time representation for $(\mathbf{r}, t)$ space:

$$\hat{C}\left(x, y, z, -i\hbar\frac{\partial}{\partial x}, -i\hbar\frac{\partial}{\partial y}, -i\hbar\frac{\partial}{\partial z}, i\hbar\frac{\partial}{\partial t}\right)\Psi(x, y, z, t) = t\,\Psi(x, y, z, t) \tag{6.96}$$

Time representation for $(\mathbf{p}, E)$ space:

$$-i\hbar\frac{\partial}{\partial E}\Psi(p_x, p_y, p_z, E) = \hat{C}\left(i\hbar\frac{\partial}{\partial p_x}, i\hbar\frac{\partial}{\partial p_y}, i\hbar\frac{\partial}{\partial p_z}, p_x, p_y, p_z, E\right)\Psi(p_x, p_y, p_z, E) \tag{6.97}$$

The time operator $\hat{C}$ describes the physically real object in time representation. $\hat{C}$ has to be developed, for example, in $(\mathbf{r}, t)$ space in terms of quantum-theoretical principles. Then the wave function $\Psi(x, y, z, t)$ can be calculated on the basis of eq. (6.96). The function $\Psi(x, y, z, t)$ has then to be inserted into eq. (6.9) and we obtain the wave function $\Psi(p_x, p_y, p_z, E)$.

On the other hand, $\hat{C}$ can also be developed in $(\mathbf{p}, E)$ space, as an equivalent procedure so to speak. Then the quantum-theoretical principles, which have also here to be applied, can probably not easily transferred from $(\mathbf{r}, t)$ space to $(\mathbf{p}, E)$

space. On the basis of $\hat{C}$, developed in $(\mathbf{p}, E)$ space, the wave function can be found as a solution of eq. (6.97). In any case, both eqs. (6.96) and (6.97) are equally relevant in the calculation of system-specific properties in time representation.

6.5.4 Alternative way

In principle, we can start from $t = t(x, y, z, p_x, p_y, p_z, E)$, that is, from relation (6.87). In Section 6.5.3, we also started from this expression but we used it in the most general form and did not express the quantum laws (6.96) and (6.97) on the basis on an analytical (mathematical) relation for $t = t(x, y, z, p_x, p_y, p_z, E)$.

No doubt, this way leads to clear statements concerning the formulation of system-specific quantum states. Nevertheless, we may go an alternative way: we can develop an analytical expression for $t = t(x, y, z, p_x, p_y, p_z, E)$ on the basis of physical principles, which are compatible with this expression in which the variables are given in classical form.

What kind of physical principles are adequate in this case? The variables in connection with $t = t(x, y, z, p_x, p_y, p_z, E)$ behave classically (Section 6.5.2), but the functional representation of the form $t(x, y, z, p_x, p_y, p_z, E)$, that is, the mathematical connection between the variables $x, y, z, p_x, p_y, p_z, E$ has to be chosen with respect to principles that are usually used in classical physics.

What is the procedure in the transition from the classical to the quantum state? We need "conclusive criteria" for the determination of a mathematical expression for $t(x, y, z, p_x, p_y, p_z, E)$. Let us briefly discuss the background of that.

The necessary and the sufficient condition

The transition from the classical, nonstatistical variables $x, y, z, p_x, p_y, p_z, E$ to the corresponding quantum operators (e.g., from $-i\hbar\partial/\partial E$ to E) is the "necessary" condition when we want to determine the quantum properties of physically real objects. But this is not "sufficient." We simply need more than that. Conditions (6.84) and (6.85) do not reflect the physical laws after which the objects behave, that is, they do not dictate the organizational form of the variables.

Conditions (6.84) and (6.85) merely change the character of the variables. Clearly, the classical variables $x, y, z, p_x, p_y, p_z, E$ behave in accordance with eqs. (6.84) and (6.85), but this does not yet lead to the "physical behavior" of the variables $x, y, z, p_x, p_y, p_z, E$, and this is expressed through the mathematical structure of the function $t(x, y, z, p_x, p_y, p_z, E)$.

For the mathematical representation of the term $t(x, y, z, p_x, p_y, p_z, E)$ we need "conclusive criteria" that are expressible in terms of the nonfluctuating, classical elements $x, y, z, p_x, p_y, p_z, E$. If this is fulfilled, we may name this mathematical expression "classical." These conclusive criteria are, for example, expressed by the effect of inertia, the principle of least action and the principle of causality. These criteria are often adapted to what we experience in everyday life.

Without these classical criteria, an analytical (mathematical) expression in terms of the classical variables cannot be formulated. We would like to assume that such a classical mathematical expression for the function $t\,(x, y, z, p_x, p_y, p_z, E)$ exists. Then instead of the general connection form $t\,(x, y, z, p_x, p_y, p_z, E)$ we get a mathematical expression that we want to indicate by the classical term (formula) $f(x, y, z, p_x, p_y, p_z, E)_{\text{classical}}$, that is, we have the transition

$$t\,(x, y, z, p_x, p_y, p_z, E) \;\rightarrow\; f(x, y, z, p_x, p_y, p_z, E)_{classical} \tag{6.98}$$

This means that the classical behavior is reflected by the variables but also through what we called "conclusive criteria."

For the alternate transformation of the variables and the operators, the basic eqs (6.8) and (6.9) have to be applied. Then the function f becomes an operator $\hat{f}$: $f \rightarrow \hat{f}$. The variables of $\hat{f}$ are operators, depending on the applied space. However, the functional structure remains in its classical form and, therefore, we have $\hat{f} = \hat{f}_{\text{classical}}$.

We need conclusive classical criteria when we want to express the general form $t\,(x, y, z, p_x, p_y, p_z, E)$ mathematically. This step is indicated through relation (6.98). These classical criteria must be expressible as a function of the nonfluctuating elements $x, y, z, p_x, p_y, p_z, E$. This does, however, not mean that quantum properties of physical objects are characterized by these classical laws alone and that the quantum aspect comes merely into existence through the exchange of the classical variables by operators. We need more than that.

The quantum object is not specified on these conclusive classical criteria alone, but extensions with respect to the quantum behavior are necessary. We need more than the conclusive classical criteria when we go from the classical state to the quantum aspect of the properties.

Further steps in the treatment

Using the rules given earlier and in Section 6.4, the following identity is obtained:

$$\begin{aligned}
&\hat{f}\left(x, y, z, \,-i\hbar\frac{\partial}{\partial x}, -i\hbar\frac{\partial}{\partial y}, -i\hbar\frac{\partial}{\partial z}, i\hbar\frac{\partial}{\partial t}\right)_{classical} \Psi(x, y, z, t) = \frac{1}{(2\pi\hbar)^2} \times \\
&\qquad \int_{-\infty}^{\infty} \hat{f}\left(p_x, p_y, p_z, E, \, i\hbar\frac{\partial}{\partial p_x}, i\hbar\frac{\partial}{\partial p_y}, i\hbar\frac{\partial}{\partial p_z}\right)_{classical} \times \\
&\qquad \Psi(p_x, p_y, p_z, E)\exp\left\{i\left[\frac{\mathbf{p}}{\hbar}\cdot\mathbf{r} - \frac{E}{\hbar}t\right]\right\}dp_x dp_y dp_z dE
\end{aligned} \tag{6.99}$$

The inverse formulation of the problem is compatible with relation (6.99):

$$\hat{f}\left(p_x, p_y, p_z, E, i\hbar\frac{\partial}{\partial p_x}, i\hbar\frac{\partial}{\partial p_y}, i\hbar\frac{\partial}{\partial p_z}\right)_{classical} \Psi(p_x, p_y, p_z, E) = \frac{1}{(2\pi\hbar)^2} \times$$
$$\int_{-\infty}^{\infty} \hat{f}\left(x, y, z, -i\hbar\frac{\partial}{\partial x}, -i\hbar\frac{\partial}{\partial y}, -i\hbar\frac{\partial}{\partial z}, i\hbar\frac{\partial}{\partial t}\right)_{classical} \times \tag{6.100}$$
$$\Psi(x, y, z, t)\, \exp\left\{-i\left[\frac{\mathbf{p}}{\hbar}\cdot\mathbf{r} - \frac{E}{\hbar}t\right]\right\} dxdydzdt$$

With the existence of eqs. (6.99) and (6.100), the classical behavior of the variables $x, y, z, p_x, p_y, p_z, E$ is eliminated through the appearance of the quantum operators. However, the conclusive criteria, the other part of the classical description, remain classical in character.

Combining eqs. (6.90) and (6.99), we obtain, in analogy to relation (6.95), the expression

$$\left\{t - \hat{f}\left(x, y, z, -i\hbar\frac{\partial}{\partial x}, -i\hbar\frac{\partial}{\partial y}, -i\hbar\frac{\partial}{\partial z}, i\hbar\frac{\partial}{\partial t}\right)_{classical}\right\}\Psi(x, y, z, t) =$$
$$\frac{1}{(2\pi\hbar)^2}\int_{-\infty}^{\infty}\left\{-i\hbar\frac{\partial}{\partial E} - \hat{f}\left(i\hbar\frac{\partial}{\partial p_x}, i\hbar\frac{\partial}{\partial p_y}, i\hbar\frac{\partial}{\partial p_z}, p_x, p_y, p_z, E\right)_{classical}\right\} \times \tag{6.101}$$
$$\Psi(p_x, p_y, p_z, E)\, \exp\left\{i\left[\frac{\mathbf{p}}{\hbar}\cdot\mathbf{r} - \frac{E}{\hbar}t\right]\right\} dp_x dp_y dp_z dE$$

The solution of eq. (6.101) is also here straightforward and allows to formulate two equations within the frame of the "time representation," one for $(\mathbf{r}, t)$ space and another for $(\mathbf{p}, E)$ space. However, we merely get quasiclassical representations in the following two forms:

Time representation for $(\mathbf{r}, t)$ space (quasiclassical case):

$$\hat{f}\left(x, y, z, -i\hbar\frac{\partial}{\partial x}, -i\hbar\frac{\partial}{\partial y}, -i\hbar\frac{\partial}{\partial z}, i\hbar\frac{\partial}{\partial t}\right)_{classical} \Psi(x, y, z, t) = t\Psi(x, y, z, t) \tag{6.102}$$

Time representation for $(\mathbf{p}, E)$ space (quasiclassical case):

$$-i\hbar\frac{\partial}{\partial E}\Psi(p_x, p_y, p_z, E) = \hat{f}\left(i\hbar\frac{\partial}{\partial p_x}, i\hbar\frac{\partial}{\partial p_y}, i\hbar\frac{\partial}{\partial p_z}, p_x, p_y, p_z, E\right)_{classical} \Psi(p_x, p_y, p_z, E) \tag{6.103}$$

If the quasiclassical expression

$$\left\{t-\hat{f}\left(x,y,z,-i\hbar\frac{\partial}{\partial x},-i\hbar\frac{\partial}{\partial y},-i\hbar\frac{\partial}{\partial z},i\hbar\frac{\partial}{\partial t}\right)_{classical}\right\}\Psi(x,y,z,t) \quad (6.104)$$

in eq. (6.101) does not fulfill completely the quantum aspect of the problem under investigation, we may extend this expression via eq. (6.101). The integral of eq. (6.101) can be written as

$$\begin{aligned} g(x,y,z,t) = {} & \frac{1}{(2\pi\hbar)^2}\times \\ & \int_{-\infty}^{\infty}\left\{\left[-i\hbar\frac{\partial}{\partial E}\right]-\hat{f}\left(i\hbar\frac{\partial}{\partial p_x},i\hbar\frac{\partial}{\partial p_y},i\hbar\frac{\partial}{\partial p_z},p_x,p_y,p_z,E\right)_{classical}\right\}\times \\ & \Psi(p_x,p_y,p_z,E)\exp\left\{i\left[\frac{\mathbf{p}}{\hbar}\cdot\mathbf{r}-\frac{E}{\hbar}t\right]\right\}dp_xdp_ydp_zdE \end{aligned} \quad (6.105)$$

Then, we get

$$\begin{aligned} & \left\{t-\hat{f}\left(x,y,z,-i\hbar\frac{\partial}{\partial x},-i\hbar\frac{\partial}{\partial y},-i\hbar\frac{\partial}{\partial z},i\hbar\frac{\partial}{\partial t}\right)_{classical}\right\}\Psi(x,y,z,t) \\ & \qquad - g(x,y,z,t)=0 \end{aligned} \quad (6.106)$$

Or in reformulated form

$$\begin{aligned} & t\Psi(x,y,z,t)- \\ & \hat{f}\left(x,y,z,-i\hbar\frac{\partial}{\partial x},-i\hbar\frac{\partial}{\partial y},-i\hbar\frac{\partial}{\partial z},i\hbar\frac{\partial}{\partial t}\right)_{quantum}\Psi(x,y,z,t)=0 \end{aligned} \quad (6.107)$$

with

$$\begin{aligned} & \hat{f}\left(x,y,z,-i\hbar\frac{\partial}{\partial x},-i\hbar\frac{\partial}{\partial y},-i\hbar\frac{\partial}{\partial z},i\hbar\frac{\partial}{\partial t}\right)_{quantum}\Psi(x,y,z,t)= \\ & \hat{f}\left(x,y,z,-i\hbar\frac{\partial}{\partial x},-i\hbar\frac{\partial}{\partial y},-i\hbar\frac{\partial}{\partial z},i\hbar\frac{\partial}{\partial t}\right)_{classical}\Psi(x,y,z,t)-g(x,y,z,t) \end{aligned} \quad (6.108)$$

The operator

$$\hat{f}\left(x,y,z,-i\hbar\frac{\partial}{\partial x},-i\hbar\frac{\partial}{\partial y},-i\hbar\frac{\partial}{\partial z},i\hbar\frac{\partial}{\partial t}\right)_{quantum}$$

which appears in eq. (6.107) has to be identified with the operator

$$\hat{C}\left(x,y,z,-i\hbar\frac{\partial}{\partial x},-i\hbar\frac{\partial}{\partial y},-i\hbar\frac{\partial}{\partial z},i\hbar\frac{\partial}{\partial t}\right)$$

see eq. (6.94).

On the other hand, we may use the inverse transformation of eq. (6.101) to complete the discussion. The inverse transformation of eq. (6.101) is given by

$$\left\{-i\hbar\frac{\partial}{\partial E}-\hat{f}\left(i\hbar\frac{\partial}{\partial p_x},i\hbar\frac{\partial}{\partial p_y},i\hbar\frac{\partial}{\partial p_z},p_x,p_y,p_z,E\right)_{classical}\right\}\Psi(p_x,p_y,p_z,E)$$
$$=\frac{1}{(2\pi\hbar)^2}\int\limits_{-\infty}^{\infty}\left\{t-\hat{f}\left(x,y,z,-i\hbar\frac{\partial}{\partial x},-i\hbar\frac{\partial}{\partial y},-i\hbar\frac{\partial}{\partial z},i\hbar\frac{\partial}{\partial t}\right)_{classical}\right\}\times$$
$$\Psi(x,y,z,t)\exp\left\{-i\left[\frac{\mathbf{p}}{\hbar}\cdot\mathbf{r}-\frac{E}{\hbar}t\right]\right\}dxdydzdt \tag{6.109}$$

If the quasiclassical expression

$$\left\{-i\hbar\frac{\partial}{\partial E}-\hat{f}\left(i\hbar\frac{\partial}{\partial p_x},i\hbar\frac{\partial}{\partial p_y},i\hbar\frac{\partial}{\partial p_z},p_x,p_y,p_z,E\right)_{classical}\right\}\Psi(p_x,p_y,p_z,E) \tag{6.110}$$

in eq. (6.109) does not fulfill completely the quantum aspect of the problem under investigation, we may extend this expression via eq. (6.109). Then, the integral of eq. (6.109) can be written as

$$g(p_x,p_y,p_z,E)=$$
$$\frac{1}{(2\pi\hbar)^2}\int\limits_{-\infty}^{\infty}\left\{t-\hat{f}\left(x,y,z,-i\hbar\frac{\partial}{\partial x},-i\hbar\frac{\partial}{\partial y},-i\hbar\frac{\partial}{\partial z},i\hbar\frac{\partial}{\partial t}\right)_{classical}\right\}\times$$
$$\Psi(x,y,z,t)\exp\left\{-i\left[\frac{\mathbf{p}}{\hbar}\cdot\mathbf{r}-\frac{E}{\hbar}t\right]\right\}dxdydzdt \tag{6.111}$$

Then, we get

$$\left\{-i\hbar\frac{\partial}{\partial E}-\hat{f}\left(i\hbar\frac{\partial}{\partial p_x},i\hbar\frac{\partial}{\partial p_y},i\hbar\frac{\partial}{\partial p_z},p_x,p_y,p_z,E\right)_{classical}\right\}\Psi(p_x,p_y,p_z,E)$$
$$-g(p_x,p_y,p_z,E)=0 \tag{6.112}$$

Or in reformulated form:

$$-i\hbar\frac{\partial}{\partial E}\Psi(p_x,p_y,p_z,E)-$$
$$\hat{f}\left(i\hbar\frac{\partial}{\partial p_x},i\hbar\frac{\partial}{\partial p_y},i\hbar\frac{\partial}{\partial p_z},p_x,p_y,p_z,E\right)_{quantum}\Psi(p_x,p_y,p_z,E)-0 \tag{6.113}$$

with

$$\hat{f}\left(i\hbar\frac{\partial}{\partial p_x},i\hbar\frac{\partial}{\partial p_y},i\hbar\frac{\partial}{\partial p_z},p_x,p_y,p_z,E\right)_{quantum}\Psi(p_x,p_y,p_z,E)=$$
$$\hat{f}\left(x,y,z,-i\hbar\frac{\partial}{\partial_x},i\hbar\frac{\partial}{\partial_y},i\hbar\frac{\partial}{\partial_z},i\hbar\frac{\partial}{\partial_t}\right)_{classical}\Psi(p_x,p_y,p_z,E)$$
$$-g(p_x,p_y,p_z,E) \tag{6.114}$$

Using eqs. (6.105) and (6.111) we immediately recognize that the functions $g(x,y,z,t)$ and $g(p_x,p_y,p_z,E)$ are connected by a Fourier transform:

$$g(p_x,p_y,p_z,E) = \frac{1}{(2\pi\hbar)^2}\int_{-\infty}^{\infty} g(x,y,z,t)\exp\left\{-i\left[\frac{\mathbf{p}}{\hbar}\cdot\mathbf{r}-\frac{E}{\hbar}t\right]\right\}dtdxdydz \tag{6.115}$$

Also $g(x,y,z,t)$ and $g(p_x,p_y,p_z,E)$ are system specific and reflect quantum-mechanical properties, which complement the classical states introduced and discussed earlier.

Comparison

The following should still be underlined: the operator

$$\hat{f}\left(i\hbar\frac{\partial}{\partial p_x}, i\hbar\frac{\partial}{\partial p_y}, i\hbar\frac{\partial}{\partial p_z}, p_x, p_y, p_z, E\right)_{quantum}$$

which appears in eq. (6.113) has to be identified with the operator

$$\hat{C}\left(i\hbar\frac{\partial}{\partial p_x}, i\hbar\frac{\partial}{\partial p_y}, i\hbar\frac{\partial}{\partial p_z}, p_x, p_y, p_z, E\right)$$

(see eq. (6.93)).

6.5.5 Energy representation

In analogy to what we developed in connection with the "time representation" of quantum phenomena, expressed by eqs. (6.96) and (6.97), we get equivalent equations in "energy representation." Also in this case the equations can be formulated within $(\mathbf{r},t)$ space as well as in $(\mathbf{p},E)$ space. That is, also here we have two equivalent equations, which can equally be used for the determination of the wave functions $\Psi(x,y,z,t)$ and $\Psi(p_x,p_y,p_z,E)$.

On the footing of what we deduced in connection with time representation, it is straightforward to show that for $(\mathbf{r},t)$ space and $(\mathbf{p},E)$ space the relations take the form

$$i\hbar\frac{\partial}{\partial t}\partial\Psi(x,y,z,t) = \hat{H}\left(x,y,z,t, -i\hbar\frac{\partial}{\partial x}, -i\hbar\frac{\partial}{\partial y}, -i\hbar\frac{\partial}{\partial z}\right)\Psi(x,y,z,t) \tag{6.116}$$

and

$$\hat{H}\left(i\hbar\frac{\partial}{\partial p_x}, i\hbar\frac{\partial}{\partial p_y}, i\hbar\frac{\partial}{\partial p_z}, p_x, p_y, p_z, -i\hbar\frac{\partial}{\partial E}\right)\Psi(p_x,p_y,p_z,E) = E\Psi(p_x,p_y,p_z,E) \tag{6.117}$$

We already know from Section 6.4 that the energy operator (Hamiltonian) $\hat{H}$ takes in $(\mathbf{r},t)$ space the form

$$\hat{H} = -\frac{\hbar^2}{2m_0}\Delta + V(x,y,z,t) \tag{6.118}$$

The four-dimensional interaction potential $V(x,y,z,t)$ is at clock time effective in the past, present and future with respect to system-specific time t. In usual quantum theory, such kind of interaction potential is not defined. Instead of the general interaction $V(x,y,z,t)$, we have here a restricted form of interaction denoted by $U(x,y,z)$ (Section 6.4), which is adapted to the experiences of human beings in everyday life. However, the physically real world is obviously more than that.

On the other hand, the operator $\hat{H}$ takes in $(\mathbf{p},E)$ space the structure

$$\hat{H} = \frac{p_x^2+p_y^2+p_z^2}{2m_0} + V\left(i\hbar\frac{\partial}{\partial p_x}, i\hbar\frac{\partial}{\partial p_y}, i\hbar\frac{\partial}{\partial p_z}, -i\hbar\frac{\partial}{\partial E}\right) \tag{6.119}$$

In this case, the interaction potential V becomes an abstract operator.

Again, the operator $\hat{H}$ is the energy operator, which is called "Hamiltonian" in the literature, in which traditional quantum theory is treated.

There is a certain similarity of eqs. (6.116) and (6.117) with Schrödinger's equations of traditional quantum theory. But this is only a formal impression. Fact is, however, that within Schrödinger's theory the system-specific time t is not defined and also not the operator for the time coordinate $-i\hbar\partial/\partial E$. More details concerning Schrödinger's equations are given in Section 6.7.

6.5.6 Discussion

Physically real objects can equally be treated in time representation (Section 6.5.3 and Section 6.5.4) as well as in energy representation. Moreover, the treatment can be done in $(\mathbf{r},t)$ space and in $(\mathbf{p},E)$ space and leads also here to equivalent information about the object under investigation. We have discussed two procedures: (1) the direct transition to the time operator and the energy operator and (2) the use of classical expressions in the transition to the quantum aspect.

1. The direct way

For the time representation and the energy representation, we started from $t = t(x,y,z,p_x,p_y,p_z,E)$ and $E = E(x,y,z,t,p_x,p_y,p_z)$ [eqs. (6.87) and (6.88)], which reflect general classical relationships. The time operator $\hat{C}$ and the energy operator $\hat{H}$ can be formulated directly on the footing of operators for the coordinates and the time (e.g., the variable t is replaced by the operator $-i\hbar\partial/\partial E$). The system-specific operators $\hat{C}$ and $\hat{H}$ are then modeled in terms of suitable quantum conceptions. We get two equivalent schemes:

$$\begin{array}{c} t\,(x,y,z,p_x,p_y,p_z,E) \rightarrow \hat{C} \\ \uparrow \\ \text{quantum conception} \end{array} \tag{6.120}$$

and

$$\begin{array}{c} E\,(x,y,z,t,p_x,p_y,p_z) \rightarrow \hat{H} \\ \uparrow \\ \text{quantum conception} \end{array} \tag{6.121}$$

Scheme (6.120) refers to Section 6.5.2 and scheme (6.121) to Section 6.5.5. Here the quantum conceptions need not be based on classical laws.

2. The start with classical arguments

Also here we start from the general classical relationships $t = t\,(x,y,z,p_x,p_y,p_z,E)$ and $E = E\,(x,y,z,t,p_x,p_y,p_z)$. We only want to discuss the time representation. There are altogether three steps in the development. An overview is given by scheme (6.122).

First, we formulate the physical problems in terms of conclusive classical criteria and these classical criteria must be expressible in terms of the nonfluctuating elements x,y,z,p_x,p_y,p_z,E. Then, we get an analytical (mathematical) expression for $t\,(x,y,z,p_x,p_y,p_z,E)$, which is still classical in character. We want to indicate this mathematical expression by $f\,(x,y,z,p_x,p_y,p_z,E)_{\text{classical}}$.

The next step is to replace the classical variables in $f_{\text{classical}}$ by operators (e.g., the variable t is replaced by the operator $-i\hbar\partial/\partial E$), which depends on the space for which the problem is treated, that is, we have to distinguish between $(\mathbf{r},t)$ space and $(\mathbf{p},E)$ space. That is, instead of $f_{\text{classical}}$ we have now an operator $\hat{f}_{\text{classical}}$, which is however still classical. The reason is obvious: the analytical form of $f_{\text{classical}}$ is based on classical criteria, but the replacement of the classical variables through operators does not change the functionality of the analytical expression and this means that the analytical expression itself still contains the classical criteria without quantum supplement. This quantum aspect can be introduced through the function g, which is expressible in $(\mathbf{r},t)$ space as well as in $(\mathbf{p},E)$ space. We finally obtain the quantum operator for the time that we indicated by $\hat{f}_{\text{quantum}}$, and $\hat{f}_{\text{quantum}}$ has to be identified with time operator $\hat{C}$. The entire scenario is summarized in scheme (6.122):

$$\begin{array}{ccccccc} t(x,y,z,p_x,p_y,p_z,E) \rightarrow & f_{classical} & \rightarrow & \hat{f}_{classical} & \rightarrow & \hat{f}_{quantum} & \leftrightarrow \hat{C} \\ & \uparrow & & \uparrow & & & \\ & \text{operators} & & g & & & \\ & (t \rightarrow -i\hbar\partial/\partial E,\text{etc.}) & & & & & \end{array} \tag{6.122}$$

This development (*the start with classical arguments*) demonstrates that the classical physical conceptions as starting point are not revealing. The direct quantum treatment of the problem (*The direct way*) should be seminal for the quantum description in terms of the time operator.

Remark

The operators for the coordinates and so on in $\hat{C}$ can be replaced by the classical variables and, in this way, we get an analytical expression but it is in part classical because the variables are classical. However, the mathematical formula itself contains the quantum-theoretical conception. Then, we have $f_{quantum}$, which is not an operator, but

$$f_{\text{quantum}} \neq f_{\text{classical}} \tag{6.123}$$

Whereas the function $f_{\text{classical}}$ contains merely the classic criteria, the function f_{quantum} contains the quantum-theoretical conception. With f_{quantum} we have the relation

$$t = t\,(x, y, z, p_x, p_y, p_z, E)_{\text{quantum}} \tag{6.124}$$

and also the relation

$$E = E\,(x, y, z, p_x, p_y, p_z, t)_{\text{quantum}} \tag{6.125}$$

Using eq. (6.125), we get the energy operator $\hat{H}$, that is, the quantum phenomenon is formulated with respect to the energy representation.

6.6 Classical behavior of the time

The quantum-theoretical treatment of a physically real object is done also within the projection theory by means of a wave function, which appears however here in an extended form since the system-specific time t has to be considered. Thus, we have for the wave function the form $\Psi(x, y, z, t)$, and $\Psi^*(x, y, z, t)$ is the corresponding complex conjugate function. First, we discuss the system-specific time and after that the simultaneous consideration of space and time.

6.6.1 System-specific time

Statistical behavior

The variables x, y, z, t are given at each time τ with a certain probability, which is proportional to the probability density $\Psi^*(x, y, z, t)\Psi(x, y, z, t)$. We consider here the situation for constant values of the coordinates, which are indicated in this case by

x_c, y_c, z_c. The probability density is assumed to be zero for $t < t_a$ and $t > t_b$ with $t_a \neq \infty$ and $t_b \neq \infty$.

At a definite clock time τ, the system-specific time t is observed with a certain probability, which is proportional to the probability density

$$\begin{gathered} \Psi^*(x_c, y_c, z_c, t)\Psi(x_c, y_c, z_c, t) \neq 0 \\ t_a \leq t \leq t_b \end{gathered} \tag{6.126}$$

Then, we obtain the sequences

$$t_1, t_2, \ldots, t_i, \ldots \tag{6.127}$$

and

$$\tau_1, \tau_2, \ldots, \tau_i, \ldots \tag{6.128}$$

The τ values increase continuously on the scale with real numbers:

$$\tau_1 < \tau_2 < \ldots < \tau_i < \ldots \tag{6.129}$$

The time t does not have property (6.129) but it varies statistically.

When the observation takes place within the range 2ε around τ, that is, at $\tau \pm \varepsilon$ not only "one" τ value is defined at $\tau \pm \varepsilon$, but the entire τ block within the range 2ε is effective. There are as many τ values within the interval 2ε as there are real numbers in 2ε, and there is an infinite number of them in 2ε. On the other hand, at each time τ there is a certain probability for the appearance of the variable t within the range between t_a and t_b, in which the probability density exists [see eq. (6.126)]. That is, there is also an infinite number of t values between t_a and t_b that belong to the τ values in 2ε. We obtain

$$\begin{gathered} \tau \to \infty \text{ in } 2\varepsilon \\ \Downarrow \\ t \to \infty \text{ in } (t_b - t_a) \end{gathered} \tag{6.130}$$

What is the condition for ε? The interval $2\varepsilon \neq 0$ contains an infinite number of τ values, but this interval can be arbitrarily small. The value for ε may even be close to zero and contains also for this case an infinite number of τ values (real numbers). That is, if we group the t interval between t_a and t_b into segments of width $2\varepsilon_t$ and if we choose for the number of the segments the value k_t, the width $2\varepsilon_t$ is given by

$$2\varepsilon_t = \frac{t_b - t_a}{k_t} \tag{6.131}$$

The value k_t is assumed to be finite ($k_t \neq \infty$). Because the interval $(t_b - t_a)$ was assumed to be finite, the segment $2\varepsilon_t$ is different from zero.

Since the number of t states within $(t_b - t_a)$ is infinity, the number of t states within $2\varepsilon_t$ must also be infinite as well ($\infty/k_t = \infty$ for $k \neq \infty$). Then we have in analogy to in (6.130) the following situation:

$$\begin{gathered} t \to \infty \text{ in } (t_b - t_a) \\ \Downarrow \\ t \to \infty \text{ in } 2\varepsilon_t \end{gathered} \tag{6.132}$$

Whereas the interval 2ε refers to the τ scale, the interval $2\varepsilon_t$ belongs to the t scale. The quantities ε and ε_t are independent of each other, but ε_t must have the same properties as ε: the interval $2\varepsilon_t \neq 0$ contains an infinite number of t values, but it can be arbitrarily small. The value for ε_t may even be close to zero and contains also for this case an infinite number of t values (real numbers).

The number of the $2\varepsilon_t$ segments is k_t and this number can be chosen arbitrarily large but different from infinity. The width $2\varepsilon_t$ decreases with increasing k_t. For a sufficiently small interval $2\varepsilon_t$, we get (almost) the complete information $\Psi^*(\mathbf{r}_c, t)\Psi(\mathbf{r}_c, t)$, $\mathbf{r}_c = (x_c, y_c, z_c)$ with any precision.

In conclusion, within the interval $2\varepsilon_t$ there must be an infinite number of t values. The number of $2\varepsilon_t$ segments within the range $(t_b - t_a)$ is $k_t = (t_b - t_a)/2\varepsilon_t$ [eq. (6.131)] and we have $k_t \neq \infty$. Since the number of t states within $(t_b - t_a)$ is infinity, the number of t states within $2\varepsilon_t$ must be infinite as well ($\infty/k_t = \infty$ for $k \neq \infty$). Within the range $(t_b - t_a)$, the number of $2\varepsilon_t$ segments is k_t with $k_t \neq \infty$ and each of these k_t intervals contains an infinite number of t values, whereby the $2\varepsilon_t$ interval can be arbitrarily small but must be different from zero.

Block behavior

The use of $\tau \pm \varepsilon$ instead of τ means that none of the τ values within the 2ε interval is soluble. In other words, the observation period of 2ε cannot be shortened. Then the observer records all what is within the segment 2ε, but he/she cannot resolve the details. He/she records all at once. The segment cannot be subdivided into parts.

In 2ε there is an infinite number of t states, but the human observer cannot resolve them but he/she experiences all at once in the form of "one block" defined through the segment of width 2ε. This block is expressed by the mean value $\bar{t}$, which is generated by all the time values within 2ε. All the τ values within the 2ε interval are defined as "one" block. This τ block acts as unified whole, and generates the mean value $\bar{t}$. This is the conception.

The system, described by t states, is observed and these states are related to the clock time τ [see eqs. (6.127) and (6.128)]. Relation here means that the system and the human observer interact with each other. This situation effectuates that the τ block overlaps with the t scale and correlates the clock time (values τ) with the t states. The overlap interval is of course also 2ε.

The τ block is for the observation of the world outside. In particular, it selects space structures. This is the task of the τ block, no less no more. The τ block can only transfer its own properties on the t scale during the process of recognition. The τ block is made for the observation of the physically real environment of the human observer. It has no other tasks. The τ block itself does not invent additional properties during the observation process. The properties of the τ block are fully transferred to the t scale of the physically real objects and are not more detailed than the properties of the τ block itself. From that the following points are relevant:

1. The overlap of the τ block with the t scale is not larger and not smaller than 2ε and defines also a block on the t scale of width $2\varepsilon_t = 2\varepsilon$. It is a t block.
2. The τ block can only resolve details of the t scale in the form of a block that is identical with the properties of the τ block, that is, the τ block acts on the t block as a unified whole and "sees" the t values within the 2ε interval all at once. This situation is expressed through the mean value $\bar{t}$ generated by all the t values within the segment $2\varepsilon_t = 2\varepsilon$.

The τ block is only able to recognize the t values in 2ε as a block and it cannot dissolve single t values. That is, the unified whole (the τ block) "sees" the t values within the 2ε interval all at once, as a unified whole so to speak; the block does not "see" single t values. In other words, the τ block records the infinite number of t values within the 2ε interval all at once and defines "one" value that expresses the mean value $\bar{t}$, which belongs to the 2ε interval. Thus, we have a t block as well. This t block has also the width 2ε and comes into existence through the overlap with the τ block. Thus, the $2\varepsilon_t$ width introduced earlier is identical with the 2ε width:

$$\varepsilon = \varepsilon_t \tag{6.133}$$

The mean value $\bar{t}$ is dependent on the position t at which the 2ε interval is just positioned. In other words, $\bar{t}$ defines the mean value of t with respect to the 2ε interval around t and of course around τ.

This entire t block with $2\varepsilon_t$ defines a mean value $\bar{t}$. The value for ε_t can be an infinitesimal small time interval; ε can be arbitrarily small but must be different from zero. Because the quantity ε can in principle be arbitrarily small, τ and $\tau \pm \varepsilon$ are practically undistinguishable: $\tau \pm \varepsilon \cong \tau$ with $\varepsilon \cong 0$.

The sequence of the mean values

The clock time τ itself moves in steps of 2ε from small to large values. Also in this case relation (6.129) is valid. Note that this is an empirical fact. For two values τ_i and τ_{i+1} we have $\tau_i < \tau_{i+1}$ with the boundary condition

$$\tau_i + \varepsilon = \tau_{i+1} - \varepsilon, \ \varepsilon \neq 0 \tag{6.134}$$

That is, the entire τ block moves in steps of 2ε and the number of mean values $\bar{t}$ in the range $t_b - t_a$ remains finite and is given by k_t. In the case of $\varepsilon = 0$, there are no mean values $\bar{t}$ definable and the number of t values is infinite in the range $t_b - t_a$. Thus, we have

$$t_i, \quad i = 1, 2, \ldots, \infty, \ (t_b - t_a) \tag{6.135}$$

$$\bar{t}_i, \quad i = 1, 2, \ldots, k_t, \ (t_b - t_a) \tag{6.136}$$

Concerning the mean value $\bar{t}$, this entity has to be interpreted as the classical time that we want to denote by t_{cl}. This point will still be discussed further.

The elementary steps

The order is to observe at clock time τ one t value with a definite probability that is proportional to $\Psi^*(\mathbf{r}_c, t)\Psi^*(\mathbf{r}_c, t)$ with $\mathbf{r}_c = (x_c, y_c, z_c)$. This situation is expressed by eqs. (6.126) and (6.127). In other words, each value within 2ε of $\tau \pm \varepsilon$ corresponds to one observational event with respect to t. The number of events within the interval 2ε of $\tau \pm \varepsilon$ is infinity. This infinite number is distributed over the entire t range, which is existent between t_a and t_b, which corresponds to the probability density $\Psi^*(\mathbf{r}_c, t)\Psi(\mathbf{r}_c, t)$. An infinite number of τ values in the range 2ε corresponds to an infinite number of t values in the range t_b–t_a.

The τ block moves on the t scale and there is an overlap with the system-specific time, which is defined between t_a and t_b [eq. (6.126)]. The movement of the τ block is in steps of 2ε, but could in principle move continuously.

This block behavior has nothing to do with a quantization but is simply a raster, which can be chosen arbitrarily and has to be different from zero. The 2ε raster can be introduced for theoretical reasons, but can also come into existence by resolutions within the frame of observations.

6.6.2 Classical time

The values for the complete t spectrum show quantum aspect at each time τ and $\varepsilon = 0$. Then the system behaves statistically and the probability density $\Psi^*(\mathbf{r}_c, t)\Psi(\mathbf{r}_c, t)$ is fully effective.

On the other hand, the system behaves classically if we admit $\varepsilon \neq 0$ for arbitrary values ε. Why? In this case, the probability densities $\Psi^*(\mathbf{r}_c, t)\Psi(\mathbf{r}_c, t)$ are eliminated and are no longer effective, that is, we have no longer a quantum characteristic. At time $\tau \pm \varepsilon$ all the k_t states for $\bar{t}$ are simultaneously occupied with "certainty." The quantum time t changes its character and we obtain a classical time t_{cl}. That is, the classical time t_{cl} is identical with the mean time $\bar{t}$:

$$t \to t_{\text{cl}} = \bar{t} \tag{6.137}$$

If the quantity ε remains sufficiently small, we obtain $\bar{t} \cong t$ and $t \cong \tau$. The differences between $\bar{t}$ and t are illustrated in Figure 6.4.

$$\tau : \textit{quantum time } t \leftarrow \Psi^*(\mathbf{r}_c,t)\Psi(\mathbf{r}_c,t)$$

$$t_i,\ i = 1,2,L\ ,\infty$$

$$\tau \pm \varepsilon : t_{i\,cl} = \bar{t}_i,\ i = 1,2,L\ ,\infty \leftarrow \textit{classical time}$$

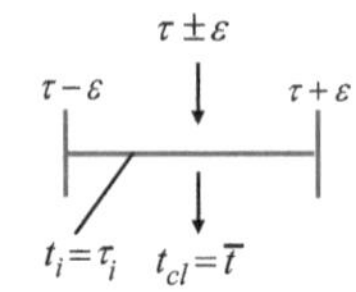

Figure 6.4: On the transition from quantum states to classical behavior within projection theory. The quantum time t is observed at clock time τ with a certain probability, which is proportional to the probability density $\Psi^*(\mathbf{r}_c,t)\,\Psi(\mathbf{r}_c,t)$ with $\mathbf{r}_c = (x_c, y_c, z_c)$. We obtain for the clock times $\tau_1, \tau_2, \ldots$the system-specific times $t_1, t_2, \ldots$. The function $\Psi^*(\mathbf{r}_c,t)\Psi^*(\mathbf{r}_c,t)$ is zero for $t < t_a$ and for $t > t_b$. On the other hand, when we admit a time interval ε around each τ, we get τ blocks as well as t blocks for $\varepsilon \neq 0$. The clock time $\tau \pm \varepsilon$ moves in steps of 2ε along the t scale, that is, the whole τ block moves and overlaps the t scale. The block behavior effectuates that only "one" value for the time is observable for each t block, and this is given by the mean time $\bar{t}$ with respect to the t interval 2ε. In the figure we have $\bar{t}_i$ at τ_i and $\bar{t}_{i+1}$ at τ_{i+1} with $\tau_{i+1} - \tau_i = 2\varepsilon$ and $\bar{t}_{i+1} - \bar{t}_i = 2\varepsilon$. This moving τ block only recognizes the t values in $2\varepsilon \neq 0$ as a block and not single t values. The τ block records the infinite number of t values within the 2ε interval all at once and defines "one" value, which expresses the mean value $\bar{t}$, which belongs to the t values within the 2ε interval. In this way, there is a perception of mean values $\bar{t}$ at τ in the interval $(t_b - t_a)$: $\bar{t}_i, i = 1, 2, \ldots, k_t$. Also the mean values $\bar{t}$ are given in steps of 2ε. The selected interval 2ε at the t scale is $2\varepsilon_t$ and is identical with 2ε. In the case of $\varepsilon = 0$ we have, in accordance with eq. (6.135), single quantum values $t_i = 1, 2, \ldots, \infty$.

As we have remarked, there are τ blocks and t blocks and they have the same interval 2ε. Each t block, all having the width $\varepsilon = \varepsilon_t$ [see eq. (6.133)], contains at each time $\tau \pm \varepsilon$ an infinite number of t states. At time $\tau \pm \varepsilon$ the number of t blocks is k_t and this is valid for each time $\tau \pm \varepsilon$. All t blocks exist simultaneously, also at each time $\tau \pm \varepsilon$. This follows directly from eqs. (6.130) and (6.132).

On the other hand, in the case of ε=0 there exists at clock time τ only one t state. However, the complete information of the system with respect to the classical time $t_{\text{cl}} = \bar{t}$ is given at $\tau \pm \varepsilon$ in the form of mean values: We have k_t mean values $\bar{t}_i, i = 1, 2, \ldots, k_t, (t_b - t_a)$, which exist simultaneously, but only that at time $\tau \pm \varepsilon$ is perceivable. This situation means that there are no longer probability densities, that is, the system behaves classical.

Remark

The classical value for the system-specific time, expressed by eq. (6.136), defines the classical behavior for $\varepsilon \neq 0$. The wave function is in this case defined but it is not effective. The object "behaves" classically.

On the other hand, the classical relation $E = E(\mathbf{p}, \mathbf{r}, t)$ is defined without wave function. The system "is" classical and a wave function is not involved and does not appear. This point has been discussed in Section 6.5. That is, relation $E = E(\mathbf{p}, \mathbf{r}, t)$ is formulated without wave function: $E = E(\mathbf{p}, \mathbf{r}, t)$ exists because it can successfully be used in the description of physically real systems, that is, it is used for the formulation of observable events. This situation belongs to the matter–mind realm and reflects the projection principle in pure culture.

The starting point for the formulation of Schrödinger's equations is Newton's classical relation, which can be expressed in short form by $E = E(\mathbf{p}, \mathbf{r})$. This corresponds to the classical relation $E = E(\mathbf{p}, \mathbf{r}, t)$ of projection theory without the system-specific time t. Similarly, a classical theory can possibly be formulated on the basis of $E = E(\mathbf{p}, \mathbf{r}, t)$.

6.6.3 The two cases

In principle, we have to distinguish between two cases: the quantum-theoretical description and the classical treatment.

Case 1: The quantum-theoretical description

Once again, in the quantum case we observe at each time τ "one" value for the system-specific time t, which is given with a certain probability and the probability density is relevant here. These are sequences $t_1, t_2, \ldots, t_i, \ldots$ and each of these values is related to a clock time: $\tau_1, \tau_2, \ldots, \tau_i, \ldots$.

Case 2: The classical treatment

We have k_t states at each $\tau \pm \varepsilon$, and all k_t segments exist simultaneously at each $\tau \pm \varepsilon$. The value for k_t can be arbitrarily large, but different from infinity. Again, we have k_t segments of width 2ε within the interval between t_a and t_a. If 2ε is sufficiently small, the number k_t of simultaneous classical events can be large and represents an almost continuous ensemble.

Each of the k_t values exist at $\tau \pm \varepsilon$ with certainty, that is, the wave function effect, which characterizes quantum effects, is eliminated. This situation defines the classical case within the frame of the projection principle.

Similarity to usual clocks

The mean values $\bar{t}_i$, $i = 1, 2, \ldots, k_t$ behave like a clock. The numbers on a usual clock correspond here to the mean values $\bar{t}_i = t_{cl}$, which behave classical. All numbers on the clock are defined simultaneously, and all values $\bar{t}_i$, $i = 1, 2, \ldots, k_t$ are defined simultaneously as well. Only one number of the clock is shown by the clock hand that is the clock time τ. The classical time structure in connection with a physically real system behaves similarly: in this case, only one $\bar{t}_i$ of the mean values is perceivable; it is $\bar{t}_i \cong \tau$.

The τ scale

There is no ensemble of τ values in analogy to the ensemble of t values at the t scale. The value τ is not a state of an existing ensemble of τ states. Only the actual value τ is existent but not values, which are smaller and larger than τ.

The reason is obvious. The wave function $\Psi(x, y, z, t)$ describes a physically real object or process in basic reality; therefore, $\Psi(x, y, z, t)$ reflects properties of basic reality. However, the wave function $\Psi(x, y, z, t)$ does not exist in this form in basic reality, and this is because space and time, that is, x, y, z, t, are not elements of basic reality.

In contrast to that, the time τ has no counterpart in basic reality; τ is an "invention" of the human being's mind and serves as a selection entity and appears in processing the basic reality information, which is given in compressed form in the mind (see Chapters 4 and 5). The pulled information from basic reality through the tools of evolution is treated and selected in terms of τ. Thus, there is no really existing τ ensemble installed in the mind, but the mind needs τ only then if there is actually something to process that comes from basic reality. The time τ serves for the representation of physically real information, that is, it is used by the mind for the representation of physical reality in proper form.

6.6.4 The transition from quantum states to classical behavior

The transition from quantum behavior to classical states is accomplished through the introduction of an interval 2ε with $\varepsilon \neq 0$. So, instead of $\tau(\varepsilon = 0)$ we have $\tau \pm \varepsilon$.

In the case of $\tau(\varepsilon = 0)$ the world behaves quantum-like: we have at each time τ one value for the system-specific time t, which is given with a definite probability that is expressed by $\Psi^*(\mathbf{r}_c, t)\Psi(\mathbf{r}_c, t)$ for constant values x_c, y_c, z_c. There is an infinite number of quantum states with respect to t within the range between t_a and t_b, where the probability density $\Psi^*(\mathbf{r}_c, t)\Psi(\mathbf{r}_c, t)$ is not zero:

$$t_i,\ i = 1, 2, \ldots, \infty \tag{6.138}$$

On the other hand, in the case of $\varepsilon \neq 0$ we have $\tau \pm \varepsilon$ and, in accordance with the statements given above, we obtain k_t mean values $\bar{t}$:

$$\bar{t}_i,\ i = 1, 2, \ldots, k_t \tag{6.139}$$

These mean values are identical with the classical time, which is outlined in Section 6.6.2.

There are two blocks with respect to time, the τ block and the t block. Both have a width of 2ε. The τ block moves in steps of 2ε along the t scale; the $\bar{t}$ value also proceeds in steps of 2ε whereby all $k_t\,\bar{t}_i$ values exist simultaneously.

In practical cases of everyday life we always work with $\tau \pm \varepsilon$ $(\varepsilon \neq 0)$, where $\varepsilon \neq 0$ reflects the observation resolution. Thus, we always perceive the physically real world outside as classical entity and, in analogy to eq. (6.12), we have

$$t_i \rightarrow t_{i\,\mathrm{cl}} = \bar{t} \tag{6.140}$$

This is in fact the behavior, which we experience in everyday life. For each resolution $\varepsilon \neq 0$, we perceive a classical world.

In Figure 6.4, the entire scenario in connection with the transition from quantum to classical behavior is summarized.

6.6.5 Quantum behavior of the reference time

Where does the τ block come from? Clearly, also the reference time τ must be based on quantum effects. Such a quantum treatment with respect to τ has been done in Ref. [81]. Then, the clock itself would be a "quantum clock" and would show a quantum reference time, which we denote here by the Greek letter κ.

The value κ has, as usual, a probability character and the probability density $\Psi^*(\kappa)\Psi(\kappa)$ is in the center [81]. This quantum clock shows, for example, the value κ' and, simultaneously, the system under investigation takes the system-specific time t' for which the probability density is given by $\Psi^*(\mathbf{r}_c, t')\Psi(\mathbf{r}_c, t')$.

The reference time τ used so far is then a mean value, defined by the variable κ which is given in the range for which the probability density $\Psi^*(\kappa)\Psi(\kappa)$ is not zero: $\Psi^*(\kappa)\Psi(\kappa) \neq 0$. This package $\Psi^*(\kappa)\Psi(\kappa) \neq 0$ moves in a strict direction from the past to the future and the mean value τ takes successively the values

$$\tau_1, \tau_2, \tau_3, \ldots, \tau_i, \ldots \tag{6.141}$$

with

$$\tau_1 < \tau_2 < \tau_3 < \ldots < \tau_i < \ldots \tag{6.142}$$

as we already outlined in Chapter 1 [eqs. (1.44) and (1.45)].

The τ block itself with a width of 2ε is defined by "one" time value, which is the mean value $\bar{\kappa}$ of κ. The κ values in the interval 2ε are distributed over the complete range of $\Psi^*(\kappa)\Psi(\kappa) \neq 0$. We exactly had the same effect with respect to the system-specific time t (see Section 6.6.4). Then, we get

$$\tau = \bar{\kappa} \tag{6.143}$$

Again, the τ block proceeds in steps of 2ε on the t scale.

6.7 Classical behavior of space and time

6.7.1 Statistical behavior in four dimensions

In Section 6.6.1, the system-specific time t has been treated for the constant space coordinates x_c, y_c, z_c. In this case, we have for the probability density the function $\Psi^*(\mathbf{r}_c, t)\Psi(\mathbf{r}_c, t)$ with $\mathbf{r}_c = (x_c, y_c, z_c)$. In general, the space variables x, y, z are not constant and behave statistically as well, like the system-specific time t.

One value of the four x, y, z, t is given at clock time τ with a definite probability that is proportional to the probability density $\Psi^*(\mathbf{r}, t)\Psi(\mathbf{r}, t)$ with $\mathbf{r} = (x, y, z)$. The probability density is assumed to be zero for

$$x < x_a, y < y_a, z < z_a, t < t_a \tag{6.144}$$

and

$$x > x_a, y > y_a, z > z_a, t > t_a, t > t_b \tag{6.145}$$

with

$$x_a \neq -\infty, y_a \neq -\infty, z_a \neq -\infty, t_a \neq -\infty \tag{6.146}$$

and

$$x_b \neq \infty, y_b \neq \infty, z_b \neq \infty, t_b \neq \infty \tag{6.147}$$

In other words, we consider the physically real objects and the entire world as finite entities. With eqs. (6.144) – (6.147) we have

$$\begin{gathered} \Psi^*(x, y, z, t)\Psi(x, y, z, t) \neq 0 \\ x_a \leq x \leq x_b,\ y_a \leq y \leq y_b,\ z_a \leq z \leq z_b,\ t_a \leq t \leq t_b \end{gathered} \tag{6.148}$$

This is exactly the situation given by eq. (6.126) with respect to time, which is here extended to space and time.

Daily life situations

In daily life, we do not observe definite values x, y, z, t at clock time τ as probability distribution, which is given by the probability density $\Psi^*(\mathbf{r}, t)\Psi(\mathbf{r}, t)$. Instead, a human being observes at τ the entire space structure of his/her environment, that is, all the details of the environment are given at the same clock time τ. Everything at once so to speak.

How is this situation described within the frame of the projection principle? The answer is relatively simple. In order to show that, we have to work with the ε effect (Section 6.1).

The ε effect in four dimensions

In the case of $\tau \pm \varepsilon$ the situation with respect to x, y, z, t is different from the case described through eq. (6.130). This is an extended form of what we have treated in connection with the time t treated in Section 6.6.1.

There is an infinite number of τ values within the interval 2ε. Thus, we have an infinite number for each of the variables x, y, z, t for which the probability densities $\Psi^*(\mathbf{r}, t)\Psi(\mathbf{r}, t)$ are cognizant. They are existent through an infinite number of probability densities $\Psi^*(\mathbf{r}, t)\Psi(\mathbf{r}, t)$ in the intervals

$$x_b - x_a,\ y_b - y_a,\ z_b - z_a,\ t_b - t_a \tag{6.149}$$

That is, the infinite number of τ values within the interval 2ε creates an infinite number for each of the variables x, y, z, t, which are distributed over the entire space-time range for which the probability density is not zero [see eq. (6.148)]. Thus, instead of eq. (6.130) we have

$$\begin{gathered} \tau \to \infty \ in\, 2\varepsilon \\ \Downarrow \\ x, y, z, t \to \infty \ in\, (x_b - x_a), (y_b - y_a), (z_b - z_a), (t_b - t_a) \end{gathered} \tag{6.150}$$

What is the condition for ε in this case? The interval $2\varepsilon \neq 0$ contains an infinite number of τ values, but it can be arbitrarily small. The value for ε may even be close to zero and contains, nevertheless, also for this case an infinite number of τ values (real numbers).

The number of states

In the next step in our analysis, we want to consider the small space–time segments

$$\varepsilon_x, \varepsilon_y, \varepsilon_z, \varepsilon_t \tag{6.151}$$

These segments refer to the variables x, y, z, t. Each variable is enclosed by intervals of widths $2\varepsilon_x, 2\varepsilon_y, 2\varepsilon_z$. In other words, in addition to time t we also treat the space coordinates x, y, z in terms of the space segments $\varepsilon_x, \varepsilon_y, \varepsilon_z$. Also the $\varepsilon_x, \varepsilon_y, \varepsilon_z, \varepsilon_t$ raster can be introduced for theoretical reasons, but can also come into existence through resolutions within the frame of observations.

As an example, let us choose the set x_i, y_i, z_i, t_i We have intervals around x_i, y_i, z_i, t_i that are given by

$$x_i \pm \varepsilon_x,\ y_i \pm \varepsilon_y,\ z_i \pm \varepsilon_z,\ t_i \pm \varepsilon_t \tag{6.152}$$

Each space variable x_i, y_i, z_i, t_i is enclosed by the intervals $2\varepsilon_x, 2\varepsilon_y, 2\varepsilon_z, 2\varepsilon_t$. We assume that the quantities $\varepsilon_x, \varepsilon_y, \varepsilon_z, \varepsilon_t$ are not zero:

$$\varepsilon_x \neq 0, \varepsilon_y \neq 0, \varepsilon_z \neq 0, \varepsilon_t \neq 0 \tag{6.153}$$

If we choose the number of intervals within the four intervals as free parameters, then the four segments $\varepsilon_x, \varepsilon_y, \varepsilon_z, \varepsilon_t$ are given by

$$\begin{aligned}(x_b - x_a)/k_x &= 2\varepsilon_x \\ (y_b - y_a)/k_y &= 2\varepsilon_y \\ (z_b - z_a)/k_z &= 2\varepsilon_z \\ (t_b - t_a)/k_t &= 2\varepsilon_t\end{aligned} \tag{6.154}$$

With eqs. (6.146), (6.147) and (6.150) we obtain

$$k_x \neq \infty, k_y \neq \infty, k_z \neq \infty, k_t \neq \infty \tag{6.155}$$

As in the case of t (Section 6.6.1), the number of x, y, z, t states within the total range is infinity. Thus, the number of states within each of the intervals $2\varepsilon_x, 2\varepsilon_y, 2\varepsilon_z, 2\varepsilon_t$ must be infinite as well (we have $\infty/k = \infty$ for $k \neq \infty$, see also Section 6.6.1). Thus, instead of eq. (6.132) we have

$$\begin{aligned}&\tau \pm \varepsilon: \\ &x, y, z, t \longrightarrow \infty \text{ in}(x_b - x_a), (y_b - y_a), (z_b - z_a), (t_b - t_a) \\ &\qquad\Downarrow \\ &x, y, z, t \longrightarrow \infty \text{ in } 2\varepsilon_x, 2\varepsilon_y, 2\varepsilon_z, 2\varepsilon_t\end{aligned} \tag{6.156}$$

What are the conditions for the segments $\varepsilon_x, \varepsilon_y, \varepsilon_z, \varepsilon_t$? Each of the four segments contains an infinite number of x, y, z, t values. This property is independent on the magnitude of the segments $\varepsilon_x, \varepsilon_y, \varepsilon_z, \varepsilon_t$; they can be arbitrarily small. The values for $\varepsilon_x, \varepsilon_y, \varepsilon_z, \varepsilon_t$ may even be close to zero and contain also for this case an infinite number of values (real numbers).

6.7.2 Classical states

If we define mean values in connection with the values within $\varepsilon_x, \varepsilon_y, \varepsilon_z, \varepsilon_t$, we get mean values $\bar{x}, \bar{y}, \bar{z}, \bar{t}$ for the x, y, z, t points that are enclosed by the segments $\varepsilon_x, \varepsilon_y, \varepsilon_z, \varepsilon_t$. These mean values are given at each time $\tau \pm \varepsilon$ with "certainty." That is, there is no longer a probability distribution because all these x, y, z, t points exist simultaneously at time $\tau \pm \varepsilon$. That is, the wave function is eliminated, as in the case of the system-specific time (Section 6.6.1). There are no longer quantum characteristics, and this situation defines the classical case. The mean values $\tau \pm \varepsilon$ express the coordinates and the time in classical form, which we want to denote by $x_{\text{cl}}, y_{\text{cl}}, z_{\text{cl}}, t_{\text{cl}}$ and we have

$$\bar{x} = x_{\text{cl}}, \bar{y} = y_{\text{cl}}, \bar{z} = z_{\text{cl}}, \bar{t} = t_{\text{cl}} \tag{6.157}$$

The case at $\tau \pm \varepsilon$ together with the segments $\varepsilon_x, \varepsilon_y, \varepsilon_z, \varepsilon_t$ always leads to classical behavior. This property is independent on the magnitude of the segments $\varepsilon, \varepsilon_x, \varepsilon_y, \varepsilon_z, \varepsilon_t$. They can be arbitrarily small. The values for $\varepsilon, \varepsilon_x, \varepsilon_y, \varepsilon_z, \varepsilon_t$ may even be close to zero.

6.7.3 Block behavior with respect to time and space distributions

In connection with the segment ε_t we have an infinite number of t values: t_i, $i=1,2,\ldots,\infty$ within $2\varepsilon_t$. For each t_i, a wave function $\Psi(x,y,z,t_i), i=1,2,\ldots,\infty$ is defined. Thus, within the segment $2\varepsilon_t$ there is an infinite number of values for the system-specific time and an infinite number of space distributions (probability densities):

$$\begin{aligned} 2\varepsilon_t \quad \rightarrow \quad & t_i,\ \Psi^*(x,y,z,t_i)\Psi(x,y,z,t_i) \\ & i=1,2,\ldots,\infty \end{aligned} \tag{6.158}$$

In Section 6.6.1, we formulated some principal remarks concerning the block behavior. Let us apply these principles also to the four-dimensional case given in connection with eq. (6.157).

Let us denote the wave function at constant t_i by $\Psi_i(x,y,z)$, i.e., we use instead of $\Psi(x,y,z,t_i)$ the notation $\Psi_i(x,y,z)$:

$$\Psi(x_i,y_i,z_i,t_i) \rightarrow \Psi_i(x,y,z) \tag{6.159}$$

In the following, we will analyze the block-behavior with respect to the system-specific time and with respect to the probability density.

Mean properties

The use of $\tau \pm \varepsilon$ instead of τ means that there is no τ value within the 2ε interval is soluble. The observational period of 2ε cannot be shortened. Then the observer records all what is within the segment 2ε, but he/she cannot resolve the details. He records all at once. This is the task of the τ block. This situation is described for the time t in Section 6.6.1, but not for the four-dimensional case.

All the τ values within the 2ε interval are defined as "one" block. This 2ε block acts as unified whole. The 2ε block overlaps with the t scale and generates a mean value $\bar{t}$ and a mean value for the probability density $\bar{\Psi}^*(x,y,z)\bar{\Psi}(x,y,z)$. These mean values refer to the t block on the t scale, which has the width of $2\varepsilon_t = 2\varepsilon$ (Section 6.6.1).

This is the conception. In other words, in $2\varepsilon_t$ there is an infinite number of $2\varepsilon_t$ states and an infinite number of probability densities [see eq. (6.158)], but the human observer cannot resolve them. However, he/she experiences all at once in the form of "blocks" defined through a segment of width $2\varepsilon_t = 2\varepsilon$.

In the four-dimensional case, there are two blocks: one block is expressed by the mean value $\bar{t}$, which is generated on the basis of all t within $2\varepsilon_t$, and the second block is expressed by the mean probability density $\bar{\Psi}_i^*(x,y,z)\bar{\Psi}_i(x,y,z)$, which is generated on the basis of all $\Psi_i(x,y,z)$, also located within $2\varepsilon_t$.

Again, in $2\varepsilon_t$ there is an infinite number of t states and an infinite number of probability densities [eq. (6.158)]. But the human observer experiences "all at once" in the form of "blocks" within the segment $2\varepsilon_t$, and there are two blocks, which are expressed by the mean values $\bar{t}$ and $\bar{\Psi}_i^*(x,y,z)\bar{\Psi}_i(x,y,z)$.

The τ block is only able to recognize the t_i values and the $\Psi^*(x,y,z,t_i)$ $\Psi(x,y,z,t_i)$ functions, localized in 2ε, as blocks and it cannot dissolve single t_i values and also not single probability densities $\Psi^*(x,y,z,t_i)\Psi(x,y,z,t_i)$. That is, the unified whole (the τ block) "sees" the t_i values and the $\Psi^*(x,y,z,t_i)\Psi(x,y,z,t_i)$ distributions all at once as mean properties. These mean properties refer to the $2\varepsilon_t$ interval, which the τ block overlaps. The τ block does not "see" single t values and single $\Psi^*(x,y,z,t_i)\Psi(x,y,z,t_i)$ distributions. The τ block acts as a unified whole, so to speak.

Distributions in space

For all t states between $\tau_i-\varepsilon$ and $\tau_i+\varepsilon$ we get an infinite number of x,y,z distributions with

$$\Psi_i^*(x,y,z)\Psi_i(x,y,z) \tag{6.160}$$

with

$$x_a \le x \le x_b,\ y_a \le y \le y_b,\ z_a \le z \le z_b \tag{6.161}$$

The block behavior effectuates the following situation: instead of an infinite number of t values we get "one" mean time value $\bar{t}$ and, furthermore, instead of an infinite number of $\bar{\Psi}^*(x,y,z)\bar{\Psi}(x,y,z)$ distributions we get "one" mean distribution function $\bar{\Psi}^*(x,y,z)\bar{\Psi}(x,y,z)$:

$$\tau_i \pm \varepsilon\colon \left\{ \begin{matrix} \bar{t}_i \\ \bar{\Psi}_i^*(x,y,z)\bar{\Psi}_i(x,y,z),\ x_a \le x \le x_b,\ y_a \le y \le y_b,\ z_a \le z \le z_b \end{matrix} \right\} \tag{6.162}$$

We have one $\bar{t}$ value for each segment $2\varepsilon_t$ and we have one probability density $\bar{\Psi}^*(x,y,z)\bar{\Psi}(x,y,z)$ for each segment $2\varepsilon_t$.

However, there is still a certain probability for the values x,y,z that exist in the ranges $x_a \le x \le x_b$, $y_a \le y \le y_b$ and $z_a \le z \le z_b$, which is described by the mean probability density $\bar{\Psi}^*(x,y,z)\bar{\Psi}(x,y,z)$. The situation is represented in Figure 6.5 and once again in clearly arranged form in Figure 6.6.

The $\bar{t}$ states outside the segment $2\varepsilon_t$ are quantum mechanical in character. This is the case for $t_a \le t < t_{a\tau_i}$ and $t_{b\tau_i} < t \le t_b$, whereby $t_{a\tau_i}$ is the lower limit of the segment $2\varepsilon_t$ and $t_{b\tau_i}$ is its upper point (see also Figure 6.5).

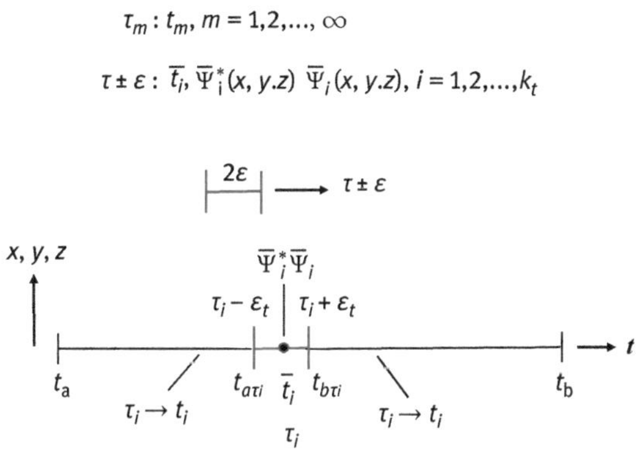

Figure 6.5: On transition from quantum states to classical behavior within the projection theory. The variables x, y, z, t of physically real objects (systems), which belong to the human observer's environment, are observed at clock time τ with a certain probability. On the other hand, when we admit a time interval ε around each τ, we get τ blocks as well as t blocks, having the widths ε and ε_t with $\varepsilon_t = \varepsilon$. The clock time $\tau \pm \varepsilon$ moves in steps of 2ε along the t scale, that is, the whole τ block moves and overlaps the t scale, but could in principle move continuously. The block behavior effectuates that only "one" value for the time and "one" function for the probability distribution is observable for each t block, and these are given by the mean value $\bar{t}$ and the mean function $\bar{\Psi}^*(\mathbf{r})\bar{\Psi}(\mathbf{r})$ with respect to the t interval $2\varepsilon_t$. In the figure, we have $\bar{t}_i$ and $\bar{\Psi}_i^*(\mathbf{r})\bar{\Psi}_i(\mathbf{r})$ at $\tau_i \pm \varepsilon$. This moving τ block only recognizes the t_i values and the functions $\Psi^*(\mathbf{r}, t_i)\Psi(\mathbf{r}, t_i)$ in $2\varepsilon \neq 0$ as a block and not in the form of single entities. In this way, we get the mean values $\bar{t}$ and $\bar{\Psi}^*(\mathbf{r})\bar{\Psi}(\mathbf{r})$, which belong to a $2\varepsilon_t$ interval. There is a perception process of mean values in the interval $(t_b - t_a)$, but only for various times $\tau \pm \varepsilon$.and the respective intervals $2\varepsilon_t$. Outside the interval $2\varepsilon_t$ the τ block is not effective and there are strict quantum states: $\tau_m \to t_m$. This is the case for $t_a \leq t < t_{a\tau_i}$ and $t_{b\tau_i} < t \leq t_b$, whereby $t_{a\tau_i}$ is the lower limit of the segment $2\varepsilon_t$ and $t_{b\tau_i}$ is its upper point. In the case of $\varepsilon = 0$ we have single quantum states $x_m, y_m, z_m, t_m,\ m = 1, 2, \ldots, \infty$, within the entire range $(t_b - t_a)$.

The τ block moves from small to large τ values, which corresponds to our clocks. The next step with respect to $\tau \pm \varepsilon$ is within the segment given by the limits $\tau_{i+1} - \varepsilon$ and $\tau_{i+1} + \varepsilon$. Here we have the mean value $\bar{t}_{i+1}$ and the mean function $\bar{\Psi}_{i+1}^*(x, y, z)\bar{\Psi}_{i+1}(x, y, z)$.

6.7.4 Block behavior in space and time

The use of $\tau \pm \varepsilon$ instead of τ means that none of the τ values within the 2ε interval is soluble. Then, the observer records all what is within the segment 2ε, but he/she cannot resolve the details. He records all at once. The segment cannot be subdivided into parts. In other words, the observational period of 2ε cannot be shortened.

So far, we have studied the block behavior with respect to time t and in connection with distribution functions. For each $2\varepsilon_t$ interval, there is a mean value $\bar{t}$ for the time and a mean probability density $\bar{\Psi}^*(x, y, z)\bar{\Psi}(x, y, z)$ for the x, y, z values. The

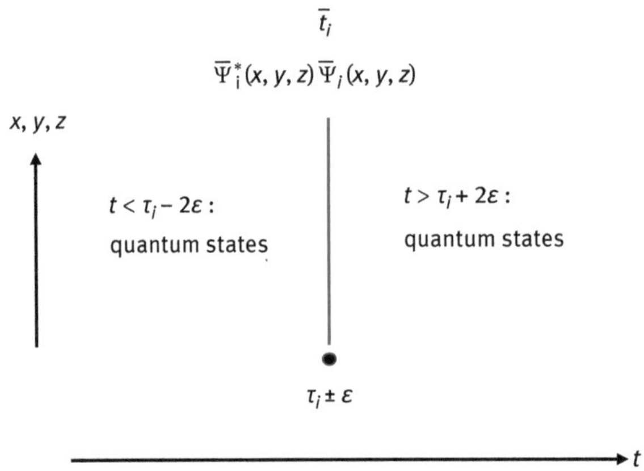

Figure 6.6: Due to the action of the τ block, we have mean values $\bar{t}$ for the time and for the probability densities $\bar{\Psi}^*(x, y, z)\bar{\Psi}(x, y, z)$. That is, there are mean probability distributions for the variables x, y, z. That is, the variables x, y, z still behave like a quantum quantity. The time exists in this case not as a quantum element but is here a classical entity. At $\tau_i \pm \varepsilon$ there is a mean value $\bar{t}_i$ for the time and a mean function $\bar{\Psi}^*(x, y, z)\bar{\Psi}(x, y, z)$. $2\varepsilon_t$. Outside the interval $2\varepsilon_t$ the τ block is not effective and there are strict quantum states.

variables x, y, z still behave like quantum entities. However, the time exists in this case not as a quantum variable.

The impact of the τ lock on the coordinates x, y, z

In the observations of daily life, we have a macroscopic world before our eyes. In this situation, not only the time behaves macroscopically, but the x, y, z values as well.

How can we go from the quantum states for the coordinates to x, y, z values that are macroscopic in character? Besides the block property $\tau \pm \varepsilon$ we introduce in addition the segments $\varepsilon_x, \varepsilon_y, \varepsilon_z$ for the x, y, z variables.

As in the case of ε and ε_t, the $\varepsilon_x, \varepsilon_y, \varepsilon_z$ raster can be introduced for theoretical reasons, but can also come into existence by resolutions within the frame of observations.

The τ block is by definition in the $\tau \pm \varepsilon$ state. The τ block is for the "observation" of physically real objects and systems, respectively. The block behavior itself is also applied on the space variables x, y, z, that is, it acts as a unified whole on the x, y, z variables as well. In other words, the τ block not only acts on the time t but also on the coordinates x, y, z.

Formulation in accordance with experiences in daily life

The following has to be considered: human beings experience their environment (trees, houses, etc.) at clock time τ, that is, all the objects in the large space around us are recorded simultaneously at time τ ($\tau \pm \varepsilon$). We have defined the coordinates x, y, z of the space through the ranges $x_a \leq x \leq x_b$, $y_a \leq y \leq y_b$, $z_a \leq z \leq z_b$ [see eq. (6.154)]. The space, experienced in daily life, can be described in this way.

Let us treat the coordinates x, y, z on the basis of the x, y, z scales of the mean probability density $\bar{\Psi}^*(x, y, z)\bar{\Psi}(x, y, z)$. This is bidden because the block behavior for the time t has already been considered in $\bar{\Psi}^*(x, y, z)\bar{\Psi}(x, y, z)$.

Treatment

These x, y, z scales of the mean probability density are subdivided into intervals of widths $2\varepsilon_x, 2\varepsilon_y, 2\varepsilon_z$. As we have outlined in Section 6.7.1, we get k_x intervals with $2\varepsilon_x$ on the x scale and k_y intervals with $2\varepsilon_y$ on the y scale, and the number of intervals with $2\varepsilon_z$ on the z scale is k_z. However, we have only one interval of width $2\varepsilon_t$ for the time t when the system is in the state $\tau \pm \varepsilon$.

Within each of these intervals $2\varepsilon_x, 2\varepsilon_y, 2\varepsilon_z$, there is an infinite number of states. But the block behavior effectuates that only one value per interval is recognizable, that is, we get for each interval one value, which is the mean value.

At each time $\tau_i \pm \varepsilon$ we get a set of mean values for the coordinates:

$$\begin{aligned} &\bar{x}_{1,i}, \bar{x}_{2,i}, \ldots, \bar{x}_{k_x,i} \\ &\bar{y}_{1,i}, \bar{y}_{2,i}, \ldots, \bar{y}_{k_y,i} \\ &\bar{z}_{1,i}, \bar{z}_{2,i}, \ldots, \bar{z}_{k_z,i} \end{aligned} \tag{6.163}$$

In addition, there is the mean time $\bar{t}$ at $\tau_i \pm \varepsilon$. All these mean values, expressed by eq. (6.163) and $\bar{t}$, exist at $\tau_i \pm \varepsilon$ simultaneously and, furthermore, they exist with certainty.

Also here, all the τ values within the 2ε interval are defined as "one" block. This τ block acts as a unified whole. The τ block overlaps with the t scale and generates a mean value $\bar{t}$ and a mean value for the probability density $\bar{\Psi}^*(x, y, z)\bar{\Psi}(x, y, z)$. This mean value and the mean function refer to the t block on the t scale, which has the width of $2\varepsilon_t = 2\varepsilon$. But, due to $2\varepsilon_x, 2\varepsilon_y, 2\varepsilon_z$, there is an additional block effect with respect to the coordinates x, y, z that is treated in connection with $\bar{\Psi}^*(x, y, z)\bar{\Psi}(x, y, z)$.

The τ block also acts as a unified whole on the space variables x, y, z and not only on the time t. Again, there are k_x intervals on the x scale and k_y intervals on the y scale, and the number of intervals on the z scale is k_z. However, for the time t only one interval $2\varepsilon_t = 2\varepsilon$ exists at $\tau \pm \varepsilon$ on which the τ block acts simultaneously.

All mean values exist simultaneously and with certainty

The τ block acts at $\tau \pm \varepsilon$ and is effective within the range $\tau - \varepsilon$ and $\tau + \varepsilon$. During this time segment, the τ block acts "simultaneously" on all the k intervals with $k = k_x\, k_y\, k_z$ and generates all the possible mean values for the coordinates. This ensemble of mean values exists simultaneously. This is due to property (6.156).

The mean values for the space coordinates, defined by eq. (6.163), and the mean value $\bar{t}$ for the time exist at $\tau \pm \varepsilon$ with "certainty" and "simultaneously." This is compatible with everyday life experiences.

Again, the x, y, z scales are subdivided into intervals of the widths $2\varepsilon_x, 2\varepsilon_y, 2\varepsilon_z$ leading to k intervals for the coordinates x, y, z, but there is one interval for the system-specific $t$scale. Within each interval of these segments, there is an infinite number of values.

It is characteristic for the behavior of the τ block that it can "see" at $\tau \pm \varepsilon$ for each interval only "one" characteristic value. This characteristic value is the mean value, that is, the τ block records the infinite number of coordinates "all at once" and generates in this way a mean value. There are $k = k_x\, k_y\, k_z$ intervals for the coordinates (Section 6.7.1) and one for the time. Therefore, there are k mean values for the coordinates and one mean value for the time. All mean values are generated simultaneously and exist with certainty.

6.7.5 Classifications

The reference time $\tau \pm \varepsilon$ moves in steps of 2ε from small to large values and selects space structures. A human being can only recognize one space structure at time $\tau \pm \varepsilon$ and this space structure varies with increasing $\tau \pm \varepsilon$:

$$\begin{gathered} \tau_1 \pm \varepsilon, \tau_2 \pm \varepsilon, \ \ldots, \tau_i \pm \varepsilon, \ \ldots \\ \tau_1 < \tau_2 < \ \ldots \ < \tau_i < \ \ldots \end{gathered} \tag{6.164}$$

For times smaller or larger than $\tau \pm \varepsilon$, the physically real world is not perceivable for human beings, but it exists.

At $\tau \pm \varepsilon$ the τ block is effective and generates mean values x, y, z, t for the coordinates and the time, which have to be interpreted as classical quantities. For times smaller and larger than $\tau \pm \varepsilon$, the τ block is not effective, and the physically real systems of the environment behave quantum mechanically.

Step i

For example, in the case of $\tau_i \pm \varepsilon$ the world is perceivable but not for $\tau_n \pm \varepsilon$ with $n = 1, 2, \ \ldots, i-1$ and $n = i+1, i+2, \ \ldots, k_t$, although the world exists simultaneously also for these times. We have three regions:

Region 1

We have a probability distribution for the variables x, y, z, t within the ranges $x_a \le x_m \le x_b$, $y_a \le y_m \le y_b$, $z_a \le z_m \le z_b$ and $t_a \le t_m < t_{a\tau_i}$, whereby the time $t_{a\tau_i}$ is lower limit of the τ block of width 2ε, that is, $t_{a\tau_i}$ is given by $t_{a\tau_i} = \tau_i - \varepsilon$. We have

$$
\begin{aligned}
&x_m, y_m, z_m, t_m \leftarrow \Psi^*(x_m, y_m, z_m, t_m)\Psi(x_m, y_m, z_m, t_m) \\
&m = 1, 2, \ldots, \infty \qquad (6.165) \\
&x_a \le x_m \le x_b, y_a \le y_m \le y_b, z_a \le z_m \le z_b, t_a \le t_m < t_{a\tau_i}
\end{aligned}
$$

In these ranges, the object or the entire environment is a strict quantum system. The τ block is not effective, that is, no observation takes place. The τ block is for the observation of the physically real world, and human beings observe the world exclusively at $\tau_i \pm \varepsilon$.

Region 2

The τ block is effective within the interval $2\varepsilon = (\tau_i + \varepsilon) - (\tau_i - \varepsilon)$, and the human being observes mean values, which are classical in character. The following scheme is valid:

$$
\begin{gathered}
\tau_i \pm \varepsilon \\
\Downarrow \\
\bar{t}_i, \overline{\Psi}^*(x,y,z)_i \overline{\Psi}(x,y,z)_i \begin{cases} \bar{x}_{1,i}, \bar{x}_{2,i}, \ldots, \bar{x}_{k_x,i} \\ \bar{y}_{1,i}, \bar{y}_{2,i}, \ldots, \bar{y}_{k_y,i} \\ \bar{z}_{1,i}, \bar{z}_{2,i}, , \ldots, \bar{z}_{k_z}, i \end{cases} \qquad (6.166)
\end{gathered}
$$

$$
\begin{aligned}
&\bar{t}_i \Leftarrow t_i - \varepsilon, t_1 + \varepsilon \\
&\bar{x}_{m,i} \Leftarrow x_{m,i} - \varepsilon_x, x_{m,i} + \varepsilon_x \\
&\bar{y}_{m,i} \Leftarrow y_{m,i} - \varepsilon_y, y_{m,i} + \varepsilon_y \\
&\bar{z}_{m,i} \Leftarrow z_{m,i} - \varepsilon_z, z_{m,i} + \varepsilon_z \\
&m = -1, 2, \ldots, k_x, k_y, k_z
\end{aligned}
$$

The observation process leads to a complete set of mean space values and one mean value for the system-specific time.

Region 3

As in the case of region 1, we have probability distribution for the variables x, y, z, t within the ranges $x_a \le x_m \le x_b$, $y_a \le y_m \le y_b$, $z_a \le z_m \le z_b$ and $t_{b\tau_i} < t_m \le t_b$, whereby the time $t_{b\tau_i}$ is the upper limit of the τ block of width 2ε, that is, $t_{b\tau_i}$ is $t_{b\tau_i} = \tau_i + \varepsilon$. We have

$$
\begin{aligned}
&x_m, y_m, z_m, t_m \leftarrow \Psi^*(x_m, y_m, z_m, t_m)\Psi(x_m, y_m, z_m, t_m) \\
&m = 1, 2, \ldots, \infty \qquad (6.167) \\
&x_a \le x_m \le x_b, y_a \le y_m \le y_b, z_a \le z_m \le z_b, t_{b\tau_i} < t_m \le t_b
\end{aligned}
$$

In these ranges, the physically real entities are strict quantum systems. The τ block is not effective, that is, no observation takes place. Again, the τ block is for the observation of the physically real world, and human beings observe the world exclusively at $\tau_i \pm \varepsilon$.

Step i + 1

When we go a step further, from $\tau_i \pm \varepsilon$ to $\tau_{i+1} \pm \varepsilon$ with $\tau_{i+1} \pm \varepsilon > \tau_i \pm \varepsilon$, exactly the same laws are valid. Instead of eqs. (6.165), (6.166) and (6.167), the following relations are relevant:

Region 1

$$\begin{aligned} &x_m, y_m, z_m, t_m \leftarrow \Psi^*(x_m, y_m, z_m, t_m)\Psi(x_m, y_m, z_m, t_m) \\ &m = 1, 2, \ldots, \infty \\ &x_a \leq x_m \leq x_b, y_a \leq y_m \leq y_b, z_a \leq z_m \leq z_b, t_a \leq t_m < t_{a\tau_{i+1}} \end{aligned} \tag{6.168}$$

The time $t_{a\tau_{i+1}}$ is a lower limit of the τ block of width 2ε, that is, $t_{a\tau_{i+1}}$ is given by $t_{a\tau_{i+1}} = \tau_{i+1} - \varepsilon_t$ $(\varepsilon_t = \varepsilon)$.

Region 2

$$\begin{gathered} \tau_{i+1} \pm \varepsilon \\ \Downarrow \\ \bar{t}_{i+1}, \overline{\Psi}^*(x, y, z)_{i+1} \overline{\Psi}(x, y, z)_{i+1} \begin{cases} \bar{x}_{1,i+1}, \bar{x}_{2,i+1}, \ldots, \bar{x}_{k_x,i+1} \\ \bar{y}_{1,i+1}, \bar{y}_{2,i+1}, \ldots, \bar{y}_{k_y,i+1} \\ \bar{z}_{1,i+1}, \bar{z}_{2,i+1}, \ldots, \bar{z}_{k_z,i+1} \end{cases} \\ \begin{aligned} &\bar{t}_{i+1} \Leftarrow t_{i+1} - \varepsilon,\ t_{i+1} + \varepsilon \\ &\bar{x}_{m,i+1} \Leftarrow x_{m,i+1} - \varepsilon_x,\ x_{m,i+1} + \varepsilon_x \\ &\bar{y}_{m,i+1} \Leftarrow y_{m,i+1} - \varepsilon_y,\ y_{m,i+1} + \varepsilon_y \\ &\bar{z}_{m,i+1} \Leftarrow z_{m,i+1} - \varepsilon_z,\ z_{m,i+1} + \varepsilon_z \\ &m = 1, 2, \ldots, k_x, k_y, k_z \end{aligned} \end{gathered} \tag{6.169}$$

Region 3

$$\begin{aligned} &x_m, y_m, z_m, t_m \leftarrow \Psi^*(x_m, y_m, z_m, t_m)\Psi(x_m, y_m, z_m, t_m) \\ &m = 1, 2, \ldots, \infty \\ &x_a \leq x_m \leq x_b, y_a \leq y_m \leq y_b, z_a \leq z_m \leq z_b, t_{b\tau_i+1} < t_m \leq t_b \end{aligned} \tag{6.170}$$

The time $t_{b\tau_{i+1}}$ is the upper limit of the τ block of width 2ε, that is, $t_{b\tau_{i+1}}$ is $t_{b\tau_{i+1}} = \tau_{i+1} + \varepsilon_t$ $(\varepsilon_t = \varepsilon)$.

The situations in connection with step i is illustrated in Figure 6.7 (a) and with respect to step $i + 1$ in Figure 6.7 (b).

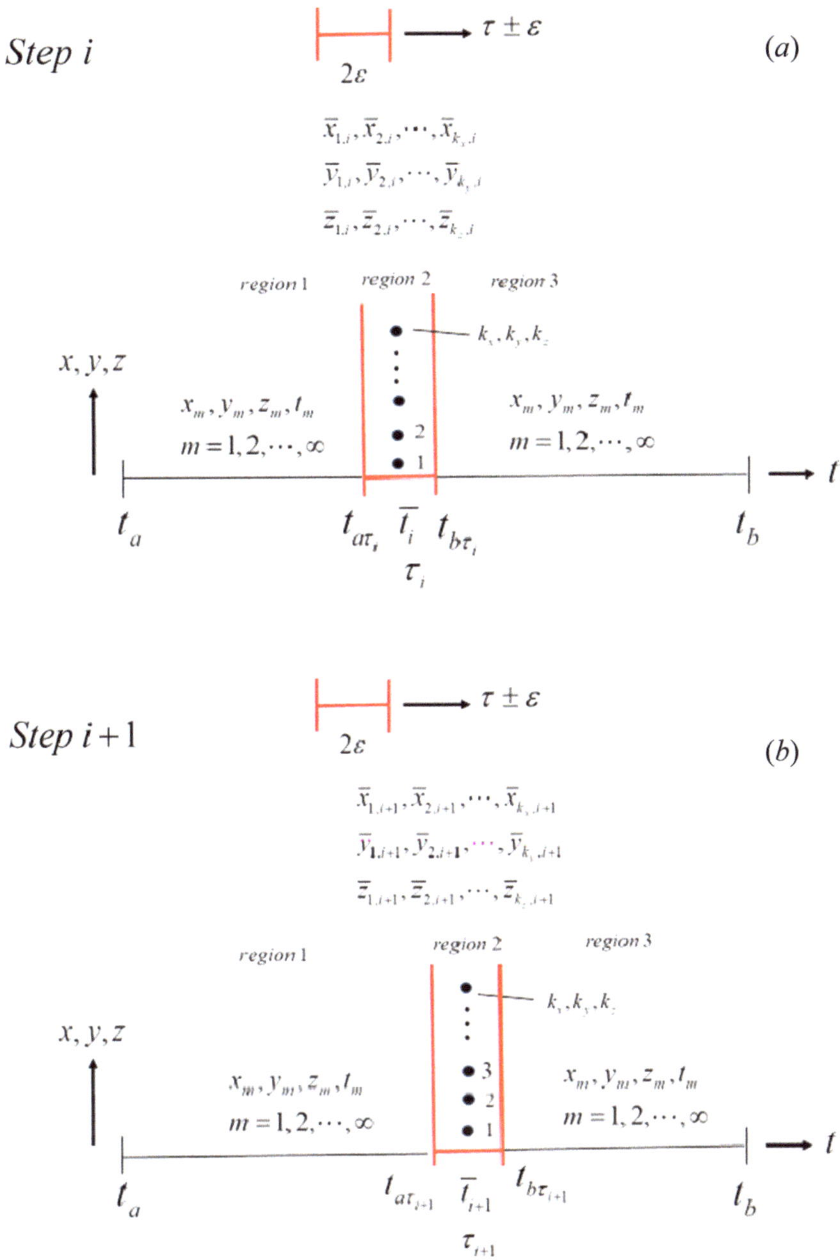

Figure 6.7: The τ block is for the observation of physically real systems. There are three regions. The τ block is not effective within two regions (regions 1 and 3); the system behaves here strictly quantum mechanically. These situations are described by eqs. (6.165) and (6.167) in the case of (a) and by eqs. (6.168) and (6.170) in the case of (b). In region 2 the τ block is effective and we get macroscopic states in the form of mean values. This is described by eq. (6.166) in (a) and by eq. (6.37) in (b).

6.8 Observational features

In everyday life, human beings observe the environment in the form of trees, houses, cars and so on. These are objects that are not observed in their quantum states but appear in macroscopic form before our eyes. From our statements in Section 6.7 directly follows that these macroscopic states are based on quantum states and become real when the τ block is effective. All objects before us appear simultaneously and this is described through the application of the τ block. That is, the macroscopic appearance comes into existence by observations. In this section, we want to deepen this view.

6.8.1 Block behavior in four dimensions: Overview

The use of $\tau \pm \varepsilon$ instead of τ means that no τ value within the 2ε interval can be dissolved. All the τ values within the 2ε interval are only defined as "one" block. This τ block acts as a unified whole consisting of an infinite number of τ values, which the 2ε interval contains.

The following points are essential in the case of the four-dimensional space–time. A brief summary is given:

1. We remarked earlier that an infinite number of τ values within the interval 2ε creates an infinite number for each of the variables x, y, z, t, which are distributed over the entire space–time range for which the probability density exists [see eq. (6.138)]:

$$\left\{\begin{array}{c} \tau-\text{values} \\ in\ 2\varepsilon \\ \downarrow \\ \infty \end{array}\right\} \Rightarrow \left\{\begin{array}{c} x,y,z,t-\text{values} \\ in\ (x_\mathrm{b}-x_\mathrm{a}),\ (y_\mathrm{b}-y_\mathrm{a}),\ (z_\mathrm{b}-z_\mathrm{a}),\ (t_\mathrm{b}-t_\mathrm{a}) \\ \downarrow \\ \infty \end{array}\right\} \tag{6.171}$$

   The ranges with respect to τ and x, y, z, t are equally occupied by an infinite number of states.
2. Since the number of x, y, z, t states within the total range is infinity, also the number of states within each of the subintervals $2\varepsilon_x, 2\varepsilon_y, 2\varepsilon_z, 2\varepsilon_t$ must be infinity as well; we have $\infty/k = \infty$ for $k \neq \infty$.
3. If the interval $2\varepsilon_x, 2\varepsilon_y, 2\varepsilon_z, 2\varepsilon_t$ are sufficiently small, the information about the system is (almost) complete, that is, the $\Psi^*(\mathbf{r},t)\Psi(\mathbf{r},t)$ distribution is closely packed with any precision; here all four dimensions x, y, z, t are concerned. Each of the intervals (the numbers are given by k_x, k_y, k_z, k_t) contains an infinite number of x, y, z, t states where the widths $2\varepsilon_x, 2\varepsilon_y, 2\varepsilon_z, 2\varepsilon_t$ may be arbitrary but different from zero.

4. The system is observed at time $\tau \pm \varepsilon \cong \tau$, but the entire x, y, z, t range is recorded at $\tau \pm \varepsilon \cong \tau$. This corresponds to what we observe in everyday life: we experience the entire space structure [(x, y, z) structure] of the environment at clock time τ ($\tau \pm \varepsilon \cong \tau$), which is identical with the mean time $\bar{t}$ and the classical time t_{cl}, respectively. Thus, we have

$$\begin{aligned} &\tau \pm \varepsilon \cong \tau: \\ &t \cong t_{\text{cl}} = \bar{t} \cong \tau \\ &\text{all } x, y, z \text{ within } (x_b - x_a),\ (y_b - y_a),\ (z_b - z) \end{aligned} \tag{6.172}$$

As we already remarked in Section 6.6.1, the τ block overlaps with the t scale and the overlap interval is $2\varepsilon_t$ and is of course identical with 2ε. The τ block is only able to recognize the t values in the interval $2\varepsilon_t$ as a block and does not see any single t value within $2\varepsilon_t$. That is, the block defines a unified whole. Thus, the τ block records the infinite number of t values within the $2\varepsilon_t$ interval "all at once" and defines "one" value which expresses the mean value $\bar{t}$, which belongs to the t values within the $2\varepsilon_t$ interval. Thus, we have a t block as well, together with the complete x, y, z structure as quoted in eq. (6.148). There are no selections with respect to space; we have therefore no reference space entities $\lambda_x, \lambda_y, \lambda_z$ (see Chapter 5) for the selection of x, y, z coordinates.

6.8.2 Objects and parts of the wave function

If the environment (physically real world) is representable as a unified whole, the wave function is expressible through the variables $\mathbf{r}, t$ with $\mathbf{r} = (x, y, z)$ and we have $\Psi(\mathbf{r}, t)$. The behavior of such kind of wave function has been investigated in Sections 6.7.1–6.7.5.

Partitions

We would like to assume that the wave function $\Psi(\mathbf{r}, t)$ is separated into N parts, and each part is symbolized together with its $\mathbf{r}, t$ ranges by $\mathbf{r}_j, t_j$, where j marks one of the N parts. Then, we have

$$\mathbf{r}_1, t_1, \mathbf{r}_2, t_2, \ \ldots, \mathbf{r}_N, t_N \tag{6.173}$$

and we formally get

$$\begin{aligned} \Psi(\mathbf{r}, t) &\to \Psi(\mathbf{r}_1, t_1, \mathbf{r}_2, t_2, \ \ldots, \mathbf{r}_N, t_N) \\ &\to \Psi_1(\mathbf{r}, t), \Psi_2(\mathbf{r}, t), \ \ldots, \Psi_N(\mathbf{r}, t) \end{aligned} \tag{6.174}$$

We have approximated the whole, expressed by $\Psi(\mathbf{r}, t)$ by N objects, which interact with each other.

We admit again the segments $2\varepsilon_x, 2\varepsilon_y, 2\varepsilon_z, 2\varepsilon_t$ for each of these N parts. The τ—block acts on all parts $\Psi_j(\mathbf{r}, t), j=1,2, \ldots, N$, simultaneously.

We have again three regions for each partition: two regions with quantum behavior (regions 1 and 3 in Figure 6.7) and one region in which the τ block acts (region 2 in Figure 6.7).

In region 2 we therefore deal with observations; the observation of all N parts takes place simultaneously and defined with certainty at each time $\tau \pm \varepsilon \cong \tau$. Thus, instead of probability distributions for the variables x, y, z, t, the mean values $\bar{x}, \bar{y}, \bar{z}, \bar{t}$ for all N parts are observed. The mean values themselves are defined in steps of $2\varepsilon_x, 2\varepsilon_y, 2\varepsilon_z, 2\varepsilon_t$ in the course of time.

6.8.3 The various steps

The τ—block acts at $\tau_1 \pm \varepsilon$, $\tau_2 \pm \varepsilon$, ..., $\tau_i \pm \varepsilon$, ... Due to ε, the clock-time moves in steps of 2ε, i.e., we for example $\tau_2 \pm \varepsilon - \tau_2 \pm \varepsilon = 2\varepsilon$. (In principle, the clock time could also move continuously.) Since there is a selection with respect to the time, at each clock-time $\tau_i \pm \varepsilon$, there is one value for the mean time, but there are many mean values for the coordinates, which are different for different parts.

For example, at clock-time $\tau_i \pm \varepsilon$, there are the following allocations for the various parts of $\Psi(\mathbf{r}, t)$:

Part 1

$$\left.\begin{array}{l} \bar{x}_{1,1,i}, \bar{x}_{1,2,i}, \ldots, \bar{x}_{1,k_{x1},i} \\ \bar{y}_{1,1,i}, \bar{y}_{1,2,i}, \ldots, \bar{y}_{1,k_{y_1},i} \\ \bar{z}_{1,1,i}, \bar{z}_{1,2,i}, \ldots, \bar{z}_{1,k_{z1},i} \end{array}\right\} \Psi_{1,\mathrm{i}}(\mathrm{r}, \bar{t}_{\mathrm{i}}) \qquad (6.175)$$
$$\bar{t}_i$$

Part 2

$$\left.\begin{array}{l} \bar{x}_{2,1,i}, \bar{x}_{2,2,i}, \ldots, \bar{x}_{2,k_{x1},i} \\ \bar{y}_{2,1,i}, \bar{y}_{2,2,i}, \ldots, \bar{y}_{2,k_{y_1},i} \\ \bar{z}_{2,1,i}, \bar{z}_{2,2,i}, \ldots, \bar{z}_{2,k_{z1},\ldots i} \\ \bar{t}_i \end{array}\right\} \Psi_{2,\mathrm{i}}(\mathrm{r}, \bar{t}_{\mathrm{i}}) \qquad (6.176)$$

$$\vdots$$

Part N

$$\left.\begin{array}{l} \bar{x}_{N,1,i}, \bar{x}_{N,2,i}, \ldots, \bar{x}_{N,k_{x1},i} \\ \bar{y}_{N,1,i}, \bar{y}_{N,2,i}, \ldots, \bar{y}_{N,k_{y_1},i} \\ \bar{z}_{N,1,i}, \bar{z}_{N,2,i}, \ldots, \bar{z}_{N,k_{z1},i} \end{array}\right\} \Psi_{N,\mathrm{i}}(\mathrm{r}, \bar{t}_{\mathrm{i}}) \qquad (6.177)$$
$$\bar{t}_i$$

At clock time $\tau_i \pm \varepsilon$ the complete information is given for each of the N parts in terms of mean values $\bar{x}, \bar{y}, \bar{z}, \bar{t}$ for the coordinates x, y, z and the system-specific time t. For example, the mean value $\bar{x}_{1,1,i}$ for the x coordinate refers to part 1 at time $\tau_i \pm \varepsilon$ and, furthermore, it refers to the first segment of the distribution within part 1. The total number of x, y, z segments within part 1 and at time $\tau_i \pm \varepsilon$ is $k_{1,i} = k_{x,1,i} k_{y,1,i} k_{z,1,i}$.

At time $\tau_i \pm \varepsilon$ for all N parts, the mean value for the time is $\bar{t}_i$. As we know, the coordinates of the system are exclusively given in the form of mean values, which are however not given as a continuous sequence of values, but in steps of $2\varepsilon_x, 2\varepsilon_y, 2\varepsilon_z$. For example, the values $\bar{x}_{N,1,i}, \bar{x}_{N,2,i}$, $\bar{y}_{N,1,i}, \bar{y}_{N,2,i}$ and $\bar{z}_{N,1,i}, \bar{z}_{N,2,i}$ are neighbors and their difference is

$$\begin{aligned} \bar{x}_{N,2,i} - \bar{x}_{N,1,i} &= 2\varepsilon_x \\ \bar{y}_{N,2,i} - \bar{y}_{N,1,i} &= 2\varepsilon_y \\ \bar{z}_{N,2,i} - \bar{z}_{N,1,i} &= 2\varepsilon_z \end{aligned} \tag{6.178}$$

Exactly, the same scheme holds when we go a step further in time, that is, from $\tau_i \pm \varepsilon$ to $\tau_{i+1} \pm \varepsilon$. Then the information for the N parts of the wave function $\Psi(\mathbf{r}, t)$ we have the following allocations:

Part 1

$$\left.\begin{array}{l} \bar{x}_{1,1,i+11}, \bar{x}_{1,2,i+1}, \ldots, \bar{x}_{1,k_{x1},i+1} \\ \bar{y}_{1,1,i+1}, \bar{y}_{1,2,i+1}, \ldots, \bar{y}_{1,k_{y1},i+1} \\ \bar{z}_{1,1,i+1}, \bar{z}_{1,2,i+1}, \ldots, \bar{z}_{1,k_{z1},i+1} \\ \bar{t}_{i+1} \end{array}\right\} \Psi_{1,i+1}(\mathbf{r}, \bar{t}_{i+1}) \tag{6.179}$$

Part 2

$$\left.\begin{array}{l} \bar{x}_{2,1,i+1}, \bar{x}_{2,2,i+1}, \ldots, \bar{x}_{2,k_{x1},i+1} \\ \bar{y}_{2,1,i+1}, \bar{y}_{2,2,i+1}, \ldots, \bar{y}_{2,k_{y1},i+1} \\ \bar{z}_{2,1,i+1}, \bar{z}_{2,2,i+1}, \ldots, \bar{z}_{2,k_{z1},i+1} \\ \bar{z}_{1,2,i+1}, \ldots, \bar{z}_{1,k_{z1},i+1} \\ \bar{t}_{i+1} \end{array}\right\} \Psi_{2,i+1}(\mathbf{r}, \bar{t}_i) \tag{6.180}$$

$\vdots$

Part N

$$\left.\begin{array}{l} \bar{x}_{N,1,i+1}, \bar{x}_{N,2,i+1}, \ldots, \bar{x}_{N,k_{x1},i+1} \\ \bar{y}_{N,1,i+1}, \bar{y}_{N,2,i+1}, \ldots, \bar{y}_{N,k_{y1},i+1} \\ \bar{z}_{N,1,i+1}, \bar{z}_{N,2,i+1}, \ldots, \bar{z}_{N,k_{z1},i+1} \\ \bar{t}_{i+1} \end{array}\right\} \Psi_{N,i+1}(\mathbf{r}, \bar{t}_i) \tag{6.181}$$

In conclusion, the macroscopic information with respect to a physically real system is given in terms of mean values.

Objects

Again, all that is valid for all the separate subsystems $\Psi_1(\mathbf{r}, t)$, $\Psi_2(\mathbf{r}, t)$, $\ldots, \Psi_N(\mathbf{r}, t)$ of the wave function $\Psi(\mathbf{r}, t)$ [see in particular eq. (6.174)]. There are k_x, k_y, k_z, k_t segments [] and, due to the action of the τ block, the following are the mean values:

$$\begin{aligned}&\bar{x}_1, \bar{x}_2, \ldots, \bar{x}_{kx}\\&\bar{y}_1, \bar{y}_2, \ldots, \bar{y}_{ky}\\&\bar{z}_1, \bar{z}_2, \ldots, \bar{z}_{kz}\\&\bar{t}_1, \bar{t}_2, \ldots, -\bar{t}_{k_t}\end{aligned} \tag{6.182}$$

All these values exist simultaneously with "certainty," that is, there are no probability distributions for them. The wave function is eliminated through the admission of the segments $2\varepsilon_x, 2\varepsilon_y, 2\varepsilon_z, 2\varepsilon_t$. In other words, the mean values are classical entities that we want to denote by $x_{\text{cl}}, y_{\text{cl}}, z_{\text{cl}}, t_{\text{cl}}$ instead of $\bar{x}, \bar{y}, \bar{z}, \bar{t}$. Then, instead of eq. (6.182) we obtain

$$\begin{aligned}&x_{\text{cl}\,1}, x_{\text{cl}\,2}, \ldots, x_{\text{cl}\,kx}\\&y_{\text{cl}\,1}, y_{\text{cl}\,2}, \ldots, y_{\text{cl}\,ky}\\&z_{\text{cl}\,1}, z_{\text{cl}\,2}, \ldots, z_{\text{cl}\,kz}\\&t_{\text{cl}\,1}, t_{\text{cl}\,2}, \ldots, t_{\text{cl}\,k_t}\end{aligned} \tag{6.183}$$

In this case, the wave functions $\Psi_1(\mathbf{r}, t), \Psi_2(\mathbf{r}, t), \ldots, \Psi_N(\mathbf{r}, t)$ have to be replaced through other characteristic functions $E(\mathbf{r}, t), F(\mathbf{r}, t), \ldots, G(\mathbf{r}, t)$:

$$\begin{aligned}&\Psi_1(\mathbf{r}, t) \to E(\mathbf{r}, t)\\&\Psi_2(\mathbf{r}, t) \to F(\mathbf{r}, t)\\&\vdots\\&\Psi_N(\mathbf{r}, t) \to G(\mathbf{r}, t)\end{aligned} \tag{6.184}$$

Each of the functions $E(\mathbf{r}, t), F(\mathbf{r}, t), G(\mathbf{r}, t)$ can be interpreted as "objects," which are in general different from each other. Their particular meaning is explained in somewhat more detail in Figures 6.8 and 6.9.

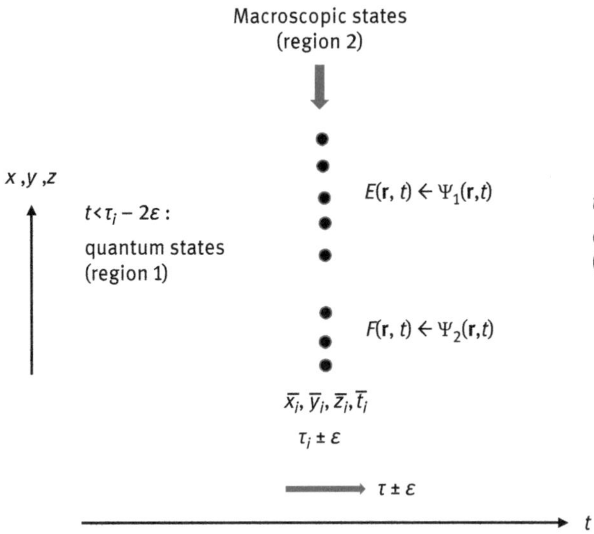

Figure 6.8: The wave function $\Psi(\mathbf{r},t)$ is separated into two parts $\Psi_1(\mathbf{r},t)$ and $\Psi_2(\mathbf{r},t)$ that are interpreted as two objects. In regions 1 and 3, the τ block is not effective and there are strict quantum-mechanical states. In region 2, the τ block creates macroscopic states expressed through mean values. All these values exist simultaneously and with certainty. The components $\Psi_1(\mathbf{r},t)$ and $\Psi_2(\mathbf{r},t)$ have to be replaced by classical functions $E(\mathbf{r},t)$ and $F(\mathbf{r},t)$ that describe the characteristics of the macroscopic state. The objects $E(\mathbf{r},t)$ and $F(\mathbf{r},t)$ have a segment structure. The wave function $\Psi(\mathbf{r},t)$ varies in the course of time $\tau\pm\varepsilon$. Thus, also the objects change their shape simultaneously together with their relative four-dimensional position.

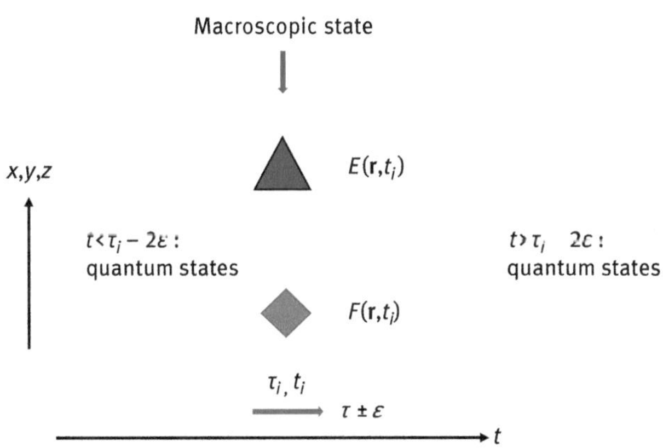

Figure 6.9: The objects $E(\mathbf{r},t)$ and $F(\mathbf{r},t)$ in Figure 6.8 have a segment structure, but human beings perceive them as "one" entity. In this way, we perceive macroscopic objects such as trees and houses. The segment structures $E(\mathbf{r},t)$ and $F(\mathbf{r},t)$ in Figure 6.8 are replaced by geometrical objects.

6.8.4 Example

In Figure 6.8 we discuss a specific example. Here the wave function $\Psi(\mathbf{r},t)$ is subdivided into two parts and objects, respectively. $\Psi(\mathbf{r},t)$ is separated into two parts $\Psi_1(\mathbf{r},t)$ and $\Psi_2(\mathbf{r},t)$.

At $\tau_i \pm \varepsilon$, the τ block acts in region 2 and creates macroscopic states expressed through mean values $\bar{x}_i, \bar{y}_i, \bar{z}_i, \bar{t}_i$, which have to be classified as classical in character. The wave function is eliminated. The components $\Psi_1(\mathbf{r},t)$ and $\Psi_2(\mathbf{r},t)$ have to be replaced by classical functions $E(\mathbf{r},t)$ and $F(\mathbf{r},t)$ that describe the characteristics of the macroscopic state of the physically real system. All these values exist simultaneously and with certainty. In regions 1 and 3, the τ block is not effective and there are strict quantum-mechanical states. In the figure, two parts of $\Psi(\mathbf{r},t)$ are shown in the macroscopic approximation $E(\mathbf{r},t)$ and $F(\mathbf{r},t)$ and can be interpreted as objects. The number of segments concerning $E(\mathbf{r},t)$ is five and concerning $F(\mathbf{r},t)$ it is three. Each segment i contains one set $\bar{x}_i, \bar{y}_i, \bar{z}_i, \bar{t}_i$. The steps between the mean values $\bar{x}, \bar{y}, \bar{z}$ are $2\varepsilon_x, 2\varepsilon_y, 2\varepsilon_z$. It is a representation for $\tau_i \pm \varepsilon$. The mean value for the time is here $\tau_i \pm \varepsilon$.

The wave function $\Psi(\mathbf{r},t)$ varies in the course of time $\tau \pm \varepsilon$. Thus, also the objects change their shape simultaneously together with their relative four-dimensional position.

The objects $E(\mathbf{r},t)$ and $F(\mathbf{r},t)$ have a segment structure and, concerning the values $\bar{x}_i, \bar{y}_i, \bar{z}_i, \bar{t}_i$, they are equivalent and exist simultaneously, that is the probability effect due to the wave functions $\Psi_1(\mathbf{r},t)$ and $\Psi_2(\mathbf{r},t)$ is eliminated. Nevertheless, the functions $E(\mathbf{r},t)$ and $F(\mathbf{r},t)$ are not only an ensemble of mean values, but they reflect the properties of the objects as well. These properties in the macroscopic state reflect quantum-mechanical features in the macroscopic approximation and are generated through the transitions $\Psi_1(\mathbf{r},t) \to E(\mathbf{r},t)$ and $\Psi_2(\mathbf{r},t) \to F(\mathbf{r},t)$ [eq. (6.184)].

The objects $E(\mathbf{r},t)$ and $F(\mathbf{r},t)$ in Figure 6.8 have a segment structure, but human beings perceive them as "one" entity. In this way, we perceive macroscopic objects such as trees and houses. The segment structures $E(\mathbf{r},t)$ and $E(\mathbf{r},t)$ in Figure 6.8 are replaced in Figure 6.9 by geometrical objects. The coordinates, describing the shape of the geometrical structures, are experienced as continuous ensemble consisting at time τ of the coordinates x, y, z. In general, the macroscopic states are characterized through the following equivalencies:

$$\begin{aligned} \bar{x} &\to x_{\text{cl}} \to x \\ \bar{y} &\to y_{\text{cl}} \to y \\ \bar{z} &\to z_{\text{cl}} \to z \\ \bar{t} &\to t_{\text{cl}} \to \tau \end{aligned} \tag{6.185}$$

The coordinates x, y, z describe the macroscopic body and they have in this case nothing to do with quantum states.

6.8.5 General statements

The functions $E(\mathbf{r},t), F(\mathbf{r},t), \ldots, G(\mathbf{r},t)$ are distributions for the description of the internal properties of the various objects. That is, the quantum-theoretical wave functions are reflected in the classical distributions of properties, which define the objects in everyday life. Again, all the objects, expressed by the functions $E(\mathbf{r},t), F(\mathbf{r},t), \ldots, G(\mathbf{r},t)$, exist at $\tau \pm \varepsilon$ with certainty. These objects can be trees, house, cars and other objects, which are observed in everyday life in the macroscopic case. In Figure 6.9, the objects $E(\mathbf{r},t)$ and $F(\mathbf{r},t)$ are represented by geometrical symbols.

The macroscopic state of the world moves through the sea of quantum states. The τ block moves of course and not the world. The τ block is for the observation of the world outside and it transforms the sea of quantum states into our usual space impressions, which we have at clock time τ before us. Clearly, these impressions are macroscopic in character.

This macroscopic world contains objects as trees, house, cars, and so on, and they appear simultaneously and with certainty. As we recognized in Section 6.7, the transition from the quantum states to the macroscopic states just generates certainty and simultaneousness.

At time τ, the quantum states are distributed over the entire x, y, z, t range that are defined by the physically real object under investigation. The τ block effectuates the transition to macroscopic properties and it selects space structures at certain t states. We observe the t state, which is just given by the clock time τ ($\tau \pm \varepsilon$). There is no selection with respect to space, that is, we observe successively the complete macroscopic x, y, z – distribution of the systems under observation at $t = \tau$.

6.8.6 Nanosystems

From the point of view of the realistic projection principle, the structure and the behavior of nanosystems have to be developed on the basis of the principles outlined in the last sections. When we enter the realm of nanoscience, we penetrate actually deeper into the details of nature and, therefore, more sophisticated description elements have to be applied.

6.9 Selection of space structures

In Sections 6.7 and 6.8, the τ block acts on the quantum states and selects incessantly x, y, z structures in the course of time τ. The block behavior is the focus in the preceding sections. In this section, we discuss the selection process without applying the τ block.

What mechanism is responsible for the selection of the wave function $\Psi(x,y,z,\tau)$ from the information $\Psi(x,y,z,t)$? $\Psi(x,y,z,t)$ is completely given at each time $\tau \pm \varepsilon$ within the interval $t_a \leq t \leq t_b$, where $\Psi(x,y,z,t)$ is not zero. The wave function $\Psi(x,y,z,t)$ is existent at each time $\tau \pm \varepsilon$, but we know that only the x,y,z structure at time τ is experienced in the case of usual observations. Thus, the part $\Psi(x,y,z,\tau)$ is in the foreground, which has to be selected from the complete information $\Psi(x,y,z,t)$. What mechanism is responsible for this selection process? In this section, we discuss this point. This situation has already been discussed in Section 5.13 without using the wave function.

6.9.1 The entire information

The system-specific time t is a quantum time and we have the following situation: at time τ only one possible t is realized with a certain probability. However, if we consider an infinitesimal time interval $\Delta\tau = \varepsilon$, an infinite number of t values are occupied, that is, the whole history (the complete past and future) of the system described within the range Δ_t is existent given within the infinitesimal time interval $\Delta\tau = \varepsilon$, where Δ_t reflects the lifetime of the system under investigation. The complete probability distribution $\Psi^*(\mathbf{r},t)\Psi(\mathbf{r},t)$ is given within the infinitesimal time interval $\Delta\tau = \varepsilon$

This is the case for any $\tau_i \pm \Delta\tau$, that is, the expression $\Psi^*(\mathbf{r},t)\,\Psi(\mathbf{r},t)$ is independent of τ; it is stationary with respect to time τ. Thus, we state the following: despite the statistical fluctuations, the whole of time structure of the physically real system, that is, its past, present and future, is laid out frozen before a human observer. This effect is also reflected in the raw information α (Section 5.13).

In other words, there is no direct connection between τ and the system-specific time t. Therefore, for the understanding of the selection process we have to construct a connection between τ and the system-specific time t because the existence of τ is a matter of fact. This connection between the reference time τ and the system-specific time t will be constructed in the next section.

Let us state once again that the time τ has primarily nothing to do with physically real systems, but only plays the role of a reference time and is realized by our clocks which we use in everyday life.

6.9.2 Caught in space and time reference time and selection processes

Principal remarks

So far, there seems to be no connection between the clock time τ and the system-specific time t. Despite the statistical fluctuations, at τ (or more precisely at $\tau \pm \varepsilon$) the whole of time t, that is, the past, present and future, is laid out frozen before us:

$$\Psi^*(x,y,z,t)\Psi(x,y,z,t)$$
$$-\infty < x,y,z < \infty, \quad -\infty < t < \infty \tag{6.186}$$

That is, at time $\tau \pm \varepsilon$ the entire information is given all at once. This is valid for all times τ. In other words, the reference time τ is not correlated to specific t values. From this point of view, it would give no sense to introduce τ within the projection theory.

Such types of processes [defined by eq. (6.186)] could not create what we call "time feeling." The system-specific time t of the system under investigation jumps statistically between the various t values without giving time a certain direction; past, present and future cannot be defined through the system.

However, we know that the entire information (6.186) is not observed at clock time $\tau \pm \varepsilon$. Instead, we always observe only a certain space structure of reality at time $\tau \pm \varepsilon$:

$$\Psi^*(x,y,z,t_0)\Psi(x,y,z,t_0)$$
$$-\infty < x,y,z < \infty \tag{6.187}$$

for $t_0 = \tau$. A usual photography represents at a certain time $\tau = t_0$ such a configuration in space,

Why does nature work in this way? There is obviously a selection processes from $\Psi^*(\mathbf{r},t)\,\Psi(\mathbf{r},t)$ to $\Psi^*(\mathbf{r},t_0)\Psi(\mathbf{r},t_0)$ with $\mathbf{r} = (x,y,z)$ [see eqs. (6.186) and (6.187)]? Why does nature choose such selection processes? The answer is given by the principles of evolution. There is an important basic principle in connection with evolution: "As little outside world as possible." Such a principle guarantees optimal chances for survival. No doubt, this principle is clearly reflected in the transition from $\Psi^*(\mathbf{r},t)\,\Psi(\mathbf{r},t)$ to $\Psi^*(\mathbf{r},t_0)\Psi(\mathbf{r},t_0)$ and reflects a certain kind of selection. The occurrence of the reference time τ is obviously one of the features for that.

6.9.3 Introduction of the reference system

How does nature organize the selection process? How comes the transition from the situation defined by eq. (6.186) to that given by eq. (6.187) into existence? This cannot be due to an internal transformation within the system alone, that is, without the influence of an external impact. Besides the physically real system, only the observer's function appears within the frame of the observation of the system. Therefore, the transition from $\Psi^*(\mathbf{r},t)\,\Psi(\mathbf{r},t)$ to $\Psi^*(\mathbf{r},t_0)\Psi(\mathbf{r},t_0)$ must be due to an interplay between the system, described by $\Psi(\mathbf{r},t)$, and the observer's influence. So far, the observer's function has been characterized only by one parameter: it is the reference time τ that is measured by our clocks in everyday life. Clearly, only one parameter (τ) is possibly not sufficient for the description of the interplay between the system and the observer's characteristics. How can we characterize the observation process more realistically?

For this purpose, let us define a reference system that is produced inside the observing human being. Furthermore, let us formally describe the reference system by the wave function $\Psi_{ref}(t)$ and the probability distribution $\Psi^*{}_{ref}(t)\Psi_{ref}(t)$, respectively. Clearly, also the time t in $\Psi_{ref}(t)$ reflects a system-specific quantity. For simplicity, we would like to assume that the reference system is not dependent on any position $\mathbf{r} = (x, y, z)$. There is in fact no section process with respect to the coordinates x, y, z.

Within the frame of the projection principle, the source for the wave function $\Psi_{ref}(t)$ are energy fluctuations in reality [$(\mathbf{p}, E)$ space] leading to an energy-dependent wave function $\Psi_{ref}(E)$ and a probability density $\Psi^*{}_{ref}(E)\Psi_{ref}(E)$. $\Psi_{ref}(E)$ and $\Psi_{ref}(t)$ are connected also here by a Fourier transform as follows (see also Ref. [12]):

$$\psi_{ref}(E) = \frac{1}{(2\pi\hbar)^{1/2}} \int_{-\infty}^{\infty} \psi_{ref}(t) \exp\left\{i\frac{E}{\hbar}t\right\} dt \tag{6.188}$$

This means that we have at clock time τ two probability distributions, $\Psi^*{}_{ref}(t)\Psi_{ref}(t)$ and $\Psi^*(x, y, z, t)\Psi(x, y, z, t)$: one for the description of the system-specific time of the reference system and the other for the description of the time structure of the system under investigation.

We may summarize that the reference system, described by $\Psi_{ref}(t)$, has essentially two tasks:

1. to describe the nature of the reference time τ more specifically, and
2. to select $\Psi^*(x, y, z, t_0)\Psi(x, y, z, t_0)$ from $\Psi^*(x, y, z, t)\Psi(x, y, z, t)$ [see eqs. (6.187) and (6.188)].

We will recognize that both items are interconnected since selection is not possible without the existence of a systematically varying reference time. First, let us discuss the feature of the reference time in more detail. After that we have we will discuss the selection process.

6.9.4 Structure of reference time

The reference time τ runs monotonically from the past to the future. However, this time feeling must also be due to a process (inside the observer) and therefore belongs also to the class of system-specific times.

However, this process might differ considerably from those treated within the frame of traditional quantum theory. The reason is simple: within projection theory, the equation for the determination of $\Psi^*(x, y, z, t)\Psi(x, y, z, t)$ [eq. (6.74)] is more general than Schrödinger's equation of traditional quantum theory because the function $V(x, y, z, t)$ is more general; it contains the system-specific time t, which is not known in usual quantum theory where the classical potential $U(\mathbf{r})$ is used (Section 6.4). In principle, $V(x, y, z, t)$ could even have an imaginary part. The appearance of the function $V(x, y, z, t)$ is a logical consequence when we work within the frame of the projection principle (see also Section 6.1).

The distribution $\Psi^*_{ref}(t)\Psi_{ref}(t)$ has been introduced in the last section. It describes the probability of finding a certain value t for the reference system in the interval Δt around t is given. The probability is given by $\Psi^*_{ref}(t)\Psi_{ref}(t)\Delta t$. In other words, the system-specific reference time t becomes uncertain, and this is because $\Psi^*_{ref}(t)\Psi_{ref}(t)$ has a certain width Δ_τ. Thus, the reference time runs no longer strictly from the past to the future as is suggested by our clocks used in everyday life. However, we have to assume that the probability distribution $\Psi^*_{ref}(t)\Psi_{ref}(t)$ for the reference time should be a relatively sharp function and, furthermore, because the time of our clocks τ runs monotonically from the past to the future, also the distribution $\Psi^*_{ref}(t)\Psi_{ref}(t)$ must run monotonically from the past to the future and we have

$$\Psi^*_{ref}(t)\Psi_{ref}(t) \to \Psi^*_{ref}(\tau - t)\Psi_{ref}(\tau - t) \tag{6.189}$$

We never measure the time τ but the quantum time t, and t is uncertain due to $\Psi^*_{ref}(\tau - t)\Psi_{ref}(\tau - t)$. However, because the distribution $\Psi^*_{ref}(\tau - t)\Psi_{ref}(\tau - t)$ can be assumed to be a relatively sharp function, the system-specific reference time t should be close to τ. Due to the existence of τ, the whole curve $\Psi^*_{ref}(\tau - t)\Psi_{ref}(\tau - t)$ moves strictly from the past to the future, but the values t for the reference time fluctuate around τ.

A simple example from macroscopic realm illustrates that. For example, a moving car, having the velocity $\mathbf{v}$, represents a strict sequence of configurations $A(\mathbf{r} - \mathbf{v}\tau)$, that is, we have for τ the strict sequence $\tau_1 < \tau_2 \ldots$. But this example is based on macroscopic observations. Therefore, in general, we have to admit a certain small width Δ_τ for $\Psi^*_{ref}(\tau - t)\Psi_{ref}(\tau - t)$.

We have outlined earlier that the source for the existence of the wave function $\Psi_{ref}(t)$ are energy fluctuations in reality, that is, within $(\mathbf{p}, E)$ space, and these fluctuations lead to $\Psi_{ref}(E)$ and $\Psi^*_{ref}(E)\Psi_{ref}(E)$, respectively. $\Psi_{ref}(E)$ and $\Psi_{ref}(t)$ are connected by the Fourier transform (6.188). However, the law for the energy fluctuations $\Psi^*_{ref}(E)\Psi_{ref}(E)$ does not change when the function $\Psi_{ref}(t)$ is shifted by τ. With

$$\phi_{ref}(E) = \frac{1}{(2\pi\hbar)^{1/2}} \int_{-\infty}^{\infty} \Psi_{ref}(\tau - t) \exp\left\{-i\frac{E}{\hbar}t\right\} dt \tag{6.190}$$

$$\phi_{ref}(E) = \Psi_{ref}(E) \exp\left\{-i\frac{E}{\hbar}\tau\right\} \tag{6.191}$$

we obtain

$$\phi^*_{ref}(E)\phi_{ref}(E) = \Psi^*_{ref}(E)\Psi_{ref}(E) \tag{6.192}$$

That is, the expression $\Psi^*_{ref}(E)\Psi_{ref}(E)$ is independent on τ. We have outlined in Ref. [81] that $\Psi^*_{ref}(E)\Psi_{ref}(E)$ is the relevant function.

In summary, at time τ we have for the description of the reference time a probability distribution of the form $\Psi^*_{ref}(\tau - t)\Psi_{ref}(\tau - t)$. In principle, also the parameter τ could behave statistically: in this case, the whole curve $\Psi^*_{ref}(\tau - t)\Psi_{ref}(\tau - t)$ would jump statistically without any systematic sequence for τ. However, this would lead to problems with respect to observations and we do not want to consider this case here.

Here, we want to assume that τ is the time of our clocks and therefore runs systematically and monotonically from the past to the future. However, at the quantum level, this time τ is not directly observable; we only observe t which is inherently uncertain, and this uncertainty is described by $\Psi^*_{ref}(\tau - t)\Psi_{ref}(\tau - t)$. We assume here that $\Psi^*_{ref}(\tau - t)\Psi_{ref}(\tau - t)$ is a relatively sharp function, and t should be close to τ. Due to the clock time τ the entire curve $\Psi^*_{ref}(\tau - t)\Psi_{ref}(\tau - t)$ moves strictly from the past to the future and, simultaneously, the values t for the reference quantum time fluctuate around τ.

6.9.5 Selections

Convolution integral

We have outlined in Section 6.9.3 that the function $\Psi_{ref}(t)$ for the reference system should also be responsible for selection processes with respect to the system-specific time for the system. In fact, the transition from $\Psi^*(x,y,z,t)\Psi(x,y,z,t)$ to $\Psi^*(x,y,z,t_0)$ $\Psi(x,y,z,t_0)$ can be explained with $\Psi_{ref}(t)$. In other words, the interplay between the two systems (the reference system described by $[\Psi_{ref}(t), \Psi_{ref}(E)]$ and, on the other hand, the system under investigation, described by the functions $[\Psi(x,y,z,t)$, $\Psi(x,y,z,E)]$ should lead to the necessary selection process. This process obviously filters out the specific space–time configuration $\Psi^*(x,y,z,t_0)\Psi(x,y,z,t_0)$ from the entire information $\Psi^*(x,y,z,t)\Psi(x,y,z,t)$.

How can we describe this selection process theoretically? The system under investigation is described in $(\mathbf{p},E)$ space by $\Psi(x,y,z,E)$ or in this case within the intermediate $(\mathbf{r},E)$ space by $\Psi(x,y,z,E)$. From $\Psi(p_x,p_y,p_z,E)$ we get the function $\Psi(x,y,z,E)$.

Through a Fourier transform (Section 6.1.3):

$$\Psi(\mathbf{r},E) = \frac{1}{(2\pi\hbar)^{3/2}} \int_{-\infty}^{\infty} \Psi(p_x,p_y,p_z,E) \exp\left\{i \left[\frac{\mathbf{p}}{\hbar}\cdot\mathbf{r}\right]\right\} dp_x\, dp_y\, dp, \qquad \mathbf{r} = (x,y,z) \tag{6.193}$$

We have two processes, one with respect to the reference system, described by $\Psi_{ref}(E)$, and another characterized by $\Psi(\mathbf{r},E)$. Both processes are independent of each other and this feature can be formulated in the form

$$\Psi_O(\mathbf{r}, E, E') = \Psi_{ref}(E)\Psi(\mathbf{r}, E') \tag{6.194}$$

However, in this case, the reference system would be decoupled from the physically real system. The coupling can, however, be introduced if we assume that the energies E and E' are strictly correlated at each time τ:

$$\tau\!: E,\ E' = E \tag{6.195}$$

Then, both systems are no longer independent of each other and can therefore be formulated in the form of

$$\Psi_O(\mathbf{r}, E) = \Psi_{ref}(E)\Psi(\mathbf{r}, E) \tag{6.196}$$

Equation (6.196) leads to the effect of observation through the registration of $\Psi_O(\mathbf{r}, E)$ and $\Psi^*{}_O(\mathbf{r}, E)\Psi_O(\mathbf{r}, E)$, respectively. The direct observations take place in $(\mathbf{r}, t)$ space. In other words, how is the information $\Psi_O(\mathbf{r}, E)$ transformed into $(\mathbf{r}, t)$ space, where our observations take place and where the selection (filtering) process from $\Psi^*(\mathbf{r}, t)\,\Psi(\mathbf{r}, t)$ to $\Psi^*(\mathbf{r}, t_O)\Psi(\mathbf{r}, t_O)$ becomes evident?

Using the Fourier transform for $\Psi_O(\mathbf{r}, E)$

$$\Psi_O(\mathbf{r}, E) = \int_{-\infty}^{\infty} \Psi_{\mathrm{CON}}(\mathbf{r}, \tau)\exp\left\{i\frac{E}{\hbar}\tau\right\}\frac{d\tau}{(2\pi\hbar)^{1/2}} \tag{6.197}$$

or its inverse transformation

$$\Psi_O(\mathbf{r}, \tau) = \int_{-\infty}^{\infty} \Psi_{\mathrm{CON}}(\mathbf{r}, E)\exp\left\{-i\frac{E}{\hbar}\tau\right\}\frac{dE}{(2\pi\hbar)^{1/2}} \tag{6.198}$$

with

$$\Psi_{\mathrm{CON}}(\mathbf{r}, \tau) \equiv \Psi_O(\mathbf{r}, \tau) \tag{6.199}$$

We assume that the function $\Psi_O(\mathbf{r}, \tau)$, which describes the coupled states between the physically real system and the human observer, is not zero for τ values between τ_a and τ_b:

$$\Psi_O(\mathbf{r}, \tau) \neq 0,\ \ \tau_a \leq \tau \leq \tau_b \tag{6.200}$$

From eq. (6.197) we immediately get a convolution integral of the following form:

$$\Psi_{\mathrm{CON}}(\mathbf{r}, \tau) = \int_{-\infty}^{\infty} \Psi_{ref}(\tau - t)\Psi(\mathbf{r}, t) - \frac{dt}{(2\pi\hbar)^{1/2}} \tag{6.201}$$

Using eqs. (6.196) and (6.201) we directly obtain the coupled term described by eq. (6.196):$\Psi_O(\mathbf{r}, E) = \Psi_{ref}(E)\Psi(\mathbf{r}, E)$.

Proof

With

$$\exp\left\{i\frac{E}{\hbar}\tau\right\}=\exp\left\{i\frac{E}{\hbar}(\tau-t)\right\}\exp\left\{i\frac{E}{\hbar}t\right\} \tag{6.202}$$

and using eqs. (6.197) and (6.201) we obtain the expression

$$\Psi_o(\mathrm{r},E)=\int\limits_{-\infty}^{\infty}\int\limits_{-\infty}^{\infty}\Psi_{ref}(\tau-t)\Psi(\mathrm{r},t)\exp\left\{i\frac{E}{\hbar}\tau\right\}\frac{dt}{(2\pi\hbar)^{1/2}}\frac{d\tau}{(2\pi\hbar)^{1/2}} \tag{6.203}$$

With the substitution $z=\tau-t$ we immediately get eq. (6.196):

$$\begin{aligned}\Psi_o(\mathbf{r},E)=&\left\{\int\limits_{-\infty}^{\infty}\Psi_{ref}(z)\exp\left\{i\frac{E}{\hbar}z\right\}\frac{dz}{(2\pi\hbar)^{1/2}}\right\}\times\\&\left\{\int\limits_{-\infty}^{\infty}\Psi(\mathbf{r},t)\exp\left\{i\frac{E}{\hbar}t\right\}\frac{dt}{(2\pi\hbar)^{1/2}}\right\}\\&=\Psi_{ref}(E)\Psi(\mathbf{r},E)\end{aligned} \tag{6.204}$$

The existence of relation (6.201) is equivalent to correlations with respect to the energies, expressed by eq. (6.196).

Model for the reference system

If we assume that the function $\Psi_{ref}(\tau-t)$ in eq. (6.201) is very sharp, we may describe it by a delta function:

$$\Psi_{ref}(\tau-t)=C\delta(\tau-t) \tag{6.205}$$

where C is a constant. The width Δ_t of the quantum-mechanical function $\Psi_{ref}(\tau-t)$ should principally not be zero but it can take any value with $\Delta_t\neq 0$. Thus, we can use the delta function with any degree of accuracy.

Using eq. (6.205) we immediately get with eq. (6.201)

$$\Psi_{CON}(\mathbf{r},\tau)=\frac{C}{(2\pi\hbar)^{1/2}}\Psi(\mathbf{r},\tau) \tag{6.206}$$

and the wave function $\Psi_O(\mathbf{r},E)$ is expressed with eq. (6.197) as follows:

$$\Psi_o(\mathbf{r},E)=\frac{C}{2\pi\hbar}\int\limits_{-\infty}^{\infty}\Psi(\mathbf{r},\tau)\exp\left\{i\frac{E}{\hbar}\tau\right\}d\tau \tag{6.207}$$

In other words, from all configurations,$\Psi(\mathbf{r},t)$, the configuration $\Psi(\mathbf{r},\tau)$ for time $t=\tau$ is filtered out, that is, it is selected. In particular, eq. (6.206) directly follows from the required filtering process, that is, the required transition from $\Psi^*(\mathbf{r},t)\Psi(r,t)$ to $\Psi^*(\mathbf{r},t_0)$ $\Psi(\mathbf{r},t_0)$ [eqs. (6.186) and (6.187)] is described by the above analysis, where $t_0=\tau$.

6.9.6 Occupations

The sensory apparatus orders the physically real effects with respect to space and time. But the senses do not record the complete x, y, z, t events, but only the x, y, z, t structure at clock time τ, that is, we have

$$t = \tau \tag{6.208}$$

The sensory apparatus selects and this tendency is in accordance with what we actually experience in daily life.

In the case of a certain τ interval ε, we have exactly the same situation. Like the τ events, distributed within ε, also the x, y, z, t events are distributed; we would like to assume that they are distributed within a definite t interval denoted by Δ_t. Here Δ_t is different from the total time range $(t_b - t_a)$ of the x, y, z, t states of the physically real system. The intervals ε and Δ_t must not be identical and also their positions on the identical time scales may be different from each other. But we want to assume that the intervals are identical and superposable:

$$\varepsilon = \Delta_t \tag{6.209}$$

Such an allocation can be made without leaving the selection condition (6.208). This feature is valid for each time τ and, therefore also for t. These correlation effects between τ and t are necessary; otherwise, there would be no connection between the physically real system and the human observer.

Correlated behavior

The reference time τ moves continuously from the past to the future and, therefore, also the interval ε moves continuously from the past to the future. When we consider eq. (6.209), we get for the values $\tau_1, \tau_2, \ldots, \tau_n, \ldots$ and t the following conditions:

$$\begin{gathered}
\left\{\begin{array}{c} \tau_1 \le \tau \le \tau_1 + \varepsilon \\ t_1 \le t \le t_1 + \Delta_t \end{array}\right\} \\
\left\{\begin{array}{c} \tau_2 \le \tau \le \tau_2 + \varepsilon \\ t_2 \le t \le t_2 + \Delta_t \end{array}\right\} \\
\vdots \\
\left\{\begin{array}{c} \tau_n \le \tau \le \tau_n + \varepsilon \\ t_n \le t \le t_n + \Delta_t \end{array}\right\} \\
\vdots
\end{gathered} \tag{6.210}$$

The values $\tau_2, \ldots, \tau_n, \ldots$ are the next real numbers after $\tau_1 + \varepsilon, \ldots, \tau_{n-1} + \varepsilon, \ldots$, and the values $t_2, \ldots, t_n, \ldots$ are the next real numbers after $t_1 + \Delta_t, \ldots, t_{n-1} + \Delta_t, \ldots$.

There is an essential difference between the clock time τ and the quantum time t: whereas τ moves within ε strictly from the past to the future, the t values within Δ_t are statistically occupied. That is, τ moves within ε from the past to the future and Δ_t moves simultaneously to τ, but the t – values within Δ_t behave statistically.

Occupations

At each time τ, defined through the intervals in eq. (6.209), there is a certain probability for the appearance of the variables x, y, z, t within the range where the probability density $\Psi^*(x, y, z, t)\Psi(x, y, z, t)$ is not zero. All values x, y, z, t of the entire range are concerned. In general, the range of is $-\infty < x, y, z, t < \infty$. These are quantum states that are occupied at a sharp value for the reference time τ. Clearly, this situation is independent of the parameter τ. Thus, we have for each time τ exactly the same information about the system under investigation:

$$\begin{aligned} &\tau: \\ &\Psi^*(x_i, y_i, z_i, t_i)\Psi(x_i, y_i, z_i, t_i) \\ &x_i, y_i, z_i, t_i \rightarrow i = 1, 2, \ldots, \infty \; in \; (x_b - x_a), (y_b - y_a), (z_b - z_a), (t_b - t_a) \end{aligned} \tag{6.211}$$

At each time τ, "one" of the four-dimensional x, y, z, t points exists, and there is an infinite number of possibilities. The intervals $(x_b - x_a)$, $(y_b - y_a)$, $(z_b - z_a)$ and $(t_b - t_a)$ define again the ranges with respect to the four-dimensional space–time (Section 6.1).

Number of events

There is an infinite number of τ values within each of the interval given in eq. (6.209). Thus, we have an infinite number for each of the variables x, y, z, t within the range $(x_b - x_a)$, $(y_b - y_a)$, $(y_b - y_a)$, $(t_b - t_a)$. Then, there is an infinite number of probability densities $\Psi^*(x, y, z, t)\Psi(x, y, z, t)$ as well.

That is, the infinite number of τ values within each of the intervals defined by eq. (6.209) create an infinite number for the variables x, y, z, t and an infinite number of probability densities, which are distributed over the entire space–time range $(x_b - x_a), (y_b - y_a), (z_b - z_a), (t_b - t_a)$ for which the probability density is not zero [see eq. (6.148)]. Thus, instead of eq. (6.130) we have

$$\begin{gathered} \tau \rightarrow \infty \text{ in } \tau_m \leq \tau \leq \tau_{m+1},\ m = 1, 2, \ldots \\ \Downarrow \\ x, y, z, t \rightarrow \infty,\ \Psi^*(x, y, z, t)\Psi(x, y, z, t) \rightarrow \infty \\ \text{in } (x_b - x_a), (y_b - y_a), (z_b - z_a), (t_b - t_a) \end{gathered} \tag{6.212}$$

As we have already outlined in Sections 6.1 and 6.6, the number of x, y, z, t states within the total range is infinity and, furthermore, we have also outlined in both sections that the number of states within the interval Δ_t must be infinite as well (we have $\infty/k = \infty$ for $k \neq \infty$, see also Section 6.6.1). Thus, in analogy to eq. (6.132), we have

$$\begin{gathered} \tau \rightarrow \text{ in } \tau_m \leq \tau \leq \tau_{m+1},\ m = 1, 2, \ldots \\ \Downarrow \\ x, y, z, t \rightarrow \infty \text{ in } (x_b - x_a), (y_b - y_a), (z_b - z_a), (t_b - t_a) \\ \Downarrow \\ x, y, z, t \rightarrow \infty \text{ in } \Delta_t \text{ and } (x_b - x_a), (y_b - y_a), (z_b - z_a) \end{gathered} \tag{6.213}$$

That is, each slice defined through Δ_t, $(x_b - x_a), (y_b - y_a), (z_b - z_a)$ contains an infinite number of events.

Furthermore, each x, y, z, t point of the slice, having an infinitesimal four-dimensional interval around this point, exists with certainty because there is also an infinite number of x, y, z, t states within this four-dimensional interval, which is arbitrarily small but different from zero. More details are discussed in the sections before.

Observation of physically real environment

In conclusion, scheme (6.81) reflects the following situation: We observe (measure) with certainty a definite space structure, expressed by the probability densities $\Psi^*(\mathbf{r}_i, t_i)\Psi(\mathbf{r}_i, t_i)$ with $\mathbf{r}_i = (x_i, y_i, z_i)$, in the form of an infinite number of events, which are quantum mechanical in character and refer to the slice, having the parameters Δ_t, $(x_b - x_a)$, $(y_b - y_a)$, $(z_b - z_a)$:

$$\begin{aligned} &\Psi^*(x_i, y_i, z_i, t_i)\Psi(x_i, y_i, z_i, t_i) \neq 0 \\ &x_a \leq x_i \leq x_b,\ y_a \leq y_i \leq y_b,\ z_a \leq z_i \leq z_b \\ &t_i \text{ in } \Delta_t \end{aligned} \tag{6.214}$$

Such situations describe what we actually experience in an unconscious way within our everyday life.

There is a correlated behavior of τ and t that is formulated by scheme (6.210). Whereas the clock time τ moves within the interval ε strictly from the past to the future, the time t behaves statistically within the time interval Δ_t.

The information about the complete system exists as a constant entity for each interval ε with $\tau_i \leq \tau \leq \tau_i + \varepsilon,\ i = 1, 2, \ldots$ and it remains constant during the τ motion. That is, the physically real system exists constantly but, in the course of time τ, a certain information is selected, from Δ_t to Δ_t, so to speak. Only those x, y, z, t events are selected, which belong to the τ interval ε and which belongs simultaneously to the t interval Δ_t, that is, the system-specific time values t occupy a time interval Δ_t, which is superposable with ε: $\Delta_t = \varepsilon$. This is reflected by scheme (6.210), which underlines that the two types of time, that is, t and τ, are strictly correlated.

No doubt, the consequences of this behavior are drastic: the physically real system is not always again created from τ to τ, but here a permanently existing system is strobed in the course of time τ.

Conclusion

Let us underline that human beings perceive (observe) the slice, defined by the space-time interval Δ_t, $(x_b - x_a), (y_b - y_a), (z_b - z_a)$ and not the complete information, which is given in the intervals $(x_b - x_a), (y_b - y_a), (z_b - z_a), (t_b - t_a)$.

The slice is selected from the whole. This is illustrated in Figure 6.10. The slice itself is recognizable through the interval ε around τ

The intervals Δ_t and ε are sufficiently small and may even be infinitesimal but have to be different from zero. Then, the situation is in accordance with that what we observe in daily life: At a certain clock time (approximately given by τ), the entire space-structure with objects is perceived.

6.9.7 Processing in the mind

The wave function, describing fictitious reality in $(\mathbf{p}, E)$ space, corresponds to the basic information α (see Section 6.2). Both entities equally characterize a physically real object. At this stage, space and time are not yet involved. The mind processes α unconsciously, and the mind also processes the function $\Psi(p_x, p_y, p_z, E)$, but consciously. Space and time come into existence through the processing in the human being's mind and they offer the frame in which the entities α and $\Psi(p_x, p_y, p_z, E)$ are represented.

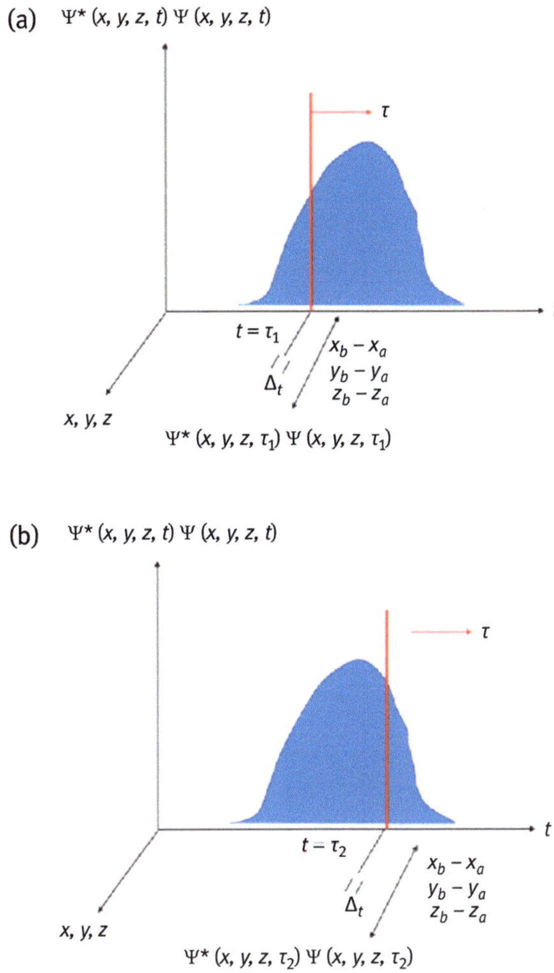

Figure 6.10: Selection of a part $\Psi(\mathbf{r},\tau)$ of the total information of the physically real system $\Psi(\mathbf{r},t)$ with $\mathbf{r}=(x,y,z)$. The selections of space structures at various clock times τ are represented in (a) for τ_1 and in (b) for τ_2 in the form of probability densities: $\Psi^*(\mathbf{r},\tau_1)\Psi(\mathbf{r},\tau_1)$, $\Psi^*(\mathbf{r},\tau_2)\Psi(\mathbf{r},\tau_2)$. In general, we have $\Psi^*(\mathbf{r},\tau_1)\Psi(\mathbf{r},\tau_1)\neq\Psi^*(\mathbf{r},\tau_2)\Psi(\mathbf{r},\tau_2)$ and this situation corresponds to what is observed in daily life. The selected information consists of events within the slice Δ_t, (x_b-x_a), (y_b-y_a), (z_b-z_a) and only this part is perceived (observable) through human beings.

In the case of α the mind produces at time τ, shown by an external clock, a space–time picture "without coordinates" in an unconscious way. It is an observation. The coordinates x,y,z at time τ are the elements of a "fictitious net" which the human being puts intellectually over the observed image in front of him/her [5].

On the other hand, the treatment of the information $\Psi(p_x,p_y,p_z,E)$ through the mind is expressed by the Fourier transform (6.8) together with the selection process

$t \rightarrow \tau$ that is analyzed in Section 6.9.5. We finally obtain a space–time structure given by $\Psi(x, y, z, \tau)$. That is, we have

$$\begin{aligned}\Psi(x, y, z, t) &= \frac{1}{(2\pi\hbar)^2} \int_{-\infty}^{\infty} \Psi(p_x, p_y, p_z, E) \exp\left\{ i \left[\frac{\mathbf{p}}{\hbar} \cdot \mathbf{r} - \frac{E}{\hbar} t \right] \right\} dp_x \, dp_y \, dp_z \, dE \\ &\downarrow \\ \Psi(x, y, z, \tau) & \end{aligned} \tag{6.215}$$

Note that space and time, that is, the elements x, y, z, t, come directly into existence by the Fourier transform (6.8), and relation (6.215) reflects a conscious action of the mind.

As we remarked earlier, the information α is processed by the mind, and α characterizes physically real entities. Therefore, the unconscious processing of α within the mind has to be performed with mind elements that are sensitive to physically real entities.

Therefore, also the selection process $t \rightarrow \tau$ and others in connection with the conscious processing through eq. (6.215) has to be performed with elements that are sensitive to physically real entities. This element is given in Section 6.9.5 by the function $\Psi_{ref}(E)$ and $\Psi_{ref}(t)$, respectively. The situation is summarized in Figure 6.11.

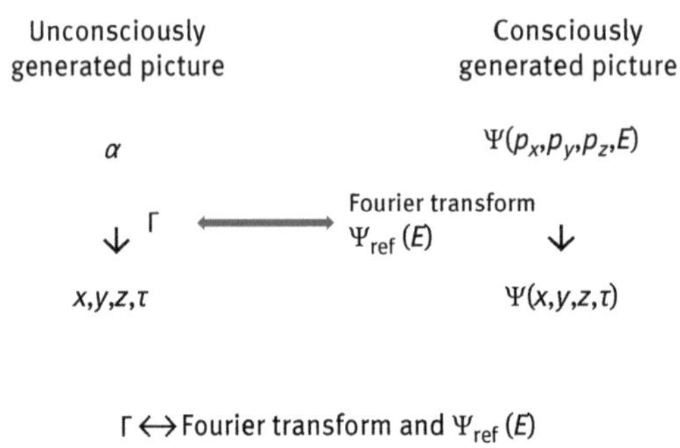

Figure 6.11: The transition from information α to the space–time picture, that is, to the x, y, z structure at clock time τ, is an unconscious process in the mind whereby the corresponding mind elements are sensitive to physically real objects. The transition itself is summarized in the figure by the letter Γ. The conscious transition, that is, the processing from $\Psi(p_x, p_y, p_z, E)$ to $\Psi(x, y, z, \tau)$, is done by a Fourier transform and the necessary selection process $t \rightarrow \tau$ by the function $\Psi_{ref}(E)$. The unconscious processes, characterized by Γ, and the conscious actions, expressed by the Fourier transform and $\Psi_{ref}(E)$, correspond each other.

In Section 6.9.5 the wave function $\Psi(x, y, z, E)$ has been used, which is however completely equivalent to $\Psi(p_x, p_y, p_z, E)$. The wave functions $\Psi(x, y, z, E)$ and $\Psi(x, y, z, E)$ are also here connected to a Fourier transform

$$\Psi(x,y,z,E) = \frac{1}{(2\pi\hbar)^2} \int_{-\infty}^{\infty} \Psi(p_x,p_y,p_z,E) \exp\left\{\frac{i}{\hbar}[\mathbf{p}\cdot\mathbf{r}]\right\} dp_x\, dp_y\, dp_z \tag{6.216}$$

Thus, the two spaces, the (**p**, *E*) space and the (**r**, *E*) space, contain exactly the same information.

6.10 Self-organizing processes

We underlined in the preceding chapters that self-organizing processes belong to the heat of nanoscience (nanotechnology). In general, we have the following situation: we put a nanosystem into a specific environment and both entities, the system and the environment, interact with each other and their states will be changed in the course of clock time τ in a self-organizing way.

6.10.1 General definition

The initial state of the nanosystem is indicated by A and its final state by B. Then the transition is characterized through

$$A \underset{\tau}{\rightarrow} B \tag{6.217}$$

The process started at time τ_A and is assumed to be finished after a certain time τ_B until stationary states are reached. The nanosystem and the environment are in stationary states for $\tau \leq \tau_A$ and $\tau \geq \tau_B$.

Such kind of processes have to be considered as "time sensitive." Just this feature is the problem when we work within traditional quantum theory. Nanosystems behave in most cases quantum mechanically but, as we underlined in the preceding chapters, in traditional physics there does not exist a quantum aspect of time. The time in traditional quantum theory is given by the external time τ, which behaves classical and is shown by our clocks.

In contrast to this traditional view, within the frame of the projection principle there is a quantum aspect of time and we have here a system-specific time t. All these have been discussed in the preceding sections.

Let us discuss some characteristics of self-organizing processes within traditional physics, where we have no quantum time. Then we will treat the problem on the basis of the more realistic projection theory, which is characterized through the system-specific quantum time.

6.10.2 Traditional physics

In most cases, nanosystems behave quantum mechanically and we have to apply Schrödinger's equations [12]. Schrödinger's equation for the stationary case is given by

$$i\hbar\frac{\partial\varphi(x,y,z,\tau)}{\partial\tau}=-\frac{\hbar^2}{2m_0}\Delta\varphi(x,y,z,\tau)+U(x,y,z)\varphi(x,y,z,\tau) \tag{6.218}$$

On the other hand, for the nonstationary case we have

$$i\hbar\frac{\partial\varphi(x,y,z,\tau)}{\partial\tau}=-\frac{\hbar^2}{2m_0}\Delta\varphi(x,y,z,\tau)+U(x,y,z,\tau)\varphi(x,y,z,\tau) \tag{6.219}$$

Once again, stationary means that the potential V is not dependent on time τ and we get eq. (6.218); if U is dependent on U we deal with the nonstationary case and eq. (6.219) is valid.

In the transition $A \underset{\tau}{\rightarrow} B$ (eq. (6.217)), the initial state A and the final state B are expressed by the following stationary equations [see in particular eq. (6.218)]:

$$i\hbar\frac{\partial\varphi_A(x,y,z,\tau)}{\partial\tau}=-\frac{\hbar^2}{2m_0}\Delta\varphi_A(x,y,z,\tau)+U_A(x,y,z)\varphi_A(x,y,z,\tau) \tag{6.220}$$

and

$$i\hbar\frac{\partial\varphi_B(x,y,z,\tau)}{\partial\tau}=-\frac{\hbar^2}{2m_0}\Delta\varphi_B(x,y,z,\tau)+U_B(x,y,z)\varphi_B(x,y,z,\tau) \tag{6.221}$$

On the other hand, for the transition from A to B, Schrödinger's nonstationary equation holds (see eq. (6.219):

$$i\hbar\frac{\partial\varphi_{AB}(x,y,z,\tau)}{\partial\tau}=-\frac{\hbar^2}{2m_0}\Delta\varphi_{AB}(x,y,z,\tau)+U_{AB}(x,y,z,\tau)\varphi_{AB}(x,y,z,\tau) \tag{6.222}$$

Here we use the τ- dependent potential $U_{AB}(x,y,z,\tau)$, but not in connection with eqs. (6.220) and (6.221). The Schrödinger theory has been discussed and analyzed in Ref. [12]. An example for a self-organizing process in traditional physics is shown in Figure 6.12 (a).

Remark

For example, in the case of the one-dimensional harmonic oscillator, the potential U is given by $1/2mx^2\omega_0^2 = U(x)$; m is the mass and ω_0 the frequency of the oscillator. The interaction of this oscillator with a specific environment can lead to the effect that the frequency ω_0 becomes τ dependent and we have $\omega_0 = \omega_0\,(\tau)$. Then, in region A the frequency is given by $\omega_A = \omega_0$ and is independent on τ; in region AB the

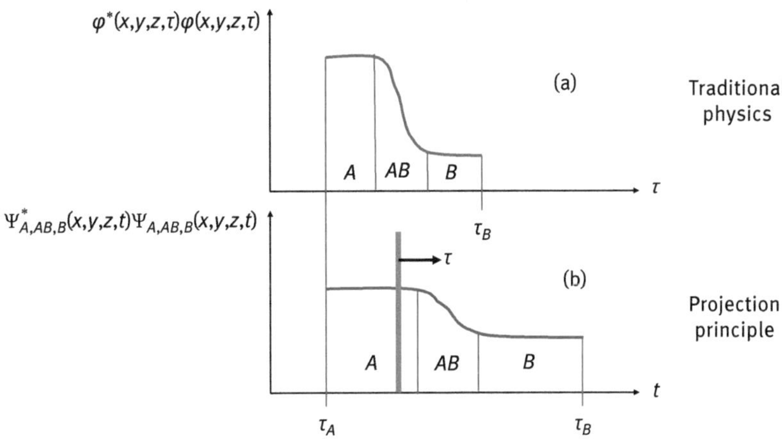

Figure 6.12: Self-organizing process described within traditional physics (a) and within the frame of the projection principle (b). For both views, the probability densities are shown. The boundary conditions at time τ_A are assumed to be identical [see eq. (6.228)]. However, the curves differ from each other. In particular, the final states, which is reached at τ_B, can differ considerably for both approaches. The characteristic three regions *A*, *AB* and *B* are located at different positions on the timescales. The scanning process is indicated in (b).

frequency is τ dependent and we get $\omega_{AB}(\tau)$; in region B the frequency is given by ω_B and is independent on τ.

The environment and the system (oscillator) influence each other and leads to changes with respect to both the oscillator and the environment. The changes, that is, the τ- dependent processes, take place during the transition in region *AB*.

6.10.3 Projection theory

Scanning information

In traditional quantum theory, the physically real systems exist at time τ and are incessantly produced in the course of time τ, but they do not exist for times $\tau' < \tau$ and $\tau' > \tau$. The view of projection theory is different. At time τ the physically real system is not produced but selected from the entire information. This has of course consequences for self-organizing processes. Let us repeat the main facts.

Despite the statistical fluctuations, at τ (or more precisely at $\tau \pm \varepsilon$), the whole of time *t*, that is, the past, present and future, is laid out frozen before us:

$$\begin{aligned} &\Psi^*(x,y,z,t)\Psi(x,y,z,t) \\ &-\infty < x,y,z < \infty, \quad -\infty < t < \infty \end{aligned} \tag{6.223}$$

That is, at time $\tau \pm \varepsilon$ the entire information is given all at once. This also means that within the range defined by τ_A and τ_B, where the self-organizing process takes place, the entire information of the wave function is given all of a piece.

The system-specific time t is defined within the range of the self-organizing process by

$$-\infty < \tau_A \leq t \leq \tau_B < \infty \tag{6.224}$$

However, we know that the entire information (6.223) is not "observed" at clock time $\tau \pm \varepsilon$. Instead, we always observe only a certain space structure of reality at time $\tau \pm \varepsilon$:

$$\begin{gathered} \Psi^*(x, y, z, t_C)\Psi(x, y, z, t_C) \\ -\infty < x, y, z < \infty \end{gathered} \tag{6.225}$$

for $t_c = \tau$. A usual photography represents at a certain time $\tau = t_c$ such a configuration in space.

The curve $\Psi^*(x, y, z, t)\Psi(x, y, z, t)$ is scanned with respect to time t and we observe merely the part at clock time t : $\Psi^*(x, y, z, \tau)\Psi(x, y, z, \tau)$. Scanning processes also take place within the frame of self-organizing processes and this is indicated in Figure 6.12b.

We may state quite generally that the information about the wave function $\Psi(x, y, z, t)$ and the probability density $\Psi^*(x, y, z, t)\Psi(x, y, z, t)$, respectively, exists completely at τ ($\tau \pm \varepsilon$) not as a static unit, but as a fluctuating quantum entity. The fluctuations cannot be dissolved by human beings in daily life and are experienced at time $\tau \pm \varepsilon$ as a compact block.

Comparison of the traditional view with the situation in projection theory

In the case of the traditional treatment of self-organizing processes, a piecewise treatment of the three regions A, AB and B is demanded and possible. Here we need for each region a specific equation, reflected by eqs. (6.220)–(6.222). The projection principle does not allow such a procedure. Within this principle, the characteristics of self-organizing processes are described by "one" relation, and this is expressed by

$$\begin{aligned} &i\hbar \frac{\partial \Psi_{A,AB,B}(x, y, z, t)}{\partial \tau} = \\ &-\frac{\hbar^2}{2m_0}\Delta\Psi_{A,AB,B}(x, y, z, t) + V_{A,AB,B}(x, y, z, t)\Psi_{A,AB,B}(x, y, z, t) \end{aligned} \tag{6.226}$$

All three regions A, AB and B are treated all of a piece. The reason is obvious: at clock time τ all quantum states t can be occupied that belong to region A, B or to region AB. This property has been discussed in the earlier section.

Nevertheless, a human observer experiences the three regions A, AB and B as a self-organizing process. He/she successively scans the curve

$$\Psi^{*}_{A,AB,B}(x,y,z,t)\Psi_{A,AB,B}(x,y,z,t) \tag{6.227}$$

with respect to t and passes through this curve in the course of clock time τ strictly from the past to the future, that is, from region A to region B over region AB. This sequence is experienced as a self-organizing process. The scanning process is indicated in Figure 6.12b.

The quantum-mechanical description of self-organizing processes in the traditional way has been excluded because the notion "time" here is not adequately treated. There is no quantum time involved, as it should be within a proper treatment of time-sensitive processes. Thus, the description of specific self-organizing processes can be utterly wrong when traditional quantum theory is applied.

At time τ_A, we may choose for the probability densities for both approaches (traditional view and projection principle) the same probability densities

$$\varphi^{*}(x,y,z,\tau_A)\varphi(x,y,z,\tau_A) = \Psi^{*}_{A,AB,B}(x,y,z,\tau_A)\Psi_{A,AB,B}(x,y,z,\tau_A) \tag{6.228}$$

Since the potential $V_{A,AB,B}(x,y,z,t)$ [eq. (6.226)] cannot be approximated by the potentials $U_A(x,y,z)$, $U_A(x,y,z)$, $U_{AB}(x,y,z,\tau)$ [eqs. (6.220), (6.221) and (6.222)], we have for $\tau > \tau_A$

$$\varphi^{*}(x,y,z,\tau)\varphi(x,y,z,\tau) \neq \Psi^{*}_{A,AB,B}(x,y,z,\tau)\Psi_{A,AB,B}(x,y,z,\tau) \tag{6.229}$$

Thus, the description of the self-organizing process with the traditional tools has only a little (or nothing) to do with the more realistic description within the frame of the projection principle. Both views are illustrated in Figure 6.12.

The prediction of self-organizing processes through the traditional way [indicated by eqs. (6.217)–(6.222)] is not realistic. Just in connection with nanosystems, a careful treatment is necessary.

6.11 Summary and final remarks

Let us summarize and complement the ideas given in this chapter through the following brief comments:

1. Within the frame of the projection principle, there are two spaces, the $(\mathbf{r},t)$ space and $(\mathbf{p},E)$ space, which are equivalent concerning their information. Both spaces are connected to each other by a Fourier transform. That is, due to $\hbar$, the variables $\mathbf{r},t,\mathbf{p},E$ are arranged in a quite different way than in classical mechanics, where $\hbar$ is not defined.

   The $(\mathbf{p},E)$ space reflects the physical reality (fictitious reality), and this reality (i.e., its information content) is completely projected onto $(\mathbf{r},t)$ space. This is consistent with the fact that we do not measure the elements x,y,z,t, but we measure the quantities p_x,p_y,p_z,E. Measurement means that an interaction process takes place and, on the other hand, interaction means an exchange of energy and momentum.

We may treat a physically real unit on the basis of the variables p_x, p_y, p_z and E "as if" they were physically real entities, but they are entities of a fictitious reality, which is projected onto space and time with x, y, z, t.

We have a (fictitious) reality, where the physically real processes are simulated. This information is projected onto space and time and we get a picture of this (fictitious) realty. This is the highest level, which human beings can take in the concrete description of the world. This is a clear and realistic point of view and we can distinguish between what is describable and observable and what not.

2. Within the frame of the projection principle, we have the following situation: Both spaces, $(\mathbf{r}, t)$ and $(\mathbf{p}, E)$, are equivalent concerning their information. Each physically real object contains in both spaces exactly the same physical information. In particular, we have the following peculiarities:
   (a) In $(\mathbf{r}, t)$ space the coordinates x, y, z and the time are numbers and the momentums and the energy are operators that are given in the form $-i\hbar\partial/\partial x$, $-i\hbar\partial/\partial y$, $-i\hbar\partial/\partial z, i\hbar\partial/\partial t$.
   (b) In $(\mathbf{p}, E)$ space, the coordinates and the time are expressed by the operators $i\hbar\partial/\partial p_x$, $i\hbar\partial/\partial p_y$, $-i\hbar\partial/\partial p_z$, $-i\hbar\partial/\partial E$. The momentums E and the energy p_x, p_y, p_z are numbers.
3. In projection theory we have a real uncertainty relation between the energy E and the system-specific time t:

$$\delta E \delta t \geq \frac{\hbar}{2}$$

Such an equation does not exist in the usual form of quantum theory. Since time is still a classical parameter within usual quantum theory, there is no uncertainty relation of type $\delta E \delta t \geq \hbar/2$ in traditional quantum theory. However, here we have an energy–time relation of the form

$$\Delta E \Delta \tau \geq \frac{\hbar}{2}$$

But the quantities ΔE and $\Delta \tau$ are differences and they have nothing to do with uncertainties. Relation $\Delta E \Delta \tau \geq \hbar/2$ is a quasiclassical equation. The value of such an equation is, however, questionable. In Mario Bunge's opinion, the energy–time relation $\Delta E \Delta \tau \geq \hbar/2$ "is a total stranger to quantum theory."
4. Physically real objects can equally be treated in time representation as well as in energy representation. Moreover, the treatment can be done in $(\mathbf{r}, t)$ space and/or in $(\mathbf{p}, E)$ space and leads also here to equivalent information about the object under investigation. We have discussed two procedures: (1) the direct transition to the time operator and the energy operator, respectively; (2) the use of classical expressions in the transition to the quantum aspect.

For the time representation and the energy representation we started from $t = t(x, y, z, p_x, p_y, p_z, E)$ and $E = E(x, y, z, t, p_x, p_y, p_z)$, which reflect general classical relationships. The time operator $\hat{C}$ and the energy operator $\hat{H}$ can be formulated directly on the footing of operators for the coordinates and the time (e.g., the variable t is replaced by the operator $-i\hbar\partial/\partial E$). The system-specific operators $\hat{C}$ and $\hat{H}$ are then modeled in terms of suitable quantum conceptions.

We have four-dimensional interaction potentials of the form $V(x, y, z, t)$, which are at clock time τ effective in the past, present and future with respect to system-specific time t. In usual quantum theory, such kind of interaction potential is not defined. Instead of the general interaction $V(x, y, z, t)$, we have here a restricted form of interaction denoted by $U(x, y, z)$.

Chapter 7
Final statements

The epistemologically value of the container model is poor. It consists at each time τ of space in which the physically real (material) objects are embedded. This is the usual conception, and the laws of theoretical physics are based on this imagination. However, the space–time itself is not physically real (Chapter 1), and this is the problem we have with the container model: The container is not physically real but the physically real objects are embedded in it.

The container model worked so far. Why? There can only be one reason: The container model reflects an approximation that is obviously "compensated" within a theoretical frame through other theory-specific approximations.

The more we penetrate into the details of nature, the more we need realistic conceptions for their description, that is, we need conceptions with increasing sophistication. In order to be able to verify the container sensitivity of the theoretical laws, we have to choose situations in which the container effect is not compensated through other approximations of a formulated theory. This is, in fact, possible when we investigate the "cosmological constant." Here, the cosmos in the form of a container is directly concerned, that is, the entire space at each time τ.

In fact, the container model effectuates a catastrophe for the numerical value of the cosmological constant [82, 83]. The expected value for it is small (almost "zero") and the predicted quantum field theoretical value is very large (not far from "infinity"). The theoretical prediction deviates from the observed value by a factor of 10^{122}. The discrepancy could not be larger; it is simply a catastrophe and is often considered as the deepest mystery of present physics.

In this connection, Leon Lederman's statement is of interest: "Einstein freed us from the either. Now we need to get rid of (today's version of either) again. We need to sweep the vacuum clean" [82]. But if there is nothing inside the container (space, vacuum), the container model is disestablished and we come inevitably to the projection principle.

In fact, this situation demands a drastic step: To drop the container model. This is not a drastic step with respect to physics, but it is a matter of habit. There is obviously no other way possible since all attempts to eliminate the large amounts of energies within space (container).

In this last chapter, we first give some statements in connection with usual quantum theory and general theory of relativity (GTR). Both theories are mutually incompatible. What could be the reason for that? Is the container model responsible for that?

https://doi.org/10.1515/9783110525595-007

7.1 Theory of relativity and quantum field theory

What can we say about the basic conceptions of traditional physics. There are two developments, on which traditional physics is essentially founded. It is the GTR and the quantum theory in its usual form. However, both theoretical blocks have been developed independently from each other. Both theories deliver excellent results, but there is a serious problem: GTR and traditional quantum theory are mutually incompatible.

7.1.1 Incompatible theories

The situation is well formulated by Brian Greene. In Ref. [84], we find the following comment:

> Through years of research, physicists have experimentally confirmed to almost unimaginable accuracy virtually all predictions made by each of these theories (General Theory of Relativity and usual quantum theory). But these same theoretical tools inexorably lead to another disturbing conclusion: As they are currently formulated, general relativity and quantum mechanics, cannot both be right. The both theories underlying the tremendous progress of physics during the last hundred years – progress that has explained the expansion of the heavens and the fundamental structure of matter – are mutually incompatible [84].

Both theories lead to sets of laws that work fantastically. However, if we put them together, we inevitably obtain irreconcilable differences. In Ref. [82], we find:

> It has been said that quantum field theory is the most accurate physical theory ever, being accurate to about one part in 10^{11}, writes Roger Penrose in The Nature of Space and Time. However, I would like to point out that general relativity has, in a clear sense, now been tested to be correct to one part in 10^{14} (and this accuracy has apparently been limited merely by the accuracy of clocks on earth).

In conclusion, both theories seem to work perfectly but, on the other hand, both theories are mutually exclusive. What might be the reason for this catastrophe? We know that both theories are formulated within the frame of the container model and this might be the reason for the fact that both theories are mutually incompatible.

From the point of view of projection theory, the application of the container model has to be done with care and can lead to a catastrophe, as we guess is the case for GTR and traditional quantum theory.

The incompatibility and, on the other hand, the excellent results, which both theories deliver, might come through the following effect: The possibly wrongly formulated theoretical background of both developments, together with the unrealistic container model, obviously compensate each other. This leads to a useful conception for both theories. Then, the theories themselves have to be considered as “recipes.”

7.1.2 Compensation effects

The incompatibility with respect to GTR and traditional quantum theory and, on the other hand, the excellent results, which both theories deliver, might come through the following effect: The theoretically formulated background of both developments and the features of the container model obviously compensate each other. This leads to a useful conception for both theories. We have to be careful with the container model, even when the background of both theories is realistic formulated.

The details of the physical laws come more and more into play when we go deeper into the nature of the things, and in such cases, we need physical laws with increasing sophistication and reliability. Compensation effects are hardly helpful in this connection. The container model reflects a hard approximation, which is obviously compensated through other approximations.

7.2 Uncertainties

As we remarked earlier, the expected value for the cosmological constant is small and the predicted quantum field theoretical value is very large. We expect a value of almost zero, but the theoretical prediction is not far from infinity. Or more precisely: The theoretical prediction deviates from the observed value by a factor of 10^{122}. The discrepancy could, in fact, not be larger. As we already remarked, this situation is often considered as the deepest mystery of present physics.

What is the reason for this discrepancy? The cause is obviously due to the fact that there is, in traditional quantum theory, no realistic uncertainty relation between the energy and the time. Let us analyze the situation.

7.2.1 Remarks on the symmetry between space and time

The symmetry between space and time is required. But traditional quantum theory does not fulfill this condition, and this is reflected in the uncertainty relations. Before we discuss the uncertainty relations with respect to this point, let us discuss some details in connection with the symmetry between space and time. We consider the special theory of relativity (STR) and the traditional form of quantum theory.

The symmetry between space and time in traditional quantum theory

Since quantum theory and STR have been developed independently from each other, the following question is of relevance: Are the features of STR fully reflected in traditional quantum theory? In other words, can quantum phenomena be treated fully in accordance with the basic laws of STR? Here, we restrict ourselves to the symmetry question. The answer is formulated as follows: Although the relativistic wave equations (e.g., Dirac's equation for the electron) are invariant under Lorentz transformation, the space coordinates x, y, z and time τ are in its physical content definitely not symmetrical to each other and that is in contrast to the fundamental results of STR. This is due to the following facts:

1. Whereas the coordinates are "statistical" quantities, the time does not behave statistically. We already mentioned in Chapter 1 that the time remains unchanged when we go from classical mechanics to quantum theory. This is clearly reflected in the fact that the coordinates can be "operators," time is always a simple "parameter." Only the clock time τ exists in all forms of traditional quantum theory.
2. The determination of the eigenfunctions and eigenvalues is restricted on space and time is not involved in this process.

The symmetry of space and time, a requirement of STR, must be reflected in all versions of traditional quantum theory, also within Schrödinger's theory. Louis de Broglie expressed this fact as follows: "*The present quantum theory in all its versions takes time as the evolution parameter and therefore destroys the symmetry between space and time*" [85].

Four-dimensional worlds

Within the framework of STR, the world has to be considered as four-dimensional, and this is because time loses its independence which it still had within Newton's mechanics. The fourth of the equations of the Lorentz transformations is given by

$$\tau' = \frac{\tau - vx/c^2}{\sqrt{1 - v^2/c^2}} \tag{7.1}$$

This equation shows that the time interval between two events in a moving frame of reference does not vanish in general. Here the frame of reference S' moves relatively to a rest system, say S, with the constant velocity v. This is even the case, when the time interval in S becomes zero. The consequence of a pure "distance in space" in S is an "interval of time" in S', and this result can directly be read from eq. (7.1). It is well known that eq. (7.1) is obviously realistic. This has been shown experimentally by the slowing down of clocks.

In conclusion, within STR, space and time are tightly interrelated and cannot be treated as independent quantities. They are, as we know, symmetrical to each other. This is however not the case in traditional quantum theory. In traditional quantum theory, space and time are independent quantities, even in the relativistic quantum case. Space and time behave differently in traditional quantum theory. Let us demonstrate that in connection with the uncertainty relations.

7.2.2 Uncertainty relations in traditional quantum theory

The symmetry between space and time is required. But traditional quantum theory does not fulfill this condition, and this is reflected in the uncertainty relations.

In usual quantum theory, there are definitely uncertainty relations [6] for the coordinates x, y, z and momentums p_x, p_y, p_y (e.g., Ref. [83]):

$$\delta p_x \delta x \geq \frac{\hbar}{2}, \quad \delta p_y \delta y \geq \frac{\hbar}{2}, \quad \delta p_z \delta z \geq \frac{\hbar}{2} \tag{7.2}$$

An analogous relation for the time τ and a quantity which has the dimensions of energy is required from the point of view of STR. Both, the energy and the time should have the features of the system (object) under investigation. But this condition is not fulfilled in the case of relation

$$\Delta E \Delta \tau \geq \frac{\hbar}{2} \tag{7.3}$$

Here the energy E is a system property but the time τ not. The time τ is, as we know, an external time (clock time) and has primarily nothing to do with the system under investigation. We can observe an object as a function of clock time τ, but its properties and its existence itself is not determined by this external clock showing the time τ. In other words, the clock is not coupled to physically real objects.

As we already outlined in Chapter 1, eq. (7.3) is often seen as uncertainty relation as well, quite in analogy to (7.2). But it is not, which is symbolically expressed by the use of Δ instead of δ.

In relations (7.2), the quantities $\delta p_x, \ldots$ and $\delta x, \ldots$ are the uncertainties in the values of the momenta and the coordinates at the same instant. As is well known, this uncertainty means that the coordinates and momenta can never have entirely definite values simultaneously.

The energy E, on the other hand, can be measured to any degree of accuracy at any instant. The quantity ΔE in eq. (7.3) is the "difference" between two "exactly measured" values of the energy at two different instants (clock times) and is not the uncertainty in the value of the energy at a given instant (see also the discussion in Refs. [6, 83]). Nevertheless, eq. (7.3) is throughout interpreted as uncertainty relation, but this is not correct. Concerning this point, Mario Bunge gave a clear statement.

7.2.3 Bunge's critique

Since time within usual quantum theory is still a classical parameter, we should consider the energy–time relation $\Delta E \Delta \tau \geq \hbar/2$ [eq. (7.3] as a quasiclassical equation. The value of such an equation is however questionable. It is, in particular, not an uncertainty relation.

In Mario Bunge's opinion, the energy-time relation $\Delta E \Delta \tau \geq \hbar/2$ (7.3) "is a total stranger to quantum theory." In particular, we find in Ref. [77]:

> This relation is made plausible by reference to some thought experiments, to radioactive decay, and to line breadths. But unlike the genuine indeterminacy relations, (7.3) has never been proved from first principles. In other words, (7.3) does not belong to quantum theory, but is just a piece of doubtful heuristics.
>
> The reason for this failure to incorporate (7.3) into quantum mechanics is the following. In this theory, as in every other known theory, time is a "c number" and, more particularly a parameter, not a dynamical variable. Moreover, τ does not "belong" (refer) to the system concerned. Even in relativistic theories the proper time, though relative to a frame of reference, is not a property of the system on the same footing as its mass or its momentum. In other words, τ does not belong to the family of operators in the Hilbert space associated to every microsystem. Therefore, τ is not a random variable and its scatter vanishes identically . . . Consequently, no matter what the scatter in the energy may be, the inequality (7.3) does not hold. Also it does not improve things to regard E, as is sometimes done, as the Hamiltonian of the system the standard deviation of the energy vanishes as well. In conclusion, the so-called fourth indeterminacy (or uncertainty) is a total stranger to the quantum theory although it can be found in works on this theory.
>
> In short, the fourth scatter relation is not deducible from the principles of the quantum theory, whether relativistic or not. But then why is it sometimes used, for example in the theory of line breadths? The reason is that it is not the same formula: although it has the same typographical form, it has a different content. In particular, "Δ" is not interpreted as a standard deviation (from what?) but as the half-life of the state ψ. But even thus reinterpreted, the formula is not deducible from the postulates of the general quantum theory. Mind, this has nothing to do with the question whether or not formula (7.3) is true under some suitable interpretation. Many other statements are true and yet they do not belong to quantum theory.

In Bunge's opinion, the formula $\Delta E \Delta \tau \geq \hbar/2$ is a total stranger to the quantum theory although it can be found in works on this theory. No doubt, Bunge is right. Nevertheless, this formula is used in connection with physically real processes. It is the basis for the incessantly particle/antiparticle production in space. The conservation law for the energy can be temporarily violated and this leads to an uncontrolled production of physically real entities in space leading to a nonacceptable (wrong) value for the cosmological constant. That is, Bunge's assessment in connection with $\Delta E \Delta \tau \geq \hbar/2$ is accompanied with a wrong value for the cosmological constant. We simply do not observe a world with such a large cosmological constant.

As in the case of the coordinates and the momentums, the energy and the time should have the features of the system (object) under investigation and we may interpret that as real uncertainties with respect to the system under investigation and this expressed by the uncertainty relations (7.2). But this condition is not fulfilled in the case of relation $\Delta E \Delta \tau \geq \hbar/2$. Equation $\Delta E \Delta \tau \geq \hbar/2$ is throughout interpreted as uncertainty relation between the energy E and the time τ. The energy E is here a quantity with respect to physically real object but not the time τ. The time is again the external clock time τ. This means that equation $\Delta E \Delta \tau \geq \hbar/2$ relates the properties of the object under investigation and an external clock. This sounds strange and is, in fact, strange. It is simply not imaginable that an object is coupled in the physically real sense to an external clock. We can observe an object as a function of time τ, but its properties and its existence itself is not determined by the external clock showing the time τ. This is hardly imaginable. It is also not imaginable that our thoughts and ideas are dependent on how fast we go.

This means that equation $\Delta E \Delta \tau \geq \hbar/2$ cannot be assessed as an uncertainty relation but it opens the door for the temporary violation of the conservation law for the energy. In other words, a doubtful equation leads to the violation of the conservation law for the energy and this is adventurous. Concerning the much too large amount of energy in the universe, we quoted Leon Lederman's statement above: "Einstein freed us from the either. Now we need to get rid of (today's version of either) again."

Remark

The conservation laws are important because the processes in nature are kept with them under control. In fact, since the energy can be (temporarily) violated in traditional quantum theory through the existence of $\Delta E \Delta \tau \geq \hbar/2$, we get an uncontrolled production of particle/antiparticle pairs, which leads to an unacceptable (wrong) value for the cosmological constant.

Relation $\Delta E \Delta \tau \geq \hbar/2$ is not an uncertainty relation. But this equation offers the possibility to violate the conservation law for the energy. But the result is a catastrophe. A genuine uncertainty relation between the energy and the (system-specific) time is required, and if such a relation could be found in traditional quantum theory, a violation of the conservation law for the energy could possibly be avoided without leaving the container model. However, such an equation could not be deduced in traditional quantum theory.

Thus, we have to go another way: We have to develop the quantum laws on the basis of a frame in which space and time are "not" physically real entities. This is fulfilled in the case of the projection principle. In Chapter 6, we have formulated the quantum laws within the frame of the projection principle. Then, in fact, the relation $\Delta E \Delta \tau \geq \hbar/2$ is replaced by the genuine uncertainty relation $\delta E \delta t \geq \hbar/2$, where t is the required system-specific time. In the case of $\delta E \delta t \geq \hbar/2$ uncontrolled energy productions are excluded. In particular, the space does not contain physically real entities like momentums and energies. The situation is summarized in Figure 7.1.

7.2.4 Consequences with respect to observations

Would the space (container) really contain all the energy, predicted by traditional quantum theory, the universe would extremely violate the observational facts. In Ref. [82], we find the following instructive example:

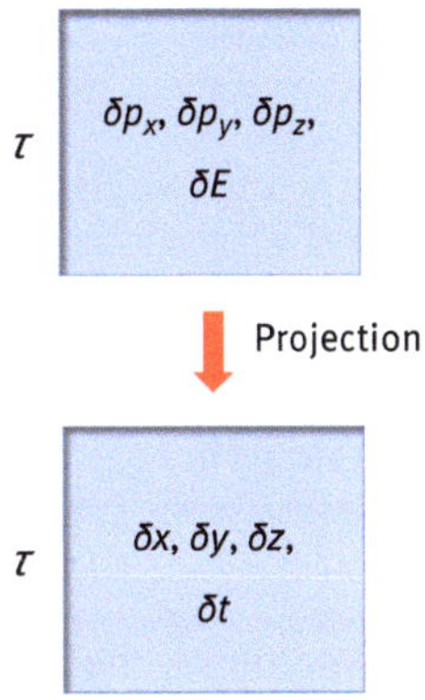

Figure 7.1: Traditional quantum theory has been developed within the frame of the container model. In this case, no system-specific time is defined and this situation leads to $\Delta E \Delta \tau \geq \hbar/2$. When we formulate the quantum laws within projection theory, the system-specific time t comes into existence and we get, in this case, the genuine uncertainty relation $\delta E \delta t \geq \hbar/2$.

> Indeed, if the vacuum contained all the energy physicists expect it to, it would be so repulsive that you wouldn't be able to see your hand in front of your nose. Even at the speed of light, the light from your hand wouldn't have time to reach your eyes before the expanding universe pulled it away. The fact that you can see anything at all, says Krauss, means that the energy of space cannot be so large.

In other words, traditional quantum theory leads to a wrong result. Due to $\Delta E \Delta \tau \geq \hbar/2$, the conservation law for the energy can be temporarily violated and this leads to an uncontrolled production of energy in space leading to a nonacceptable (wrong) value for the cosmological constant and predicts observational facts that have nothing to do with that what we really observe. The value of relation $\Delta E \Delta \tau \geq \hbar/2$ has to be judged negatively, which we have outlined in Sections 7.2.1–7.2.3.

Even when the cosmological constant could be reduced, the container model with its physically real space–time elements would remain an unrealistic conception. We know that the space–time elements are "not" physically real; they can simply not be observed and measured, respectively. The container model reflects a too narrow frame for the understanding of the material world and it has not the capability for the introduction of a system-specific time. Let us repeat that Schrödinger tried to construct an operator for the time but without success. Schrödinger did it within the frame of traditional quantum theory, which is based on the container model.

7.2.5 Two faces of the energy

Each pair of particles that pops in and out is due to the existence of $\Delta E \Delta \tau \geq \hbar/2$, and this formula allows that the conservation law for the energy can be temporarily violated and this leads to an uncontrolled production of particle/antiparticle pairs in space. However, gravity does not see them, and this is strange. In Refs. [82, 83], the following instructive remarks:

> If something has matter or energy, gravity will respond by warping; gravity is the warping; if there's no warping, there's no matter or energy. So, what does gravity "see" when it looks out into this fast writhing vacuum? Nothing or almost nothing. For some unknown reason, all the activity of the vacuum is completely invisible to gravity. And yet, each jitter of the vacuum, each pair of particles that pops in and out, should be creating a small gravity well around it, just like a rock or a planet. The combined warping of all the gravity wells should be huge. You would think, said Strominger, that the whole universe would collapse because of the gravitational attraction of this sea of virtual particles. [82]

In connection with this strange situation (not to say impossible situation), the following note is important: the physically real space energy (equivalent with the existence of the cosmological constant) does not reflect a certain kind of antigravity. However, the repulsive force in the universe is not due to matter and energy but comes entirely from empty space itself.

In conclusion, the energy that comes out of nothing behaves quite different from that coming out of something and this is strange. We have two faces of energy! In Ref. [82], we find that the following comment reflects the present situation: "I don't know a good heuristic argument about why it should be so crazy," said Lawrence Krauss. "Why the energy comes out of nothing is so different from the energy that comes out of something."

All this seems to be artificial and is not convincing at all; this situation strongly suggests eliminating this picture of space-filling quantum fields and quantum space energy in order to get the space free of unacceptable peculiarities. In other words, the container model has to be eliminated.

Let us highlight in this connection the following point: We have two types of energy, although the notion "energy" has not been redefined. Two sorts of energy have to be assumed without being able to know why and without redefining the original notion "energy." Why the energy comes out of nothing is so different from the energy that comes out of something? Within this view, the energy has two "faces" without being able to define the faces. There is, in fact, no scientific law for the differentiation between the two energy faces.

This reflects rather a misconstruction and not a real situation, which is essentially based on the doubtful relation $\Delta E \Delta \tau \geq \hbar/2$. All this seems to be artificial and is not convincing.

7.2.6 Some principal questions

No doubt, the container model is problematic. However, for certain aspects, the container model is able to work, but not when we enter the region, where the true laws are disclosed. Here, instead of the container model, the projection principle is adequate because it is more realistic. What can we say about basic conceptions and the absolute truth?

Traditional quantum theory and classical mechanics have their own area of validity. But both approaches are based, on the other hand, on a common principle, which must not be final truth.

The container model is such a common conception on which quantum theory and classical mechanics are equally based. But the container model is obviously too rudimentary for the solution of basic questions.

In fact, the more we penetrate into the details of the physically real world, the more we need realistic conceptions for their description, that is, we need conceptions with increasing reliability. That is, in such cases the basic laws will be more and more disclosed. This often means that the complete theoretical frame has to be renewed from the very first. The use of a modified version of the space–time is such step in the case of modern physics, where we enter the realm of cosmology as well as that of nanoscience. In both cases, we expect the existence of a system-specific quantum time.

Then, we must eliminate the weak points of this specific basic conception, in this case the weak points of the container model. To modify or replace the container model, we have to use a more realistic space–time conception in which space and time are no longer physically real in the sense of Mach. As we know, within the container model, space and time are treated as physically real entities. This has to be seen as the reason why this conception is not fully applicable not only in cosmology (the cosmological constant turned to be wrong) but also not in connection with nanoscience because a system-specific quantum time is not contained within the fame of the container model. The projection principle goes really a step further, that is, it penetrates deeper into the nature of the various phenomena in the world. It should not be considered as the final basic truth but is a way in this direction,

Within the container model, a system-specific time is not definable. This view exclusively refers to the external time τ. The basic experience dictates this situation: We observe the world outside at each time τ as space structure; that is, the objects are directly before our eyes at τ, and no other type of time comes here into play. The door for the introduction of a system-specific time is not open. But it is a view in connection with the container model, and this conception is throughout seen as the absolute truth. But a system-specific quantum time is only definable in connection within the frame of the projection principle where the information is given in compressed form (Chapter 5). Projection theory works on the basis of more information than the container model and it is not restricted on the immediate space–time impressions alone. There is still the basic reality which is outside of space and time. A system-specific time becomes possible.

What is the consequence? Once again, Schrödinger wanted to expand quantum theory on the basis of the already "existing frame." But this direction does not work. The already existing frame is the problem. This frame is burdened with the container model in which space and time play the role of physically real entities and this is against Mach's principle.

However, we have to be careful. The more we penetrate into the details of nature, the more we disclose the true things of the world and we need theoretical conceptions with increasing sophistication. That is, in such cases, the basic laws will be more and more disclosed. This often means that the complete theoretical frame has to be renewed from the very first. The use of a modified version of the space–time is such a step.

7.2.7 What situations in physics are container sensitive?

The container model works well in many cases, although it is epistemologically not acceptable at all. In Section 7.1, we discussed GTR and traditional quantum theory. There are obviously cases where the results are not container sensitive. It is, however, more probable that a theoretically formulated theory and the features of the container model compensate each other. This can obviously lead to useful conceptions and laws as a function of space and time. We cannot exclude that GTR and traditional quantum field theory belong to this category.

For testing the container model more directly, we need an effect which encompasses the "entire" space–time. The following question arises: how is the whole space–time influenced through matter, which is embedded in it and which occupy each point in the space–time? That is, we need effects that are not superimposed by other effects and assumptions, respectively.

7.2.8 A genuine uncertainty between energy and time

Concerning relation $\Delta E \Delta \tau \geq \hbar/2$, two points have to be highlighted:

1. It is not an uncertainty relation.
2. It is even not a strict quantum equation. The time τ is a strict classical element.

In the development of quantum theory, an uncertainty relation between the energy and the time is required when we base our considerations on STR. Relation (7.3), that is, $\Delta E \Delta \tau \geq \hbar/2$, is however not the required uncertainty relation, but it is a "stranger," as Mario Bunge remarked. Equation (7.3) is an outcome of traditional quantum theory and, therefore, we have to consider traditional quantum theory with care because it leads to doubtful results. The reason for that has to be seen in the container model. Relation $\Delta E \Delta \tau \geq \hbar/2$ belongs exclusively to traditional quantum and it follows from this version of traditional quantum theory.

The elements x, y, z and p_x, p_y, p_y are genuine uncertain in traditional quantum theory and the uncertainty relations (7.2) are definable for them. The variables x, y, z and p_x, p_y, p_y belong to the system (object) under investigation and they characterize it. In $\Delta E \Delta \tau \geq \hbar/2$, only the energy E is a system variable but not the time τ. The time τ

is an observer index, external in character, and it behaves strictly classically. The clock time τ has nothing to with do the system (object) under investigation. In other words, the time τ, which appears in $\Delta E \Delta \tau \geq \hbar/2$, is not a system-specific quantity.

Again, in traditional quantum theory a quantum-time does not exist (Chapter 1), and the phenomenon "time" is here merely expressed through the external parameter τ. This is in fact a disadvantage of traditional quantum theory.

The quantum treatment within the frame of the container model leads to wrong results when we consider cosmological aspects. We do not observe the outcome predicted by the theory, but we observe another world. The theory delivers a wrong result for the cosmological constant. The much too large value for the cosmological constant is often judged in this way, and there is in fact no other assessment possible. The critical statements by Mario Bunge and Leon Lederman undoubtedly point in this direction.

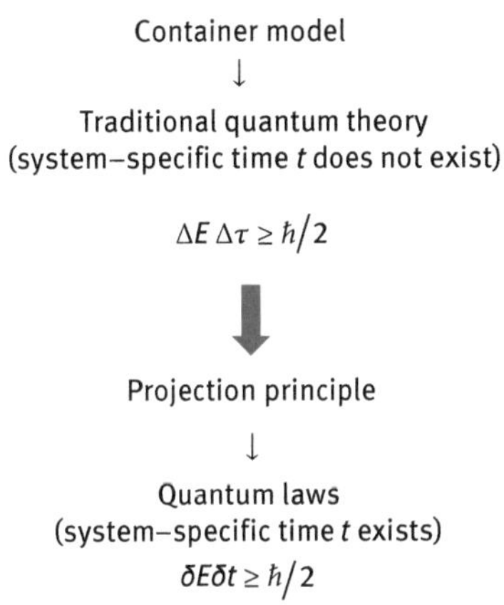

Figure 7.2: Uncertainties in projection theory. The information $\Psi(\mathbf{p}, E)$ is projected from $(\mathbf{p}, E)$ space onto $(\mathbf{r}, t)$ space and we get the information $\Psi(\mathbf{r}, t)$. The values p_x, p_y, p_z, E of $(\mathbf{p}, E)$ space are at clock time τ inherently uncertain and we have the uncertainties $\delta p_x, \delta p_y, \delta p_z, \delta E$ here (see Fig. 7.1). On the other hand, the values x, y, z, t of $(\mathbf{r}, t)$ space are, also at the same clock time τ, inherently uncertain and we have here the uncertainties $\delta x, \delta y, \delta z, \delta t$ (see Fig. 7.1). t is the system-specific time, which is not known in traditional physics. Due to the coupling of $(\mathbf{p}, E)$ space with $(\mathbf{r}, t)$ space, the uncertainties of $(\mathbf{p}, E)$ space are related to the uncertainties of $(\mathbf{r}, t)$ space and we get uncertainty relations as, for example, $\delta E \delta t \geq \hbar/2$, which is quite different from relation $\Delta E \Delta \tau \geq \hbar/2$ of traditional quantum theory (here, the elements ΔE and $\Delta \tau$ are not uncertainties).

Traditional quantum theory works within the container model, and within this view, space and time are considered as physically real entities. In particular, traditional quantum theory leads to the doubtful relation $\Delta E \Delta \tau \geq \hbar/2$. As we remarked several times, it is obviously not possible to define a system-specific time when we work within the frame of the container model. Erwin Schrödinger demonstrated that.

However, within projection theory a system-specific time t exists and, instead of relation $\Delta E\Delta\tau \geq \hbar/2$, we get a genuine uncertainty relation between the energy E and the time t: $\delta E\delta t \geq \hbar/2$ (see Chapter 6) where t is the required system-specific time. This means that we have to replace relation $\Delta E\Delta\tau \geq \hbar/2$ by $\delta E\delta t \geq \hbar/2$:

$$\begin{gathered}\Delta E\Delta\tau \geq \hbar/2\\ \Downarrow\\ \delta E\delta t \geq \hbar/2\end{gathered} \tag{7.4}$$

Then, the requirement of STR is fulfilled. In projection theory, we have genuine uncertainties not only for the coordinates and momentums but for the energy and time as well. More details are given in Figure 7.2 and Chapter 6.

7.3 No physically real space–time effects

According to Mach, space–time can never be the source for physically real effects, as, for example, inertia. Space and time are not physically real and, therefore, they cannot be the source for physically real effects.

But this condition is also not fulfilled at all within GTR. This has been demonstrated by de Sitter. In Ref. [86], we find the following text:

> In 1917, an eminent Dutch astronomer, Willem de Sitter, pointed out to Einstein that there was a finite valued solution of his field equations that gave the inertial mass of a particle even it was the only one in the universe. In this case, the curved space-time of general relativity would be flat, that is the geodesic line passing through the particle would be straight. The lone particle would be guided along this geodesic line as if it was made of inertial matter. Einstein initially argued strongly against this solution. However eventually he conceded that his interpretation of inertia could not therefore be due to other matter, as required by Mach's principle, because there was no other matter around in the de Sitter's example.

In other words, the field equations of GTR lead to a physically real space–time. This is not surprising because GTR works within a four-dimensional container (space–time).

7.3.1 Container with material objects

Mach's principle is obviously not accomplishable within the frame of the container model. The entity space or space–time, which defines the container, has to be physically real if we embed physically real (material objects) into the container.

Interaction of the coordinates with matter

The definition of the container with physically real objects means that there is a relation between the container and the material objects, where the container itself is at each time τ defined through the space. Let us consider at time τ_n the coordinates of the space, and let us denote them by

$$(x_i, y_i, z_i)_{\text{space}}, \quad i=1,2, \ldots, \infty \tag{7.5}$$

The space is a continuum and the number of space points is, therefore, infinity. We assume that at the same time τ_n there are N material bodies (objects) of mass m_k embedded in this space. Thus, we have

$$m_k, \quad k=1,2, \ldots, N \tag{7.6}$$

At each clock time τ_n, there has to be a "coupling" between the objects and the space. Otherwise, the objects cannot be related to the space, that is, the space and the objects would be independent of each other. If they are independent of each other, we cannot talk of a container filled with matter. Then, the space knows, at each time τ_n, nothing about the objects and vice versa. In this case, it makes no sense to express the object-properties as a function of the space coordinates. For example, the wave function Ψ is not expressible in terms of x_i, y_i, z_i.

In other words, there is a condition: The space point $(x_i, y_i, z_i)_{\text{space}}$ and the objects of mass m_k must be coupled. Coupling means that they interact with each other, which we symbolize by

$$(x_i, y_i, z_i)_{\text{space}} \Leftrightarrow m_k \tag{7.7}$$

Clearly, there is, in general, also an interaction between the objects that we express here by

$$m_k \leftrightarrow m_m, \; m \neq k \tag{7.8}$$

However, both interaction types, formulated by eqs. (7.7) and (7.8), are principally different from each other. In particular, interaction (7.7) may exist without interaction (7.8).

The necessary coupling between the space and the material objects comes into existence through single events of type (7.7). This situation defines the basis of the container model.

Objections

There are two objections against this view. The interaction process (7.7) provides that single, isolated space points are at each time τ observable, but this is not the case, that is, single space points are not defined. They are not physically real. The following two points are relevant (see Chapter 1):

1. We can only say something about distances in connection with masses, and time intervals in connection with physical processes [5, 83].

2. Such kind of physically real interaction processes provide that the space is at each time τ physically real, which is however not the case.

In conclusion, also from the analysis given in this section follows that the container model has to be excluded.

7.3.2 The situation within projection theory

In the case of the projection principle, we have a picture of the physically real (material) objects. The objects themselves appear at time τ as geometrical positions within space. The coupling between the geometrical positions and the space is given automatically because a geometrical position is a point in space.

The geometrical position is created together with the space in the observer's mind. The raw information α is processed through the processing characteristic η (see Chapter 4) and α is transformed under the simultaneous creating of space and time. The raw data α are at time τ inbuilt in this space and appears as geometrical position in space directly before the eyes of the observing human being. The raw data α and space and time do not exist separately from each other, but α is processed while space and time are created simultaneously. η reflects "one" process creating space and time and couples the transformed data α to space and time.

Space and time are created through the mind because the observer needs a representation of the physical real object, which is embedded in basic reality. The raw data α and the space–time are coalesced, that is, they are coupled. The space–time is then created if there is a physically real object to represent. This defines the coupling between the object and space at clock time τ.

Distances and time intervals

Let us deepen the facts, which appear within the frame of the projection principle, just in connection with the appearances in space and time.

In projection theory, we have at least "two" objects, the object and its counterpart (Section 6.1.7); the objects themselves are geometrical structures in space and time (denoted by $(\mathbf{r}, t)$ space). Because there are at least two objects (geometrical positions) at each clock time τ, there can only be at least two space–time position. Therefore, there are at clock time τ only space–time distances $\Delta x, \Delta y, \Delta z, \Delta t$ but never isolated space–time points x, y, z, t. The space–time distances are arbitrary in the case of distance-independent interactions (correlations) and systematically in the distance-dependent case.

The fact that only space–time distances $\Delta x, \Delta y, \Delta z, \Delta t$ of objects (geometrical objects in space and time) are definable is in accord with the fundamental observations (Chapter 1): We can only say something about "distances" in connection with masses, and "time intervals" in connection with physical processes. The four numbers of single, isolated x, y, z, t points are not expressible. Thus, single isolated objects cannot exist in space and time. The projection principle fulfills this condition. That is, the projection principle gives automatically that what a human being actually observes.

Object/antiobject pairs in projection theory

Again, within the frame of the projection principle, at least two objects exist at each clock time τ, and this is because any object can only exist in connection with its counterpart, which is a certain kind of antiobject (Chapter 6). That is, in projection theory we have object/antiobject pairs. The existence of antiobjects is necessary because the conservation laws of energy and momentum must be fulfilled.

In traditional quantum theory, just the opposite is the case: There exist particle/antiparticle pairs (Section 7.2) because, due to $\Delta E \Delta \tau \geq \hbar/2$, the conservation law for the energy is temporarily violated.

7.4 Summary and final remarks

1. We first discussed the situation in GTR and in the traditional form of quantum field theory. Both theories deliver excellent results, but these theories are mutually incompatible. This is a problem and we tried to explain this unsatisfactory situation as follows: The container model, on which both frames are based, reflects an approximation, which is obviously compensated within both theoretical frames through other theory-specific approximations.
2. The critical analysis with respect to the container model points on effects in nanoscience and cosmology. The developments in both realms need a system-specific time, which is not accomplishable within the container view, but a more sophisticated conception is required. The projection principle is here in the foreground.
3. We discussed the cosmological constant. The expected value for it is small and the predicted quantum field theoretical value is very large. The theoretical prediction deviates from the observed value by a factor of 10^{122}. The discrepancy could in fact hardly be larger. This situation is often considered as the deepest mystery of present physics. The reason is obviously due the fact that there is no genuine uncertainty relation between the energy and the time in the traditional form of quantum theory.
4. Space and time are in traditional quantum theory definitely not symmetrical to each other and this is in contrast to the fundamental results of STR. In this connection the following facts are relevant: Whereas the coordinates are "statistical" quantities, the time does not behave statistically. The time remains unchanged

when we go from classical mechanics to quantum theory. This is clearly reflected in the fact that the coordinates can be "operators," time is always a simple "parameter." Only the clock time τ exists in all versions of traditional quantum theory. Within projection theory clock time τ is a reference time, but we have here system-specific times.

5. The more we penetrate into the details of nature, the more we need realistic conceptions for their description. Thus, we need scientific conceptions with increasing sophistication. In such cases, the basic laws will be more and more disclosed. This often means that the complete theoretical frame has to be renewed. The use of a modified version of the space–time is such step, where we enter the realm of "cosmology" as well as that of "nanoscience." In both cases, we expect the existence of a system-specific quantum time for the understanding of self-organizing processes in nanoscience and for the description of the cosmological constant.
6. The conservation laws are important because the processes in nature are kept with them under control. In fact, since the energy can be (temporarily) violated in traditional quantum theory through the existence of $\Delta E \Delta \tau \geq \hbar/2$, we get an uncontrolled production of particle/antiparticle pairs, which leads to an unacceptable (wrong) value for the cosmological constant. Within the frame of the projection principle, the conservation law of energy is valid and we get in fact instead of relation $\Delta E \Delta \tau \geq \hbar/2$ the genuine uncertainty relation $\delta E \delta t \geq \hbar/2$, where t is the required system-specific time.

References

[1] Ratner, D, and Ratner, M. Nanotechnology and Homeland Security, Upper Saddle River, NJ 07458, Prentice Hall International, 2004.
[2] Wolfram Schommers, Nanodesign, World Scientific, 2014.
[3] Prigogine, I. From Being to Becoming (Time and Complexity in the Physical Sciences), Freeman, W.H., and Francisco, S, 1979.
[4] Schommers, W. Space-Time and Quantum Phenomena, Quantum Theory and Pictures of Reality, editor, Schommers, W. Springer-Verlag, Heidelberg, New York, 1989.
[5] Schommers, W. Mind and Reality, World Scientific, 2015.
[6] Landau, L.D, and Lifschitz, E.M. Quantum Mechanics, Pergamon, Oxford, 1965.
[7] Schommers, W. Basic Physics of Nanoscience, Elsevier, 2019.
[8] Jung, C.G. Synchronizität, Akausalität und Okkultismus, Deutscher Taschenbuch Verlag GmbH & Co, KG, München, 1990.
[9] Brückner, R. Das schielende Kind, Schwabe Verlag, Basel, Stuttgart, 1977.
[10] Thomas, S.K. The Structure of Scientific Revolution, University of Chicago Press, Chicago, 1970.
[11] Fraser, J. Die Zeit, Deutscher Taschenbuch Verlag, München, 1991.
[12] Schommers, W. The Visible and the Invisible, World Scientific, New Jersey, London, Singapore, 2011 and Wolfram Schommers, Quantum Processes, World Scientific, New Jersey, London, Singapore, 2011.
[13] John, D. B. The Artful Universe, Little, Brown and Company, Boston, New York, Toronto, London, 1995.
[14] Schommers, W. Handbook of Theoretical and Computational Nanotechnology, 1, 137, American Scientific Publishers, Los Angeles, 2006.
[15] Rieth, M, and Schommers, W. Handbook of Theoretical and Computational Nanotechnology, 1, 15, American Scientific Publishers, Los Angeles, 2006.
[16] Schommers, W. Nanostructured Materials 9, 693 (1997) and Michael Rieth and Wolfram Schommers, J. Comput. Theor. Nanosci., 1, 40, 2004.
[17] Broderick, D. THE SPIKE. How Our Lives Are Being Transformed By Rapidly Advancing Technologies, Tom Doherty Associates, LLC, New York, 2001.
[18] Gödel, K. Monatsh, Math. Phys., 38, 173, 1931.
[19] Pastalkova, E, Serrano, P, Pinkhasova, D, Fallace, E, Fenton, A.A, and Sacktor, T.C. Science, 313, 1141, 2006.
[20] Kasabov, N, and Beniskova, L. J. Comput. Theor. Nanosci., 1, 47, 2004.
[21] Lees, A.W., and Edwards S.F. J. Phys., C5, 1921, 1972.
[22] Gosling, E.M., and McDonald, I.R. Mol. Phys., 26, 1475, 1973.
[23] Ashurst, W.T., and Hoover, W.G. Phys. Rev. Lett., 31, 206, 1973.
[24] Car, R., and Parinello, M. Phys. Rev. Lett., 55, 2471, 1985.
[25] Seifert, G. Physik der Nanostrukturen, Schriften des Forschungszentrums Jülich, Jülich, 1998.
[26] Car, R., and Parinello, M. Simple Molecular Systems at Very High Densities, NATO Advanced Study Institute, Series Plenum Press, New York, 1988, 455.
[27] Mermin, N.D. Phys. Rev. A, 137, 1441, 1965.
[28] Milburn, J. J. Comput. Theor. Nanosci., 2, 161, 2005.
[29] Jammer, M. The Philosophy of Quantum Mechanics, Wiley, New York, 1974.
[30] Gribbin, J. Schrödinger's Kittens and the Search for Reality, Little, Brown and Company, Boston, New York, London, 1995.
[31] Home Ghose, D., and Agarwal, G.S. Phys. Lett. A, 153, 403, 1991.
[32] Ghose, P. Testing Quantum Mechanics on a New Ground, Cambridge University Press, 1999.

https://doi.org/10.1515/9783110525595-008

[33] Ghose, P., Testing Quantum Mechanics on a New Ground, Cambridge University Press, 1999.
[34] Mizobuchi, Y., and Othake, Y. Phys. Lett. A, 168, 1, 1992.
[35] Unnikrishnan, C.S., and Murphy, S.A. Phys. Lett. A, 221, 1, 1996.
[36] Brida, G., Genovese M., Gramegna M., and Predazzi, E. Phys. Lett. A, 328, 313, 1996.
[37] Schommers, W. Structure and Dynamics of Many-Particle Systems: Big Data Sets and Data Analysis, Big Data for Remote Sensing: Visualization, Analysis and Interpretation, Editors, Dey, N, Bhatt, C, and Ashour, A. S. 61–97, Springer, 2019.
[38] Schommers, W. Phys. Rev. Lett., 38, 1536, 1977 W. Schommers, Phys. Rev. B 21, 847 (1980).
[39] Morse, P.M. Phys. Rev., 34, 57, 1929.
[40] Mohammed, K., Shukla, M.M., Milstein, F., and Merz, J.L. Phys. Rev. B, 29, 3117, 1984.
[41] Eichler, M., and Peyzl, M. Phys. Stat. Sol., 35, 333, 1969.
[42] Diener, H.D., Heinrich, R., and Schellenberger W. Phys. Stat. Sol. B, 44, 403, 1971.
[43] Vitzk, V., Perrin, C.R., and Bown D.K. Philos. Mag., 21, 1049, 1970.
[44] Gehlen, C., Rosenfeld, A.R., and Hahn, G.T. J. Appl. Phys., 39, 5346, 1968.
[45] Chang, R. Philos. Mag., 16, 1021, 1967.
[46] Cotterill, R.M.J., and Doyama M. Phys. Rev., 145, 465, 1966.
[47] Doyama, M., and Cotterill R.M.J. Phys. Rev., 137, A994, 1965.
[48] Johnson, R.A. Phys. Rev., 134, A1329, 1964.
[49] Gififalco, L.A., and Weizer, V.G. J. Phys. Chem. Solids, 12, 260, 1960.
[50] Wilson, W.D., and Bisson C.L. Phys. Rev. B, 3, 2979, 1971.
[51] Anderman, A., and Gehman W.G. Phys. Stat. Sol., 30, 283, 1968.
[52] Furth, R. Proc. R. Soc. (London), 183, 87, 1944.
[53] Gififalco, L.A., and Weizer, V.G. Phys. Rev., 114, 687, 1959.
[54] Lincoln, R.G., Koliwad, K.M., and Ghate P.B. Phys. Rev., 157, 463, 1967.
[55] Goodman, F.O. Phys. Rev., 164, 1113, 1967.
[56] Moody, M., Ray, J.R., and Rahman, A. Phys. Rev. B, 35, 557, 1987.
[57] Strachan, A., Cagin, T., and Goddard W.A. Phys. Fev. B, 60, 15084, 1999.
[58] Azizian, S., and Gobal F. React. Kinet. Catal. L, 76, 43, 2002.
[59] Kao, C.L., Weaver, J.F., and Madix, R.J. Surf. Sci., 505, 115, 2002.
[60] Fang, T.H., Wenig, C.I., and Chang J.G. Surf. Sci., 501, 138, 2002.
[61] Kao, C.L., Carlson A., and Madix R.J. Surf. Sci., 497, 356, 2002.
[62] Vijay, A, and Metiu, H. J. Chem. Phys., 116, 60, 2002.
[63] Torrens, M. Interatomic Potentials, Academic Press, New York and London, 1972.
[64] Schommers, W. Int. J. Modern Phys., B8, 3411, 1994.
[65] Barker, J.A., Watts R.O., Lee, J.K, Schafer, T.P., and Lee, Y.T. J. Chem. Phys., 61, 3081, 1974.
[66] Romahn, U., Göbel H., Schommers, W., and von Blanckenhagen, P. Mod. Phys. Lett., B5, 1873, 1991.
[67] Wolf, D., Wang J., Phillpot, S.R., and Gleiter H. Phys. Rev. Lett., 74, 4686, 1995.
[68] Axilrod, B.M., and Teller, E. J. Chem. Phys., 11, 229, 1943.
[69] Schommers, W. Phys. Rev., 22, 2855, 1980.
[70] Schommers, W., in "Structure and Dynamics of Surfaces I", ed. by Schommers W., and von Blanckenhagen P. Springer-Verlag, Berlin, Heidelberg, New York, 1987.
[71] Schommers, W., in "Structure and Dynamics of Surfaces II", ed. by Schommers W., and von Blanckenhagen P. Springer-Verlag, Berlin, Heidelberg, New York, 1987.
[72] Rieth, M, and Schommers, W. Encyclopedia of Nanoscience and Technology, edited by, Nalwa, H.S., American Scientific Publishers, Volume, 2, 51–103, 2004.
[73] von Ditfurth, H. Der Geist fiel nicht vom Himmel, Deutscher Taschenbuch Verlag.
[74] Ziman, J. Reliable Knowledge, Cambridge University Press, Cambridge, 1978.
[75] Schneider, W. Hypothese, Experiment, Theorie, Göschen, S. Walther de Gruyter, Berlin, 1978.

[76] Richard, P.F. The Character of Physical Law, M.I.T. Press Cambridge, MA and London, 1977.
[77] Bunge, M. Can. J. Phys, 48, 1410, 1970.
[78] van Hove, L. Phys. Rev, 95, 249, 1954.
[79] Schommers, W. Phys. Rev, A16, 327, 1977.
[80] von Baeyer, Hans Christian. Information. The Language of Science, Harvard University Press, Cambridge, Massachusetts, 2004.
[81] Wolfram Schommers, J. Comp. Theor. Nanosci., 4, 992, 2007.
[82] Cole, K.C. The Hole in the Universe, A Harvest Book, HARTCOURT Inc, San Diego, New York, London, 2001.
[83] Schommers, W, Advanced Science Letters 1, 59 2008.
[84] Greene, B. The elegant Universe, Vintage Books, A Division of Random House Inc, New York, 1999.
[85] de Broglie, L. Die Elementarteilchen, Goverts, Hamburg, 1943.
[86] Graneau, P, and Graneau, N. In the Grip of the Distance Universe, World Scientific, New Jersey, London, Singapore, 2006.

Index

https://doi.org/10.1515/9783110525595-009

Bei Fragen zur Produktsicherheit wenden Sie sich bitte an:
If you have any questions regarding product safety,
please contact:

Walter de Gruyter GmbH
Genthiner Straße 13
10785 Berlin
productsafety@degruyterbrill.com